Computer Graphics – Systems and Applications

Managing Editor: J. L. Encarnação

Editors: K. Bø J. D. Foley R. A. Guedj
P. J. W. ten Hagen F. R. A. Hopgood M. Hosaka
M. Lucas A. G. Requicha

M. Hosaka

Modeling of Curves and Surfaces in CAD/CAM

With 90 Figures

Springer-Verlag
Berlin Heidelberg New York
London Paris Tokyo
HongKong Barcelona
Budapest

Mamoru Hosaka
Tokyo Denki University
Kanda-Nishikicho 2-2
Chiyoda-ku
Tokyo 101, Japan

ISBN-13:978-3-642-76600-8 e-ISBN-13:978-3-642-76598-8
DOI: 10.1007/978-3-642-76598-8

Library of Congress Cataloging-in-Publication Data
Hosaka, M. (Mamoru)
Modeling of curves and surfaces in CAD/CAM/M. Hosaka. p. cm. – (Computer graphics –systems and applications) Includes bibliographical references and index.
ISBN-13:978-3-642-76600-8
1. Computer graphics. I. Title. II. Series: Symbolic computations. Computer graphics – systems and applications. T385.H6958 1992
670.285'66 – dc20 91-43266 CIP

Softcover reprint of the hardcover 1st edition 1992

Typesetting: Camera ready by author
45/3140-543210 – Printed on acid-free paper

Preface

1 Aims and Features of This Book

The contents of this book were originally planned to be included in a book entitled *Geometric Modeling and CAD/CAM* to be written by M. Hosaka and F. Kimura, but since the draft of my part of the book was finished much earlier than Kimura's, we decided to publish this part separately at first. In it, geometrically oriented basic methods and tools used for analysis and synthesis of curves and surfaces used in CAD/CAM, various expressions and manipulations of free-form surface patches and their connection, interference as well as their quality evaluation are treated. They are important elements and procedures of geometric models. And construction and utilization of geometric models which include free-form surfaces are explained in the application examples, in which the methods and the techniques described in this book were used. In the succeeding book which Kimura is to write, advanced topics such as data structures of geometric models, non-manifold models, geometric inference as well as tolerance problems and product models, process planning and so on are to be included. Consequently, the title of this book is changed to *Modeling of Curves and Surfaces in CAD/CAM.*

Features of this book are the following. Though there are excellent text books in the same field such as G. Farin's *Curves and Surfaces for CAD/CAM*[1] and C.M. Hoffmann's *Geometric and Solid Modeling*[2], this book differs from them in the methods of analysis, interpretation and range of the problems to be treated. Though most of the theories and the methods treated in this book were developed from academic points, they were applied to real problems which were performed for some Japanese companies during the period from mid-1960 to 1987. Since before the mid-1970s the literature in this field outside Japan was hardly available for the author and the necessity of the development was urgent, the theories and the methods were developed independently of those from outside Japan. Since most of them appeared only in Japanese journals and proceedings and the application results have been company assets, some of the contents of this book may be new in the English language. They have been successfully applied in the companies and the intuitive approach of the analysis has enabled the practical engineers there to understand the essence of the methods and to become interested in their application. Moreover the contents treat current and future problems of importance, so the author believes the contents are worth publishing.

Nowadays, there are many commercial software systems for CAD/CAM.

They are useful for ordinary products, but they do not seem to be adequate for application to products with special features attractive to customers, because there have still been companies which asked the author of this book for advice as to the application of the system to their special product developments. The author thinks the contents of the book include some hints on how to respond to these questions.

2 Structure and Contents of the Book

This book is written for practical engineers who need knowledge about theories of curves and surfaces and their processing methods for the design and manufacture of engineering objects of high quality. The author has tried to make the description as easily understandable and applicable as possible. Accordingly, though the various methods are systematically formalized, they are also graphically interpreted so that one can analyze problems just using pencil and paper with the help of the drawings. This improves one's understanding of the problem he is trying to solve.

The contents of the book are divided into the following six parts.

- Chapters 1–4. The basic mathematical tools and linear and quadratic curves and surfaces.
- Chapters 5–8. The fundamental theory of curves and surfaces extracted from the theory of differential geometry and its application. Also the curve passing through a given point sequence.
- Chapters 9–12. Curves and surfaces with control points and their regular connection.
- Chapters 13–15. Non-regular connection of surface patches and triangular patches.
- Chapter 16 and Appendix. Interference of surfaces. Tracing methods of intersection curves by numerical calculation.
- Chapter 17. Geometric models and engineering drawings. Feature input, construction of geometric models which include free-form surfaces, and their utilization.

In Chap. 1, the theory of vectors and matrices together with their notations which are used in this book are concisely summarized with application examples. In Chap. 2, coordinate transformations and movement of a solid body are treated. Since it is rather difficult for a person to grasp intuitively the three-dimensional movement of a body, we formalize its description. And for its intuitive understanding, we introduce mirror reflections to represent rotations and translations. In Chap. 3, objects which are represented by linear equations are treated. Since this is related to the control polygons and nets, we introduce Menelaus' theorem, which is a classic theorem of geometry. Its variants are used in various places afterwards. A topology of faces, edges and vertices of a solid body is explained simply.

In Chap. 4, conics and quadrics are treated, because they are represented by second degree equations and their properties are well investigated. Since the shape of a free-form surface around a point on it can be expressed approximately by a quadric surface and also the shape of intersection curves of surfaces near their singular point is conic, knowledge of quadrics and conics is important. And many mechanical parts have shapes consisting of quadrics and planes.

Chapters 5 and 6 are excerpts from the differential geometry to be adapted to analysis and interpretation of curves and surfaces described in this book. Geometric quantities such as tangent and normal vectors, curvature and torsion are introduced together with the formulas of Frenet-Serret. The shape of a curve near a point on it is described approximately with these values. When an elevation and a plan of a space curve to be designed are given independently, a problem of determining an expression of the space curve is treated as a practical example. In the last part, a method of converting a parametric expression of a plane curve into an implicit form is described. Conics are represented by a rational Bézier curve of degree two, which is treated in Chap. 12.

For a surface, its basic vectors, fundamental magnitudes, curvatures and lines of curvature are explained, which determine the shape around a point on the surface. Weingarten's formula and Rodrigues' equation, which describe the behavior of the normal vector, are introduced and used in Chap. 7, where various curves on a surface are defined to express its characteristics. These curves are classified according to their features: curves intrinsic to its shape, those dependent on direction of interest, those related to its environment and the direction, and those defined to search for special points on the surface. Umbilics, lines of curvature, equi-brightness curves, silhouette patterns, highlight curves, ridge valley curves, and extremum search curves and others are discussed. The fundamental magnitudes and curvatures of offset surfaces and ruled surfaces are also determined, which are to be used afterwards.

In Chap. 8, curves which pass through a given point sequence are treated. Lagrange's formula which uses a polynomial is explained together with a method of calculating interpolated or extrapolated points. Its rational equivalent is also introduced. Though these methods are not directly used in CAD/CAM operations, they are included in an excellent differential equation solver used in Chap. 16 and explained in the Appendix. Then curve segments expressed by parametric polynomials are described. The point sequence can be connected smoothly by the curve segments with the pseudo-elastic energy stored in all the segments being kept minimum. In these connections of parametric curves, the importance of scale ratios which relate to relative real magnitudes of the curve segments is shown by examples of connected curve shapes. So far, fundamental knowledge on curves and surfaces has been explained, then in Chaps. 9–15, that which is used in CAD/CAM is treated.

In Chap. 9, so-called Bézier curves and surface patches are defined from the special characters of control points. Here the shifting operators for subscripts of control points are introduced for simplicity of the expressions as well as of their manipulation. Geometric properties of a Bézier curve are intuitively obtained from a Bézier polygon which consists of the control points for the segment. Division and degree elevation of the curve segment correspond to those of the Bézier polygon. A Bézier net for a surface patch is obtained by the tensor product of two Bézier polygons for curve segments. The net determines geometric properties of the patch.

In Chap. 10, first, the meaning of degree of freedom for the connection of Bézier polygons is explained, then a Connection Defining Polygon (CDP) which determines connection of the adjacent curves in $C^{(n-1)}$ is defined. When continuity of derivatives at the junction of two curves of degree n is guaranteed up to the $(n-1)$-th derivative, the condition of connection is called $C^{(n-1)}$. Since at each junction of connecting Bézier polygons in $C^{(n-1)}$, there exists a CDP, we explain that the adjacent CDP are related by another polygon called the spline polygon. Inversely, from a given spline polygon we can deduce a sequence of $C^{(n-1)}$ connected Bézier polygons. We explain practical methods of determining these Bézier polygons from an arbitrary spline polygon whose shape is a rough picture of the shape of the generated connected curve. The spline polygon is considered to control the shape of connected curves of degree n and the $C^{(n-1)}$ condition means that the shape control is local. In a Bézier polygon, the control is global.

Then we explain a method of division of a spline polygon which is equivalent to increase of vertices of the spline polygon. Next, for a spline polygon of degree three we show a method to add more freedom for shape control by changing the connecting condition from $C^{(2)}$ to $G^{(2)}$, which only guarantees continuity of the curvature vector at the junctions instead of the second derivative. In the last part of this chapter, effects of the scale ratios on the shape of the generated shape of the curve and that of the radius of curvature profile, whose shape is used as an aesthetic criterion by designers, are explained.

In the first section of Chap. 11, by the method of tensor product Bézier nets are determined from a given spline net. In the succeeding sections, proofs of the methods of determining Bézier polygons from a spline polygon are given. Here Menelaus' theorem is extensively used.

In Chap. 12, rational Bézier and rational spline curves are treated. First comes the relation between linear division of a line segment and rational division of the same segment with a different weight value attached to each of its end points. The geometric properties of a Bézier polygon can easily be transferred to the rational Bézier polygon and vice versa. A simple graphical mapping method between the respective polygons is given. In the similar mapping from a rational spline polygon we obtain rational Bézier polygons connected in $C^{(n-1)}$. We give examples of effects of the weights on the shape of the generated curve and surface.

Since a rational Bézier of degree two is converted to an implicit form which is the same as a conic, we discuss the effect of the weight on the class of conics. In the last sections we introduce the use of negative values of a weight and a curve fitting method by conics. We show that the optimum number of segments for fitting a given point sequence is obtained.

In Chaps. 13–15, shapes which cannot be expressed by spline surfaces are treated. Such shapes are often needed in special regions of surface expressed by regular connection of patches. In Chap. 13 it is shown that regions of convex corner and convex-concave corners can be rounded by using only non-regular connection of Bézier patches. In Chap. 14, methods to correct a mismatch between freedoms of given boundary conditions and that of a patch to be defined are treated. At first the classic Coons' method is explained, then a patch consisting of a weighted mean of patches, each of which partially satisfies boundary conditions, and a patch with double twist vectors or double inner control points are introduced.

Next, a method of correction patches which compensate the mismatch of cross-boundary tangents and higher derivative vectors is described with various examples. In the last section the method of a rolling ball for smooth connection of surfaces is explained.

In Chap. 15, a three-sided patch in operator form is treated, which has the similar characters as those of a usual Bézier patch. But their connection is much more constrained because of the lesser number of control points. The correction patch similar to that for the four-sided patch is explained for their flexible connection. Division and degree elevation of the patch can be performed in a similar way.

In Chap. 16, interference of surfaces of various types is treated. Intersection of free-form surfaces is a difficult and important problem in the real application of surface design. This chapter was completely revised from the original draft, because a new method which has proved to give good results was found during the writing of the draft. Differential equations which describe the intersection curves of surfaces of various types, including offset surfaces, are given and numerical methods of solving these differential equations with adaptive stepsize are successfully applied for tracing the intersection curves.

All the starting points of the curve are determined by other methods described in this section, and truncation errors in the integration process are kept within allowable tolerances. We are freed from the problem of determining stepsize to obtain a next point and we can use the method even in the very near region of a singular point where the normal vectors of both surfaces are coincident. Around a singular point we can investigate the behavior of the intersection curve by considering only higher degree terms of the differential equations, which are usually neglected. Various examples of solved results are given.

In Chap. 17, main features of two examples of the successful integrated CAD/CAM systems are described using data from Toyota Motor Corporation and Sony Corporation. In motor car body design and manufacture, there are two groups of persons who are least adaptable to computerization. One is a group of sensitive style designers, who are artists, and the other is the group of experts on die-face design of stamping presses, whose accumulated knowledge on stamping press processes is valuable for successful die-making. If human interfaces and interaction procedures which these persons would agree to use willingly can be provided, a successful integrated CAD/CAM system can be constructed. Its other various and highly automated operations can be implemented technically even if they are complicated. We describe such parts of the successful system. The other feature is that the master models which have been used in various stages of manufacturing are replaced by the numerical master model from which various standard and control information is extracted for use in manufacturing the stamping dies and inspection of the manufactured objects and the welding fixtures and jigs.

The next example is a very small commercial product such as an in-the-ear headphone and its case. Though the size of the objects is very small, similar design and evaluation techniques are used, and we explain them briefly. In both examples the theories and methods which are described in this book were applied.

In the Appendix, numerical methods of differential equation solving with adaptive stepsize, which are used in this book for tracing or drawing various curves on a surface including intersection curves, are explained. We have tried to make the description understandable without special knowledge of numerical integration.

3 Short Historical Note

Many of the materials for this book were taken from the author's research work performed from the mid-1960s to the late 1980s in Japan. A short historical note on this research and development is described to clarify the background.

The author's practical experience on design task dates back to the period of 1943-1945, when he engaged in the structural design of airplanes. After 1947 he engaged in research and development work at the Research Institute of National Railways of Japan. Among various works performed there the most famous one was his proposal for and completion of the on-line real time seat reservation system for trains [3], which began its successful field operation from the beginning of 1960. With this system he learned a great deal about computer system design and the importance of man-machine interaction.

After his transfer to the University of Tokyo in 1959, he had an interest in graphics and designed graphics hardwares devices which at that time were not available in Japan. These included graphical input and output devices and a unique Digital Differential Analyzer [4]. In late 1964, he read papers by Sutherland and Coons [5] in the proceedings of 1963 SJCC, and was greatly shocked especially by the article by S.Coons in which a revolutionary intelligent machine for interactive designing would be completed in five years. Stimulated by Sutherland's SKETCHPAD, he designed a graphic system whose hardware was constructed by Hitachi Ltd. in Japan and which surprised people who knew nothing about the tricks of programming of graphics [6].

With these experiences he knew that the prediction by Coons was only a dream and Coons' surface expressions were not adequate for high quality surface description. So, he developed his own theory of free-form surfaces [7], for which two large motor car makers in Japan approached him, because only his theory among all others passed their tests, as each of them informed him separately. At that time he knew nothing about car industry. One company wanted to apply his theory to drawings in car body design. The other company, Toyota, wanted to apply it to NC machining of stamping dies for car bodies. For the latter he proposed the use of a data structure for describing the shape of a part of a car body, which would be an internal model of the real object. This was an idea of geometric modeling.

At that time, concerning information creation and communication and transfer to downstream of the design tasks, he felt the necessity for an information description which can be treated directly by the computer and replace the conventional engineering drawings. With this idea and his surface theory, Toyota developed a system called TINCA (Toyota Integrated Numerical Control Approach), completed around 1972 [8], which produced cutter paths for machining stamping dies from the internal model.

While the model was used in the company, his idea on modeling led to his design of an interactive language called GIL (Generalized Interactive Language) which operated on the laboratory-made minicomputer at the University of Tokyo. With these facilities, creation, manipulation and display of geometric models of polyhedra became possible [9]. This work was performed independently without knowledge of works of I.Braid and M.E.Engli on geometric modeling [10]. An English paper on this work was published [11]. In 1974 the author published a book on computer graphics and its application based on his research work during 1964-1972 [12].

The work of geometric modeling was further developed by F.Kimura into the system GEOMAP (GEOmetric Modeling And Processing) at the ETL (Electric Technical Laboratory) of the Japanese Government, where powerful computers were available. The GEOMAP system supported a unique two-dimensional input method of data and instruction with pen and paper. Works concerning GEOMAP were published in English [13][14].

The author of this book learned about the Bézier curve for the first time while he was in Europe attending IFIP Congress '74. He was interested in it and soon found a shift operator could be used to simplify its expression and manipulation [15]. This expression has been used ever since then [15][16].

Around 1976, the Toyota Motor Corporation again asked the author to help with their new project, which was to assist style designers' activities using a computer to shorten the design time and to make the design shape information a direct input to the various downstream tasks performed by the computers such as the structure design system of car bodies and the stamping die design and manufacturing system (TINCA system). This project seemed very attractive but it was difficult to attain its objective in a short time.

At that time Toyota had a system called NTDFB (New Toyota Data File system for Body engineering) which took in measured data from a clay model made by designers and constructed a wire frame model of a car body with interpolation by Coons' patches. But the quality of the surface was not satisfactory and it was an enormous task to correct and to convert the data for use by TINCA system for manufacturing stamping dies.

For the new project, at Toyota the designers' conventional methods were systematically scrutinized and analyzed. According to this information new methods of surface synthesis and evaluation were developed with the collaboration of the style designers. The new methods were a kind of simulation of designers' activities during their creative works of constructing a clay model which was the expression of their intentions. Accordingly, the system had to provide various methods and tools for construction and evaluation of the model in order for the designers and modelers to feel they were using their accustomed methods.

Through the experiments, armed with the theoretical works and experience on a prototype and after more than three years development with much effort and investment by Toyota, the system was completed and has been used since 1981 and an epoch-making enhancement of productivity of the design as well as high product quality of the final products have been attained. By means of this system and downstream systems integrated with it, all the physically constructed master models were abolished and the number of tryouts of the stamping press with real models was extremely diminished [18].

During cooperation on the development of the system, the author found many problems to be solved, but owing to the limited developing time they were not all solved satisfactorily. These were problems related to the evaluation, interference and connection of free-form surfaces, some of which were analyzed more elegantly afterwards [17][19][20]; these topics are treated in this book. In 1981 the author left the University of Tokyo according to the age-limit rule there and transferred to Tokyo Denki University, where he served as the director of the university research center. After the term of office in 1988, he had time to make his efforts to write this book in which most of his practically useful work on free-form curve and surfaces are included in a systematic way, although some of the complicated analyses were omitted owing to the introductory character of the book.

After the successful development work of the first Seat Reservation System of Japanese National Railways, the author had to be engaged in consulting work on the development of the large on-line stock information and transaction system of the Tokyo Stock Exchange. They both have quite different characters from the

CAD/CAM system of Toyota, but the common factor was that they involved expert persons without whom the systems could not be operated successfully. The author has realized that human considerations and good interfaces with the machine were among the most important factors for the success of computerized systems together with their high reliability and high performance.

4 Acknowledgements

The author owes his research work in this field to many people who gave problems to him and discussed with him and tested research results. Especially, he thanks very much the people at Toyota Motor Corporation and Prof. emeritus T. Sata of the University of Tokyo, who first introduced me to this field, and Prof. F. Kimura of the University of Tokyo, who was a student of my laboratory in the same university and afterwards has been my colleague in research work in this field.

For materials in Chap. 17, Mr. M. Ohara and Dr. M. Higashi (now at Toyota Technological Institute) of Toyota Motor Corporation and Dr. T. Kuragano of Sony Corporation permitted me free use of various results of the developments performed there. All the figures except those in Chap. 17, were produced by Mr. T. Saitoh of Tokyo Denki University and Mr. T. Kushimoto and Mr. H. Kobayasi of Stanley Electric Corporation by calculating equations in this book. Their research works with the author are also included in the book.

By the effort of Mr. J.Andrew Ross of Springer-Verlag, many errors, typographical and grammatical, of the original draft were corrected; I thank him very much for his contribution.

Tokyo, August 1991 Mamoru Hosaka

Contents

1 Excerpts from Vector and Matrix Theory

1.1 Introduction

In treating geometric objects or quantities, we use vector and matrix notations and operations throughout this book to simplify their description and also to make the reader's geometric understanding easier. We show in this chapter only the essence of vector and matrix operations with their application examples, which will be referred to from various places in this book.

1.2 Notations of Vectors and Vector Arithmetic

A vector is a quantity which has magnitude and direction. In diagrams, a vector is expressed by a line segment with an arrow head: the line length represents its magnitude and the arrow head indicates its direction. Usually its absolute position need not be considered except in the case of a position vector.

In a three-dimensional space, the degree of freedom of a vector is three: the direction and the magnitude or the three components in a coordinate system in the space. Coordinate systems are not necessarily orthogonal, but for simplicity, at first we use an orthogonal coordinate system and afterwards we treat an oblique coordinate system. Three unit vectors, each of which corresponds to one axis of the coordinate system, are used as the basis. Let $\mathbf{i}$, $\mathbf{j}$ and $\mathbf{k}$ be the unit vectors. Their expressions are given by their components values on the coordinate axes:

$$\mathbf{i} = (1,0,0),\ \mathbf{j} = (0,1,0),\ \mathbf{k} = (0,0,1), \tag{1.1}$$

and if we let components of an arbitrary vector $\mathbf{a}$ be a_x, a_y and a_z, then $\mathbf{a}$ is expressed as

$$\mathbf{a} = (a_x, a_y, a_z). \tag{1.2}$$

Multiplication of a vector $\mathbf{a}$ by a constant c, addition and subtraction of two vectors $\mathbf{a}$ and $\mathbf{b}$ are defined as the same operations between their components:

$$\mathbf{c} = c\mathbf{a} = (ca_x, ca_y, ca_z), \tag{1.3}$$

$$\mathbf{c} = \mathbf{a} \pm \mathbf{b} = (a_x \pm b_x, a_y \pm b_y, a_z \pm b_z). \tag{1.4}$$

In Fig. 1.1, the above operations are shown graphically. Using equations (1.1), (1.2), (1.3) and (1.4), a vector $\mathbf{a}$ can be written in a form:

$$\mathbf{a} = a_x\mathbf{i} + a_y\mathbf{j} + a_z\mathbf{k}. \tag{1.5}$$

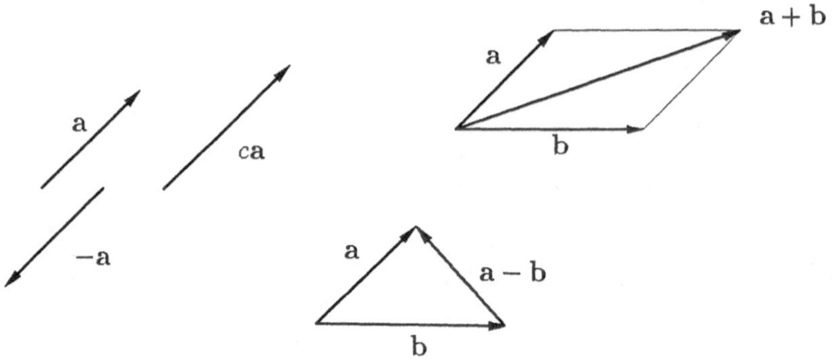

Figure 1.1 Graphic representation of vectors and their arithmetic

Since the position of a point P in the space is determined by its distance and the direction from the origin of the coordinate system, it is a vector. Usually when a vector represents a position, the tail of the vector **p** is fixed at the origin; its component values are not explicitly subtracted by the corresponding value (0,0,0) of the origin. Therefore, the coordinates values (p_x, p_y, p_z) of the position P are the components of its position vector **p**:

$$\mathbf{p} = (p_x,\ p_y,\ p_z). \tag{1.6}$$

To express a vector, its components are written horizontally and enclosed by a pair of parentheses. The other form is to write its components vertically enclosed by a pair of large parentheses thus:

$$\begin{pmatrix} a_x \\ a_y \\ a_z \end{pmatrix}. \tag{1.7}$$

Since this latter form occupies a large space and is not suitable for usual typewriting, when this form is needed, the following alternative is frequently used:

$$\mathbf{a}^T = (a_x, a_y, a_z)^T. \tag{1.8}$$

The superscript ^{T} indicates transposition.

When a reference point of a position vector **p** is at a position **q**, a displacement vector **d** is defined by

$$\mathbf{d} = \mathbf{p} - \mathbf{q} = (p_x - q_x, p_y - q_y, p_z - q_z). \tag{1.9}$$

Its magnitude is the distance or the length between the points **p** and **q**, which is denoted by

$$|\mathbf{d}| = |\mathbf{p} - \mathbf{q}| = \sqrt{(p_x - q_x)^2 + (p_y - q_y)^2 + (p_z - q_z)^2}. \tag{1.10}$$

The magnitude of a vector is the square root of the sum of the squares of its components. It is also called the *absolute value* of a vector. A vector $\hat{\mathbf{a}}$ whose magnitude is 1 and whose direction is same as that of $\mathbf{a}$, is called a *unit vector* in the direction of $\mathbf{a}$. It is derived from a vector $\mathbf{a}$ by dividing by its absolute value $|\mathbf{a}|$:

$$\hat{\mathbf{a}} = \frac{\mathbf{a}}{|\mathbf{a}|}. \tag{1.11}$$

The unit vectors which have the directions of the coordinates axes x, y and z are $\mathbf{i}$, $\mathbf{j}$ and $\mathbf{k}$ as used above.

1.3 Product of Vectors

1.3.1 Inner Product of Vectors

Next, an inner product $\mathbf{a}\cdot\mathbf{b}$ of two vectors $\mathbf{a}$ and $\mathbf{b}$ is defined. The inner product is a scalar value and is equal to the product of the absolute values of two vectors multiplied by the cosine of the angle θ between them:

$$\mathbf{a}\cdot\mathbf{b} = |\mathbf{a}|\cdot|\mathbf{b}|\cos\theta. \tag{1.12}$$

An inner product $\mathbf{a}\cdot\mathbf{a}$, which is written $\mathbf{a}^2$, is the same as the square of the absolute value of $\mathbf{a}$. Hence,

$$|\mathbf{a}| = \sqrt{\mathbf{a}\cdot\mathbf{a}}\,. \tag{1.13}$$

If non-zero vectors $\mathbf{a}$ and $\mathbf{b}$ are orthogonal, $\mathbf{a}\cdot\mathbf{b}$ is equal to zero. Naturally, we have $\mathbf{i}^2 = \mathbf{j}^2 = \mathbf{k}^2 = 1$ and $\mathbf{i}\cdot\mathbf{j} = \mathbf{j}\cdot\mathbf{k} = \mathbf{k}\cdot\mathbf{i} = 0$ by definition. Since $\mathbf{a}$ and $\mathbf{b}$ can be expressed in the form (1.5), $\mathbf{a}\cdot\mathbf{b}$ is given by the sum of the products of corresponding components of the two vectors in the orthogonal coordinate system:

$$\mathbf{a}\cdot\mathbf{b} = a_x b_x + a_y b_y + a_z b_z. \tag{1.14}$$

If the inner product of two non-zero vectors is positive, or zero or negative, then the angle between them is smaller than, or equal to, or greater than 90 degrees. The inner product does not depend on the order of multiplication.

1.3.2 Vector Product

A vector product $\mathbf{a}\times\mathbf{b}$ of two vectors $\mathbf{a}$ and $\mathbf{b}$ is a vector. Let it be $\mathbf{c}$, whose magnitude is equal to the product of the absolute values of the two vectors $\mathbf{a}$ and $\mathbf{b}$ multiplied by absolute value of the sine of the angle θ between them:

$$|\mathbf{c}| = |\mathbf{a}|\cdot|\mathbf{b}||\sin\theta|. \tag{1.15}$$

When the tails of $\mathbf{a}$ and $\mathbf{b}$ are coincident, $|c|$ gives the area of the parallelogram made by them. The direction of $\mathbf{c}$ is equal to that of the right screw in a rotation of $\mathbf{a}$ towards $\mathbf{b}$ by the angle θ. Hence, the vector $\mathbf{c}$ is orthogonal both to $\mathbf{a}$ and

b. When the order of multiplication is reversed, the direction of the resulting vector is also reversed:

$$\mathbf{a} \times \mathbf{b} = -\mathbf{b} \times \mathbf{a}. \tag{1.16}$$

If a vector product of two non-zero vectors is zero, they are parallel. The normal direction of a plane is determined by the vector product of two non-parallel vectors in the plane. The front side and the rear side of a plane are defined by the direction of its normal vector.

When there is a relation

$$\mathbf{c} = \mathbf{a} \times \mathbf{b}, \tag{1.17}$$

the components of the product $\mathbf{c}$ can be expressed by those of $\mathbf{a}$ and $\mathbf{b}$. Let their components be

$$\left.\begin{array}{rcl} \mathbf{a} &=& (a_x, a_y, a_z), \\ \mathbf{b} &=& (b_x, b_y, b_z), \\ \mathbf{c} &=& (c_x, c_y, c_z). \end{array}\right\} \tag{1.18}$$

By the two orthogonal conditions $\mathbf{a} \cdot \mathbf{c}=0$ and $\mathbf{b} \cdot \mathbf{c}=0$, we have

$$\left.\begin{array}{rcl} a_x c_x + a_y c_y + a_z c_z &=& 0, \\ b_x c_x + b_y c_y + b_z c_z &=& 0. \end{array}\right\} \tag{1.19}$$

From the above two linear equations for the three variables c_x, c_y and c_z, we obtain the ratios among the three components of $\mathbf{c}$,

$$c_x = \gamma \begin{vmatrix} a_y & a_z \\ b_y & b_z \end{vmatrix}, \quad c_y = \gamma \begin{vmatrix} a_z & a_x \\ b_z & b_x \end{vmatrix}, \quad c_z = \gamma \begin{vmatrix} a_x & a_y \\ b_x & b_y \end{vmatrix}, \tag{1.20}$$

where γ is a constant independent of $\mathbf{a}$ and $\mathbf{b}$. To determine the value of the constant γ, we choose $\mathbf{a} = (1,0,0)$ and $\mathbf{b} = (0,1,0)$, then $\mathbf{c}$ is determined by the definition of the vector product as (0,0,1). Comparing the left-hand side and the right-hand side of eq. (1.20), we can fix the value of γ equal to 1. It follows that the components of the result vector of $\mathbf{a} \times \mathbf{b}$ are given by eq. (1.20) with $\gamma = 1$. They can collectively be written:

$$\mathbf{c} = c_x\mathbf{i} + c_y\mathbf{j} + c_z\mathbf{k} = \begin{vmatrix} \mathbf{i} & \mathbf{j} & \mathbf{k} \\ a_x & a_y & a_z \\ b_x & b_y & b_z \end{vmatrix}. \tag{1.21}$$

1.4 Triple Products

1.4.1 Scalar Triple Product

The *scalar triple product* is a kind of inner product of two vectors, one of which is represented by a vector product of another two vectors. Suppose the tails of three vectors $\mathbf{a}$, $\mathbf{b}$ and $\mathbf{c}$ coincide. See Fig. 1.2. The value of $|\mathbf{a} \times \mathbf{b}|$ is equal to

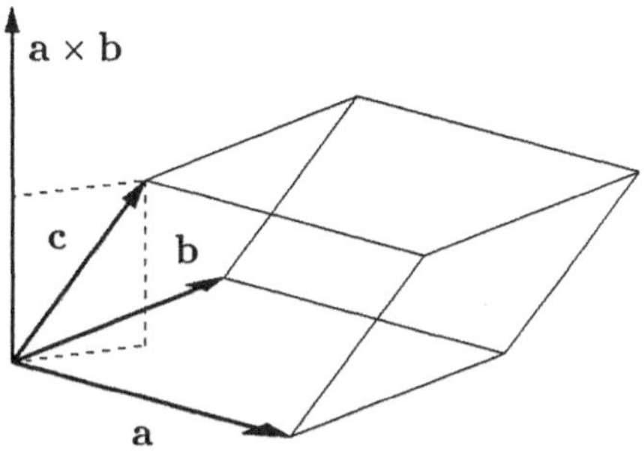

Figure 1.2 Scalar triple product

that of the area of a parallelogram whose adjoining sides are the vectors $\mathbf{a}$ and $\mathbf{b}$. The unit normal $\mathbf{n}$ of the parallelogram is given by

$$\mathbf{n} = \frac{(\mathbf{a} \times \mathbf{b})}{|\mathbf{a} \times \mathbf{b}|}. \tag{1.22}$$

Since the inner product $(\mathbf{n} \cdot \mathbf{c})$ is the normal component of $\mathbf{c}$, the expression $|\mathbf{a} \times \mathbf{b}|(\mathbf{n} \cdot \mathbf{c})$, which represents a product of the bottom area and the height of the parallelogram, is equal to the volume of the parallelpiped whose adjoining three edges are $\mathbf{a}$, $\mathbf{b}$ and $\mathbf{c}$. If its value is zero, the three vectors are coplanar. If it is positive, the vector $\mathbf{c}$ is in the front side of a plane determined by $\mathbf{a} \times \mathbf{b}$, if negative, it is in the rear side of the plane.

In the evaluation of the volume of the parallelpiped, the area defined by $\mathbf{a}$ and $\mathbf{b}$ are considered to be its bottom face. But as any one of the areas defined by two vectors of the three can equally well be taken as a bottom face, its volume is given also by $(\mathbf{b} \times \mathbf{c}) \cdot \mathbf{a}$ or by $(\mathbf{c} \times \mathbf{a}) \cdot \mathbf{b}$ as well as by $(\mathbf{a} \times \mathbf{b}) \cdot \mathbf{c}$. They are all denoted by $[\mathbf{a}, \mathbf{b}, \mathbf{c}]$ and are called the *scalar triple product* of the vectors and have the relations:

$$[\mathbf{a}, \mathbf{b}, \mathbf{c}] = [\mathbf{b}, \mathbf{c}, \mathbf{a}] = [\mathbf{c}, \mathbf{a}, \mathbf{b}] = -[\mathbf{b}, \mathbf{a}, \mathbf{c}]. \tag{1.23}$$

When any two element vectors of a scalar triple product have the same or opposite direction or when its three element vectors are coplanar, the value of the scalar triple product is zero. Its sign is used to determine in which side of a plane a given point exists.

1.4.2 Vector Triple Product

A *vector triple product* has forms $(\mathbf{a} \times \mathbf{b}) \times \mathbf{c}$ or $\mathbf{a} \times (\mathbf{b} \times \mathbf{c})$. These forms appear frequently in manipulation of vectors in geometric modelling. The direction of a vector $(\mathbf{a} \times \mathbf{b})$ is normal to the plane determined by $\mathbf{a}$ and $\mathbf{b}$. Therefore the direction of a vector $(\mathbf{a} \times \mathbf{b}) \times \mathbf{c}$ is normal to $(\mathbf{a} \times \mathbf{b})$, that is, it is in the plane defined by $\mathbf{a}$ and $\mathbf{b}$. Therefore, it is expressed by a linear combination of $\mathbf{a}$ and $\mathbf{b}$ such as

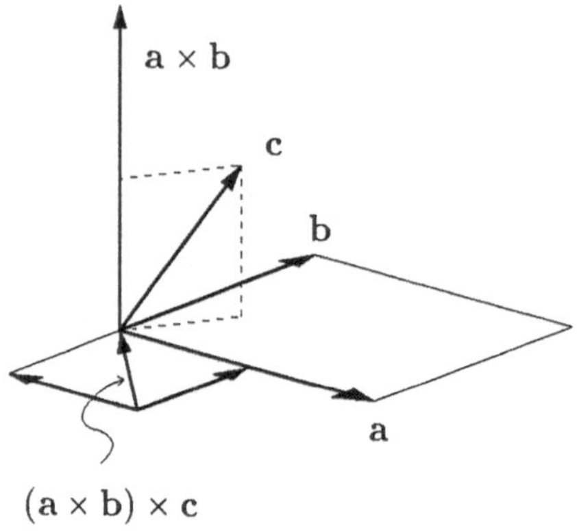

Figure 1.3 Vector triple product

$$(\mathbf{a} \times \mathbf{b}) \times \mathbf{c} = \lambda \mathbf{a} + \mu \mathbf{b}, \tag{1.24}$$

where λ and μ are scalar constants to be determined. See Fig. 3. By multiplying both sides of eq. (1.24) with the vector $\mathbf{c}$, the left-hand side becomes $[\mathbf{c}, \mathbf{a} \times \mathbf{b}, \mathbf{c}]$, which is zero, because of the two identical elements.

The right-hand side becomes

$$\lambda(\mathbf{a} \cdot \mathbf{c}) + \mu(\mathbf{b} \cdot \mathbf{c}) = 0. \tag{1.25}$$

The following relations satisfy the above equation:

$$\lambda = -\gamma(\mathbf{b} \cdot \mathbf{c}), \;\; \mu = \gamma(\mathbf{a} \cdot \mathbf{c}),$$

where γ is a constant independent of $\mathbf{a}$, $\mathbf{b}$ and $\mathbf{c}$. To determine γ, we set $\mathbf{a}$ and $\mathbf{b}$ orthogonal and $\mathbf{b}$ equal to $\mathbf{c}$, then we evaluate both sides of eq. (1.24) using the relations $\lambda = -\gamma \mathbf{b}^2$ and $\mu = 0$ which are obtained from the above setting, and determine that the constant γ is equal to 1. It follows that eq. (1.24) becomes

$$(\mathbf{a} \times \mathbf{b}) \times \mathbf{c} = (\mathbf{a} \cdot \mathbf{c})\mathbf{b} - (\mathbf{b} \cdot \mathbf{c})\mathbf{a}. \tag{1.26}$$

By changing the multiplication order and then exchanging the names of variables, we get

$$\mathbf{a} \times (\mathbf{b} \times \mathbf{c}) = (\mathbf{a} \cdot \mathbf{c})\mathbf{b} - (\mathbf{a} \cdot \mathbf{b})\mathbf{c}. \tag{1.27}$$

Another vector product formula may be obtained. Multiplying both sides of eq. (1.26) by a vector $\mathbf{d}$, we have

$$\{(\mathbf{a} \times \mathbf{b}) \times \mathbf{c}\} \cdot \mathbf{d} = (\mathbf{a} \cdot \mathbf{c})(\mathbf{b} \cdot \mathbf{d}) - (\mathbf{b} \cdot \mathbf{c})(\mathbf{a} \cdot \mathbf{d}).$$

Its left-hand side is a scalar triple product $[(\mathbf{a} \times \mathbf{b}), \mathbf{c}, \mathbf{d}]$. By applying eq. (1.23), it is equal to $(\mathbf{a} \times \mathbf{b}) \cdot (\mathbf{c} \times \mathbf{d})$. Therefore, the formula

$$(\mathbf{a} \times \mathbf{b}) \cdot (\mathbf{c} \times \mathbf{d}) = (\mathbf{a} \cdot \mathbf{c})(\mathbf{b} \cdot \mathbf{d}) - (\mathbf{b} \cdot \mathbf{c})(\mathbf{a} \cdot \mathbf{d}) \tag{1.28}$$

is obtained.

1.4.3 Application Examples

Example 1. Construction of a Coordinate System

The vector triple product is used to construct a coordinate system in a space. Let $\mathbf{n}$ be a normal unit vector to a plane determined by the vectors $\mathbf{a}$ and $\mathbf{b}$:

$$\mathbf{n} = \frac{\mathbf{a} \times \mathbf{b}}{|\mathbf{a} \times \mathbf{b}|}. \tag{1.29}$$

Then writing $\hat{\zeta} = \mathbf{n}$, $\hat{\xi} = \hat{\mathbf{a}}$, we make $\hat{\eta}$ orthogonal to $\hat{\zeta}$ and $\hat{\xi}$:

$$\hat{\eta} = \hat{\zeta} \times \hat{\xi} = \frac{(\mathbf{a} \times \mathbf{b}) \times \mathbf{a}}{|\mathbf{a} \times \mathbf{b}| \cdot |\mathbf{a}|}. \tag{1.30}$$

Using eq. (1.26), the numerator of the right-hand side of the above equation is changed to

$$(\mathbf{a} \times \mathbf{b}) \times \mathbf{a} = (\mathbf{a} \cdot \mathbf{a})\mathbf{b} - (\mathbf{b} \cdot \mathbf{a})\mathbf{a} = \mathbf{a}^2\mathbf{b} - (\mathbf{b} \cdot \mathbf{a})\mathbf{a},$$

so the expression for $\hat{\eta}$ is

$$\hat{\eta} = \frac{\hat{\mathbf{b}} - (\hat{\mathbf{b}} \cdot \hat{\mathbf{a}})\hat{\mathbf{a}}}{|\hat{\mathbf{a}} \times \hat{\mathbf{b}}|}.$$

The orthogonal unit vectors $\hat{\xi}$, $\hat{\eta}$ and $\hat{\zeta}$ thus determined make an orthogonal coordinate system.

Example 2. Use of Eq. (1.28)

Let there be three vectors $\mathbf{a}$, $\mathbf{b}$ and $\mathbf{c}$, whose tails coincide, let the plane defined by $\mathbf{a}$ and $\mathbf{b}$ be orthogonal to the plane defined by $\mathbf{b}$ and $\mathbf{c}$, and let the angles between $\mathbf{a}$ and $\mathbf{b}$, between $\mathbf{b}$ and $\mathbf{c}$ and between $\mathbf{a}$ and $\mathbf{c}$ be θ_1, θ_2 and θ_3 respectively. Then the following relation holds:

$$\cos\theta_3 = \cos\theta_1 \cdot \cos\theta_2.$$

To prove this, set $\mathbf{d} = \mathbf{b}$ in eq. (1.28); since $\mathbf{a} \times \mathbf{b}$ and $\mathbf{b} \times \mathbf{c}$ are orthogonal, the left side of eq. (1.28) is zero. By rewriting the inner product with absolute values and the cosines of the angles between the vectors concerned, the above relation is obtained.

1.4.4 Oblique Coordinate System

A vector is usually resolved into its components in an orthogonal coordinate system, but sometimes it is required to be resolved in an oblique coordinate system. Let $\mathbf{e}_1, \mathbf{e}_2$ and $\mathbf{e}_3$ be the unit vectors of axes of the coordinate system, let the angle between $\mathbf{e}_1$ and $\mathbf{e}_2$ be ω, and that between $\mathbf{e}_3$ and the normal of the plane made by $\mathbf{e}_1$ and $\mathbf{e}_2$ be θ.

Now we are to express an arbitrary unit vector $\mathbf{a}$ by its components in the oblique coordinate system. Since the vector $\mathbf{a}$ can be expressed as

$$\mathbf{a} = \lambda\mathbf{e}_1 + \mu\mathbf{e}_2 + \nu\mathbf{e}_3, \tag{1.31}$$

we determine λ, μ and ν in terms of $\mathbf{e}_1$, $\mathbf{e}_2$, $\mathbf{e}_3$, $\mathbf{a}$, θ and ω.

Multiplying both sides of eq. (1.31) by $\mathbf{e}_2 \times \mathbf{e}_3$, we obtain

$$\mathbf{a} \cdot (\mathbf{e}_2 \times \mathbf{e}_3) = \lambda\mathbf{e}_1 \cdot (\mathbf{e}_2 \times \mathbf{e}_3) = \lambda[\mathbf{e}_1, \mathbf{e}_2, \mathbf{e}_3],$$

and similarly

$$\begin{aligned} \mathbf{a} \cdot (\mathbf{e}_3 \times \mathbf{e}_1) &= \mu\mathbf{e}_2 \cdot (\mathbf{e}_3 \times \mathbf{e}_1) = \mu[\mathbf{e}_1, \mathbf{e}_2, \mathbf{e}_3], \\ \mathbf{a} \cdot (\mathbf{e}_1 \times \mathbf{e}_2) &= \nu\mathbf{e}_3 \cdot (\mathbf{e}_1 \times \mathbf{e}_2) = \nu[\mathbf{e}_1, \mathbf{e}_2, \mathbf{e}_3]. \end{aligned}$$

Since the triple product $[\mathbf{e}_1, \mathbf{e}_2, \mathbf{e}_3]$ is the volume of a parallelpiped of unit length edges, which is equal to $\sin\omega\cos\theta$, we have

$$\left.\begin{aligned} \lambda &= [\mathbf{a}, \mathbf{e}_2, \mathbf{e}_3]/\sin\omega\cos\theta, \\ \mu &= [\mathbf{a}, \mathbf{e}_3, \mathbf{e}_1]/\sin\omega\cos\theta, \\ \nu &= [\mathbf{a}, \mathbf{e}_1, \mathbf{e}_2]/\sin\omega\cos\theta. \end{aligned}\right\} \tag{1.32}$$

If $\omega = 90$ degrees and $\theta = 0$, then the unit vectors $\mathbf{e}_1$, $\mathbf{e}_2$ and $\mathbf{e}_3$ are mutually orthogonal. Accordingly, the above equations are reduced to

$$\lambda = \mathbf{a} \cdot \mathbf{e}_1, \ \ \mu = \mathbf{a} \cdot \mathbf{e}_2, \ \ \nu = \mathbf{a} \cdot \mathbf{e}_3.$$

1.5 Differentiation of Vectors

When a vector is a function of parameters, it is often necessary to differentiate the vector with respect to the parameters. For example, when $\mathbf{r}(t)$ represents a curve, $d\mathbf{r}(t)/dt$ is a vector expressing the tangent direction of the curve. This is obvious from the definition of differentiation.

When differentiating a vector product or an inner product of two vectors, the process is the same as differentiation of a product of scalar variables. But the order of differentiation must not be changed for the vector product, and we have

$$\frac{d(\mathbf{a} \cdot \mathbf{b})}{dt} = (\frac{d\mathbf{a}}{dt}) \cdot \mathbf{b} + \mathbf{a} \cdot (\frac{d\mathbf{b}}{dt}), \tag{1.33}$$

$$\frac{d(\mathbf{a} \times \mathbf{b})}{dt} = (\frac{d\mathbf{a}}{dt}) \times \mathbf{b} + \mathbf{a} \times (\frac{d\mathbf{b}}{dt}). \tag{1.34}$$

Examples.

Let $\mathbf{t}$, $\mathbf{b}$ and $\mathbf{n}$ be unit vectors orthogonal to each other and functions of a parameter s, which is an arc length, and let $\mathbf{b}$ be defined by $\mathbf{b} = \mathbf{t} \times \mathbf{n}$. Since

$\mathbf{t}^2 = 1$, we have $\mathbf{t}(d\mathbf{t}/ds) = 0$ which indicates that $d\mathbf{t}/ds$ is orthogonal to $\mathbf{t}$. Introducing a constant κ, we define $\mathbf{n}$ such that $d\mathbf{t}/ds = \kappa\mathbf{n}$. By differentiating the relations $\mathbf{b} \cdot \mathbf{b} = 1$ and $\mathbf{t} \cdot \mathbf{b} = 0$ with respect to s, we obtain

$$\mathbf{b} \cdot \frac{d\mathbf{b}}{ds} = 0, \quad \frac{d\mathbf{t}}{ds} \cdot \mathbf{b} + \mathbf{t} \cdot \frac{d\mathbf{b}}{ds} = 0.$$

As $d\mathbf{t}/ds = \kappa\mathbf{n}$, we have the relation $(d\mathbf{t}/ds) \cdot \mathbf{b} = 0$ and hence we get $\mathbf{t} \cdot (d\mathbf{b}/ds) = 0$. This means $(d\mathbf{b}/ds)$ is orthogonal to $\mathbf{t}$ as well as to $\mathbf{b}$, so it has a component of $\mathbf{n}$ only. Introducing a proportional constant $-\tau$, we can express it as

$$\frac{d\mathbf{b}}{ds} = -\tau\mathbf{n}.$$

Making the vector product of $\mathbf{t}$ with both sides of $\mathbf{b} = \mathbf{t} \times \mathbf{n}$, and using eq. (1.26), we get a relation $\mathbf{n} = \mathbf{b} \times \mathbf{t}$. Differentiating this by s, we obtain

$$\frac{d\mathbf{n}}{ds} = \frac{d\mathbf{b}}{ds} \times \mathbf{t} + \mathbf{b} \times \frac{d\mathbf{t}}{ds},$$

which is converted to

$$\frac{d\mathbf{n}}{ds} = -\tau\mathbf{n} \times \mathbf{t} + \kappa\mathbf{b} \times \mathbf{n} = \tau\mathbf{b} - \kappa\mathbf{t}.$$

This equation is called Serret-Frenet's formula. See Sect. 5.3.

1.6 Matrix Notations and Simple Arithmetic of Matrices

Matrix notations and operations are used in this book for coordinate transformations, displacements of solid objects, and for expressions denoting objects. The reason is that they are concise and systematic as well as visually understandable. So, we explain their notations and some of operations in this section.

Letting m and n be integers, $m \times n$ real numbers or vectors or symbols of the same attribute are arranged in m rows and in n columns enclosed by a pair of brackets as shown below:

$$A = \begin{bmatrix} a_{11} & a_{12} & \cdots & a_{1n} \\ a_{21} & a_{22} & \cdots & a_{2n} \\ \vdots & \vdots & \ddots & \vdots \\ a_{m1} & a_{m2} & \cdots & a_{mn} \end{bmatrix}. \tag{1.35}$$

The right-hand side arrangement is called a matrix of size $m \times n$ and a_{ij} are elements of the matrix. Matrices can have names: the name of the above matrix is A. For simplicity, an notation $A = [a_{ij}]$ may be used.

A vector of m components can be thought of a matrix of size $1 \times m$, or its transpose is a matrix of size $m \times 1$.

Let the i-th row of the matrix A be denoted by $\mathbf{a}_i$:

$$\mathbf{a}_i = \begin{pmatrix} a_{i1} & a_{i2} & \dots & a_{in} \end{pmatrix}, \quad (1 \le i \le m). \tag{1.36}$$

Then the matrix A can be written:

$$A = \begin{bmatrix} \mathbf{a}_1 \\ \mathbf{a}_2 \\ \vdots \\ \mathbf{a}_m \end{bmatrix}. \tag{1.37}$$

But to save space and to type it easily, the following expression is frequently used:

$$A = \begin{bmatrix} \mathbf{a}_1 & \mathbf{a}_2 & \dots & \mathbf{a}_m \end{bmatrix}^T. \tag{1.38}$$

Let each column of the matrix A be a transposed vector or matrix of $m \times 1$:

$$\mathbf{a}_j'^T = \begin{pmatrix} a_{1j} & a_{2j} & \dots & a_{mj} \end{pmatrix}^T, \quad (1 \le j \le n). \tag{1.39}$$

Then the matrix A can be denoted by

$$A = [\ \mathbf{a}_1'^T \quad \mathbf{a}_2'^T \quad \dots \quad \mathbf{a}_n'^T\ ]. \tag{1.40}$$

When the size of matrices A, B and C are the same, the following arithmetic operations:

$$C = A \pm B, \quad C = cA, \tag{1.41}$$

are defined in such a way that the elements of C are equal to the arithmetic operations between the corresponding elements of the matrices A and B,

$$c_{ij} = a_{ij} \pm b_{ij}, \quad c_{ij} = ca_{ij}\ , \tag{1.42}$$

where c is a scalar constant.

1.7 Products of Matrices

1.7.1 Multiplication of a Vector and a Matrix

Multiplying a vector $\mathbf{b}$ of m components by a matrix A of $m \times n$, we get a vector $\mathbf{c}$ of n components,

$$\mathbf{c} = (c_1, c_2, c_3, \cdots c_n) = \mathbf{b}A. \tag{1.43}$$

A component of $\mathbf{c}$ corresponds to an inner product of the vector $\mathbf{b}$ and the corresponding column vector of the expression eq. (1.40) of A.

Since they are matrices of size $1 \times m$ and size $m \times 1$, their inner product is denoted by $\mathbf{b}\mathbf{a}_j'^T$. Using this notation and eq. (1.40), we can express eq. (1.43) in the following form,

$$\mathbf{c} = \mathbf{b}A = \begin{pmatrix} \mathbf{b}\mathbf{a}_1'^T & \mathbf{b}\mathbf{a}_2'^T & \cdots & \mathbf{b}\mathbf{a}_n'^T \end{pmatrix}. \tag{1.44}$$

Comparing eq. (1.43) and (1.44), we have

$$c_i = \mathbf{b}\mathbf{a}_i'^T, \quad (1 \le i \le n). \tag{1.45}$$

Example. Coordinate transformation

When a position vector $\mathbf{p}$ is defined on the coordinate system C1 and it is required to indicate this fact explicitly, $\mathbf{p}$ and its components are written with a superscript indicating the coordinate system C1, such as

$$\mathbf{p}^{(1)} = (p_x,\ p_y,\ p_z)^{(1)}.$$

Usually the superscript is omitted when there is no confusion. If the same point $\mathbf{p}$ is to be described by the coordinate system C2, there arise the necessity of coordinate transformation from the system C1 to the system C2. The vector $\mathbf{p}^{(1)}$ is transformed to $\mathbf{p}^{(2)}$ by multiplying the transformation matrix T_{12} with $\mathbf{p}^{(1)}$,

$$\mathbf{p}^{(2)} = \mathbf{p}^{(1)} T_{12}. \tag{1.46}$$

The detailed description on the transformation is given in the next chapter.

1.7.2 Product of Matrices

The product of a matrix B of size $k \times m$ and a matrix A of size $m \times n$ produces a matrix of size $k \times n$. For $k = 1$, the multiplication is explained in the previous section. This is extended to any integer k. Let the product matrix be C. Then we use the notation:

$$C = BA. \tag{1.47}$$

As in eq. (1.38), B is composed of k row vectors $\mathbf{b}_j$ of m elements and C of k row vectors $\mathbf{c}_j$ of n elements, $(1 \le j \le k)$:

$$B = [\ \mathbf{b}_1\ \ \mathbf{b}_2\ \ \cdots\ \ \mathbf{b}_k\ ]^T, \quad C = [\ \mathbf{c}_1\ \ \mathbf{c}_2\ \ \cdots\ \ \mathbf{c}_k\ ]^T. \tag{1.48}$$

Since BA is $[\ \mathbf{b}_1\ \ \mathbf{b}_2\ \ \cdots\ \ \mathbf{b}_k\ ]^T A$, the product C becomes

$$C = [\ \mathbf{b}_1 A\ \ \mathbf{b}_2 A\ \ \cdots\ \ \mathbf{b}_k A\ ]^T. \tag{1.49}$$

Hence, the i-th row of C is

$$\mathbf{c}_i = \mathbf{b}_i A. \tag{1.50}$$

The j-th component c_{ij} of $\mathbf{c}_i$ is equal to an inner product of a vector $\mathbf{b}_i$ and the j-th column vector of A.

An element c_{ij} of C, which is the product of two matrices A and B, is constructed from the inner product of i-th row vector of the matrix B and j-th column vector of the matrix A.

Example.

Let a matrix T_{12} be the coordinate transformation matrix from the system C1 to the system C2, and T_{23} be the coordinate transformation matrix from the system C2 to the system C3. The coordinate transformation matrix T_{13} from the system C1 to the system C3 is the product of T_{12} and T_{23}, such as

$$T_{13} = T_{12} T_{23}. \tag{1.51}$$

1.8 Square Matrix, Inverse Matrix and Other Related Matrices

A matrix whose numbers of rows and columns are equal is called a *square matrix*. A unit matrix is one whose diagonal elements are all 1 and the other elements are all zero. The unit matrix I has the following properties,

$$\mathbf{b}I = \mathbf{b}, \quad AI = IA = A. \tag{1.52}$$

The unit matrix I corresponds to 1 in algebraic multiplication.

When product of two square matrices A and B is the unit matrix, A is called the *inverse matrix* of B or vice versa. A matrix whose rows and columns are interchanged from the original matrix A is called the *transposed matrix* of A, and is denoted by A^T. Considering a vector $\mathbf{b}$ as $1 \times n$ matrix, $\mathbf{b}^T$ is an $n \times 1$ matrix or a column vector.

The transposed matrix of a product AB is equal to the product of the transposed matrices B^T and A^T, that is,

$$(AB)^T = B^T A^T. \tag{1.53}$$

Multiplying a vector $\mathbf{b}$ with a matrix B generates a vector:

$$\mathbf{c} = \mathbf{b}B. \tag{1.54}$$

This is the form of the post-multiplication of a matrix.

When the pre-multiplication form is required, transposition of both sides of eq. (1.54) gives

$$\mathbf{c}^T = (\mathbf{b}B)^T = B^T\mathbf{b}^T = C\mathbf{b}^T, \tag{1.55}$$

where B^T is denoted by C.

Example 1. Matrix expressions of conics

The general expression for a conic is given by the following implicit equation:

$$ax^2 + 2hxy + by^2 + 2gx + 2fy + d = 0, \tag{1.56}$$

where a, b, d, g and h are constants. This can be expressed in matrix form as

$$(x\ y\ 1) \cdot \begin{bmatrix} a & h & g \\ h & b & f \\ g & f & d \end{bmatrix} \cdot (x\ y\ 1)^T = 0. \tag{1.57}$$

Example 2. An expression for a *tensor product surface* patch

Let each element $\mathbf{p}_{ij}$ of a $n \times n$ square matrix C represent a characteristic value or a vector, such as a control point, of a surface patch. By multiplying $\mathbf{p}_{ij}$ by a weighting function $f_i(u)f_j(v)$ and taking the sum of all of them, we obtain a surface patch expression, where u and v are the parameters. To simplify the expression and its numerical calculation, instead of the double summation $\sum\sum f_i(u)f_j(v)\mathbf{p}_{ij}$, the equivalent matrix representation is frequently used. Let $\mathbf{b}(u)$ be a vector whose components are $f_i(u),(i = 1\cdots n)$ and $\mathbf{b}(v)$ be a vector whose components are $f_j(v),(j = 1\cdots n)$, then the expression of the surface patch is formally written as

$$\begin{aligned} \mathbf{s}(u,v) &= \mathbf{b}(v)\cdot C\cdot \mathbf{b}(u)^T = \mathbf{b}(u)\cdot C\cdot \mathbf{b}(v)^T, \\ \text{where} \quad C &= [\mathbf{p}_{ij}]. \end{aligned} \tag{1.58}$$

This is one of expressions for a tensor product surface patch. The derivation of eq. (1.58) is explained in Chap. 9.

The *inverse matrix* of a square matrix is a square matrix which produces the unit matrix I by its multiplication with the original square matrix. If the relation

$$AB = BA = I \tag{1.59}$$

holds, a matrix A is called the inverse of a matrix B, or matrix B is the inverse of matrix A. They are denoted by

$$B = A^{-1}, \quad A = B^{-1}. \tag{1.60}$$

In this case, the matrices A and B are non-singular. The inverse matrix does not always exist: if value of determinant of a square matrix is zero, the matrix is called singular.

When a vector $\mathbf{b}$ is unknown, a vector $\mathbf{c}$ and a matrix B are known, and there exists a relation:

$$\mathbf{c} = \mathbf{b}B, \tag{1.61}$$

the unknown vector $\mathbf{b}$ can be obtained by post-multiplying the inverse matrix B^{-1}, if it exists, with both sides of eq. (1.61):

$$\mathbf{c}B^{-1} = \mathbf{b}BB^{-1} = \mathbf{b}I = \mathbf{b}. \tag{1.62}$$

This is the solution of eq. (1.61).

The inverse of a product of two matrices is equal to

$$(BA)^{-1} = A^{-1}B^{-1}. \tag{1.63}$$

When the inverse and the transposed matrices of a given matrix are the same, then the given matrix is called the *orthogonal matrix*. A coordinate transformation matrix of rotation about any axis is the orthogonal matrix. For example, let $T_{1,2}$ be the transformation matrix of rotation around the z axis, which is given by

$$T_{12} = \begin{bmatrix} \cos\theta & -\sin\theta & 0 \\ \sin\theta & \cos\theta & 0 \\ 0 & 0 & 1 \end{bmatrix}, \tag{1.64}$$

then the relation $T_{12}^{-1} = T_{21} = T_{12}^T$ holds. Since a rotation matrix about any axis is expressed by a product of three orthogonal matrices of rotation around the respective coordinate axes, its inverse is equal to its transpose because of equations (1.53) and (1.63).

1.9 Principal Directions and Eigenvalues

If a product of a vector **e** and a square matrix A gives a vector of the same direction as that of **e**, this is called the principal direction and the vector is called the eigenvector of the matrix A. It is determined from the equation given by $\mathbf{e}A = \lambda\mathbf{e}$, that is,

$$\mathbf{e}(A - \lambda I) = 0, \tag{1.65}$$

where λ is the constant to be determined. For the existence of a non-null vector **e** , an equation

$$det(A - \lambda I) = 0 \tag{1.66}$$

must hold. This is an algebraic equation for λ: the characteristic equation of the matrix A.

When A is symmetric, the equation (1.65) has real roots which are called the *eigenvalues* of A. For each eigenvalue, the equation (1.32) gives the ratio of the components of **e**, from which the principal directions, the *eigenvectors*, are determined. The eigenvectors are proved to be orthogonal each other.

Example.

$$A = \begin{bmatrix} a & h \\ h & b \end{bmatrix},$$

and $\mathbf{e} = (e_x, e_y)$. Then $\mathbf{e}(A - \lambda I) = 0$ is resolved into its components:

$$(a - \lambda)e_x + he_y = 0, \quad he_x + (b - \lambda)e_y = 0. \tag{1.67}$$

If these simultaneous equations have solutions other than $e_x = e_y = 0$,

$$det(A - \lambda I) = \begin{vmatrix} (a - \lambda) & h \\ h & (b - \lambda) \end{vmatrix} = \lambda^2 - (a + b)\lambda + ab - h^2 = 0 \tag{1.68}$$

must hold. The above equation is a quadratic equation for λ and its roots λ_1,λ_2 are given by

$$\lambda_1, \lambda_2 = \{a + b \pm \sqrt{(a - b)^2 + 4h^2}\}/2. \tag{1.69}$$

Corresponding to these λ_1, λ_2, the directions of the eigenvectors $\mathbf{e}_1(e_{x1}, e_{y1})$ and $\mathbf{e}_2(e_{x2}, e_{y2})$ can be determined from eq. (1.67) by

$$\frac{e_{yi}}{e_{xi}} = \frac{(\lambda_i - a)}{h} = \frac{h}{(\lambda_i - b)}, \quad i = 1, 2. \tag{1.70}$$

From equations (1.69) and (1.70), we obtain

$$\frac{e_{y1}}{e_{x1}}\frac{e_{y2}}{e_{x2}} = \frac{(\lambda_1 - a)}{(\lambda_2 - b)} = -1. \tag{1.71}$$

Accordingly, the inner product of $\mathbf{e}_1$ and $\mathbf{e}_2$ becomes zero:

$$e_{x1} \cdot e_{x2} + e_{y1} \cdot e_{y2} = 0. \tag{1.72}$$

Therefore $\mathbf{e}_1$ and $\mathbf{e}_2$ are orthogonal.

2 Coordinate Transformations and Displacements

2.1 Introduction

In CAD and CAM, objects to be designed and manufactured have to be modelled in the computer. In dealing with such models, geometric relations among the objects have to be described and their calculations systematically performed. Vector and matrix operations are very useful in these tasks. Among their applications, coordinate transformation and the movements of solid bodies are the fundamental tools in treating 3D objects mathematically. We explain them in this chapter. A position vector **p** of a point is denoted by $\mathbf{p}^{(1)}$ to indicate that it is seen from the coordinate system C1. The same is applied for its components:

$$\mathbf{p}^{(1)} = (x, y, z)^{(1)}. \tag{2.1}$$

When this point is to be seen from the coordinate system C2, it must be written in the same form as eq. (2.1):

$$\mathbf{p}^{(2)} = (x, y, z)^{(2)}. \tag{2.2}$$

We have to establish the relation between $\mathbf{p}^{(1)}$ and $\mathbf{p}^{(2)}$. When a body moves to other location changing its orientation, a point $\mathbf{p}^{(1)}$ of the body moves to $\mathbf{q}^{(1)}$ seen from the same coordinate system. We have to also establish their relation.

Recognition and manipulation of the geometrical relations of objects in 3D space are rather difficult tasks for a person without using real models or other suitable visualizing methods. So, we formalize the mathematics used in three-dimensional movements of bodies and coordinate transformations in a systematic way so that they may easily be applied for computer processing.

In the last section of this chapter, simulation of the displacement of objects by mirror reflections is treated, because sometimes movements of a body in space are easily understood by analogy with mirror reflections.

2.2 Coordinate Transformation Matrix 1

Suppose the coordinate system C2 is set by rotating the coordinate system C1 around a line through its origin by a given angle. A case of translation with the rotation of the system C2 is treated afterwards.

Since there is no translation, the components of $\mathbf{p}^{(2)}$ can be expressed by a linear combination of the components of $\mathbf{p}^{(1)}$, that is, the former is obtained from the latter by multiplying with a 3×3 matrix, which is called the coordinate transformation matrix T_{12}. So we have a relation:

$$\mathbf{p}^{(2)} = \mathbf{p}^{(1)} T_{12}. \tag{2.3}$$

The position $\mathbf{p}^{(1)}$ must be derived by formally applying T_{21} to $\mathbf{p}^{(2)}$:

$$\mathbf{p}^{(1)} = \mathbf{p}^{(2)} T_{21}. \tag{2.4}$$

Replacing $\mathbf{p}^{(1)}$ of eq. (2.3) by eq. (2.4), we get

$$\mathbf{p}^{(2)} = \mathbf{p}^{(2)} T_{21} T_{12}. \tag{2.5}$$

From this the following relation must hold,

$$T_{21} T_{12} = I, \tag{2.6}$$

where I is the unit matrix of size 3×3. The transformation matrix T_{21} is the inverse of T_{12}:

$$T_{21} = T_{12}^{-1}. \tag{2.7}$$

Seeing the same point $\mathbf{p}$ from the coordinate system C3, we have the following relation:

$$\mathbf{p}^{(3)} = \mathbf{p}^{(1)} T_{13} = \mathbf{p}^{(2)} T_{23}. \tag{2.8}$$

Replacing $\mathbf{p}^{(2)}$ in the right-hand side of eq. (2.8) by eq. (2.3), we get

$$\mathbf{p}^{(1)} T_{13} = \mathbf{p}^{(1)} T_{12} T_{23}, \tag{2.9}$$

from which we obtain a law of multiplication of the transformation matrices,

$$T_{13} = T_{12} T_{23}. \tag{2.10}$$

Taking its inverse and applying the relation (2.7), we have

$$T_{13}^{-1} = T_{23}^{-1} T_{12}^{-1}, \quad T_{31} = T_{32} T_{21}. \tag{2.11}$$

This is the same relation as eq. (2.10). We can state that in the sequence of coordinate transformations, the intermediate coordinates have no effect. So, its general form is

$$T_{ij} = T_{i(k_1)} T_{(k_1)(k_2)} \cdots T_{(k_{n-1})(k_n)} T_{(k_n)j}. \tag{2.12}$$

2.3 Calculation of Transformation Matrix

The size of a transformation matrix T_{12} is 3×3. Let the matrix T_{12} be

$$T_{12} = \begin{bmatrix} t_{11} & t_{12} & t_{13} \\ t_{21} & t_{22} & t_{23} \\ t_{31} & t_{32} & t_{33} \end{bmatrix}. \tag{2.13}$$

To determine its elements t_{ij}, we set $\mathbf{p}^{(1)} = \mathbf{i}_1^{(1)} = (1,0,0)^{(1)}$. Then using eq. (2.3), we obtain the first row of eq. (2.13):

$$\begin{aligned} \mathbf{i}_1^{(2)} &= (\ t_{11}\ \ t_{12}\ \ t_{13}\ ) \\ &= (\ (\mathbf{i}_1 \cdot \mathbf{i}_2)\ \ (\mathbf{i}_1 \cdot \mathbf{j}_2)\ \ (\mathbf{i}_1 \cdot \mathbf{k}_2)\ ). \end{aligned}$$

Similarly, for the second and third rows:

$$\begin{aligned} \mathbf{j}_1^{(2)} &= (\ (\mathbf{j}_1 \cdot \mathbf{i}_2)\ \ (\mathbf{j}_1 \cdot \mathbf{j}_2)\ \ (\mathbf{j}_1 \cdot \mathbf{k}_2)\ ), \\ \mathbf{k}_1^{(2)} &= (\ (\mathbf{k}_1 \cdot \mathbf{i}_2)\ \ (\mathbf{k}_1 \cdot \mathbf{j}_2)\ \ (\mathbf{k}_1 \cdot \mathbf{k}_2)\ ). \end{aligned}$$

Thus the matrix T_{12} can be written :

$$T_{12} = [\ \mathbf{i}_1^{(2)}\ \ \mathbf{j}_1^{(2)}\ \ \mathbf{k}_1^{(2)}\ ]^T = [\ \mathbf{i}\ \ \mathbf{j}\ \ \mathbf{k}\ ]_1^{(2)T}. \tag{2.14}$$

This expression means that values of elements of the transformation matrix T_{12} are obtained from those of a transposed matrix of the basis vectors of the system C1, which are seen from the system C2. Similarly, T_{21} is given by

$$T_{21} = [\ \mathbf{i}\ \ \mathbf{j}\ \ \mathbf{k}\ ]_2^{(1)T}. \tag{2.15}$$

This matrix T_{21} is the inverse of T_{12}, and also the transpose of T_{12}. The transformation matrix, therefore, is the orthogonal matrix.

2.4 Coordinate Transformation Matrix 2

We treat a case where the origins of the systems C1 and C2 do not coincide. The origin $\mathbf{o}_1$ of C1 has non-zero components seen from the system C2:

$$\mathbf{o}_1^{(2)} = (\ t_{41}\ \ t_{42}\ \ t_{43}\ )_1^{(2)}. \tag{2.16}$$

In order to express the transformation in the same form as eq. (2.3), equations (2.1), (2.2) and (2.16) are extended to the following forms,

$$\mathbf{p}^{(1)} = (\ x\ \ y\ \ z\ \ 1\ )^{(1)}, \tag{2.17}$$

$$\mathbf{p}^{(2)} = (\ x\ \ y\ \ z\ \ 1\ )^{(2)}, \tag{2.18}$$

$$\mathbf{o}_1^{(2)} = (\ t_{41}\ \ t_{42}\ \ t_{43}\ \ 1\ )_1^{(2)}, \tag{2.19}$$

then the transformation matrix T_{12} takes the form:

$$T_{12} = \begin{bmatrix} t_{11} & t_{12} & t_{13} & 0 \\ t_{21} & t_{22} & t_{23} & 0 \\ t_{31} & t_{32} & t_{33} & 0 \\ t_{41} & t_{42} & t_{43} & 1 \end{bmatrix}. \tag{2.20}$$

The relations (2.5) – (2.12) hold without modification, equations (2.14) and (2.16) are modified to

$$T_{12} = [\ \mathbf{i}\ \ \mathbf{j}\ \ \mathbf{k}\ \ \mathbf{o}\ ]_1^{(2)T}, \tag{2.21}$$

$$T_{21} = [\ \mathbf{i}\ \ \mathbf{j}\ \ \mathbf{k}\ \ \mathbf{o}\ ]_2^{(1)T}. \tag{2.22}$$

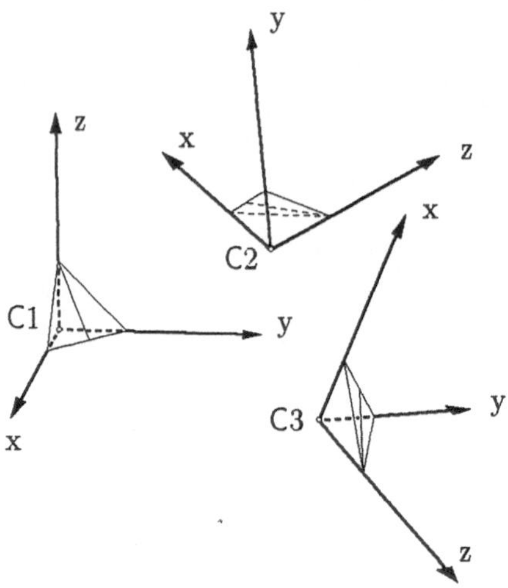

Figure 2.1 Displacement of a body

2.5 Movement and Coordinate Transformations

In the previous section a point is fixed, but the coordinate systems C1 and C2 are different, or the coordinate system C1 moves to a different position by rotation and translation to become the system C2. In this section, a coordinate system C1 is fixed, but a body B moves to another position by rotation and translation. We are to formulate its movement.

A coordinate system C2 fixed in a body B becomes the coordinate system C3 when the body B moves to another location. See Fig. 2.1. Any point of the body seen from the body coordinate systems C2 and C3 does not change though its position $\mathbf{p}$ changes to $\mathbf{q}$ if seen from the coordinate system C1. So, we can express

$$\mathbf{p}^{(2)} = \mathbf{q}^{(3)}. \tag{2.23}$$

If we see $\mathbf{q}$ from the system C2, we consider $\mathbf{q}^{(2)}$ has moved from $\mathbf{p}^{(2)}$ and we describe this fact by using the movement matrix S,

$$\mathbf{q}^{(2)} = \mathbf{p}^{(2)}S. \tag{2.24}$$

On the other hand, multiplying T_{32} with eq. (2.23) to get $\mathbf{q}^{(2)}$, we have

$$\mathbf{q}^{(2)} = \mathbf{q}^{(3)}T_{32} = \mathbf{p}^{(2)}T_{32}. \tag{2.25}$$

Comparing the above two equations (2.24) and (2.25), we have

$$S = T_{32}. \tag{2.26}$$

We can state that the value of S, which is the movement matrix of a body n its initial coordinate system C2, is the same as that of the coordinate transormation matrix from C3 to C2.

If we use another coordinate system C1 in eq. (2.24), we can change eq. (2.25) o

$$\mathbf{q}^{(1)}T_{12} = \mathbf{p}^{(1)}T_{12}S, \tag{2.27}$$

rom which we have

$$\mathbf{q}^{(1)} = \mathbf{p}^{(1)}T_{12} \cdot S \cdot T_{21}. \tag{2.28}$$

This is the formula of the movement matrix in the system C1. We use $S^{(1)}$ in lace of $T_{12} \cdot S \cdot T_{21}$, then the above equation becomes

$$\mathbf{q}^{(1)} = \mathbf{p}^{(1)}S^{(1)}. \tag{2.29}$$

ince S introduced in eq. (2.24) is the movement matrix in the system C2, it hould be written $S^{(2)}$, and the coordinate transformation of S from the system C2 to C1 becomes

$$S^{(1)} = T_{12} \cdot S^{(2)} \cdot T_{21}. \tag{2.30}$$

When there are successive movements of a body, which include rotations, this an be expressed mathematically by the successive multiplication of relevant S natrices. As it is rather difficult to grasp this kind of movement visually, the ther approach which uses the mirror (reflection) transformation is described in Sect. 2.7.

2.6 Application Examples

2.6.1 Successive Rotations in Space

n rotation of a body in space, the coordinate system attached to the body also otates, so human understanding of the motion of the body becomes difficult. But rotation around a coordinate axis is easier to grasp, and also construction of a mechanism of rotation around a fixed axis is easier, so in most cases general otation is resolved into successive application of single axis rotations.

With the coordinate system attached to a body, at first the body rotates around its z axis by an angle ψ in the direction from the x to the y axes, then it otates around its y axis by an angle θ in the direction from the z to the x axes, inally it rotates around its x axis by an angle ϕ from the y to the z axes.

Let consider the coordinate systems C0, C1, C2 and C3, the origins of which coincide and their definitions are (refer to Fig. 2.2).

C0: the initial coordinate system fixed in the space.
C1: the system defined by a rotation of C0 around its z axis by an angle ψ.
C2: the system defined by a rotation of C1 around its y axis by an angle θ.
C3: the system defined by a rotation of C2 around its x axis by an angle ϕ.

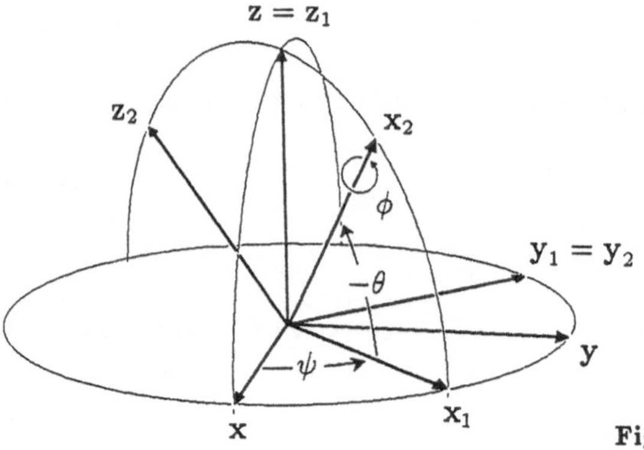

Figure 2.2 Rotation of coordinates

Then the transformation matrices from one system to the next are constructed using eq. (2.14):

$$T_{01}(\psi : z) = \begin{bmatrix} \cos\psi & -\sin\psi & 0 \\ \sin\psi & \cos\psi & 0 \\ 0 & 0 & 1 \end{bmatrix}, \tag{2.31}$$

$$T_{12}(\theta : y) = \begin{bmatrix} \cos\theta & 0 & \sin\theta \\ 0 & 1 & 0 \\ -\sin\theta & 0 & \cos\theta \end{bmatrix}, \tag{2.32}$$

$$T_{23}(\phi : x) = \begin{bmatrix} 1 & 0 & 0 \\ 0 & \cos\phi & -\sin\phi \\ 0 & \sin\phi & \cos\phi \end{bmatrix}. \tag{2.33}$$

The transformation matrix of the synthesized rotations is given by

$$T_{03} = T_{01}(\psi : z)T_{12}(\theta : y)T_{23}(\phi : x). \tag{2.34}$$

For example, for $\psi = 45°, \theta = 60°, \phi = 30°$,

$$T_{03} = \begin{bmatrix} 1/2\sqrt{2} & -\sqrt{3}/4\sqrt{2} & 5/4\sqrt{2} \\ 1/2\sqrt{2} & 3\sqrt{3}/4\sqrt{2} & 1/4\sqrt{2} \\ -\sqrt{3}/2 & 1/4 & \sqrt{3}/4 \end{bmatrix}. \tag{2.35}$$

Its transpose is equal to the inverse, because it is the orthogonal matrix. The squared sum of each row or column of the above matrix is equal to 1, because each row or column of the above matrices is the unit vector as are shown in eq. (2.14). When the origins of C0 and C1 do not coincide, T_{03} becomes a 4×4 matrix, whose 4th row is the vector of the origin of C0 seen from C3 and the 4th column is the vector $(0\ 0\ 0\ 1)^T$.

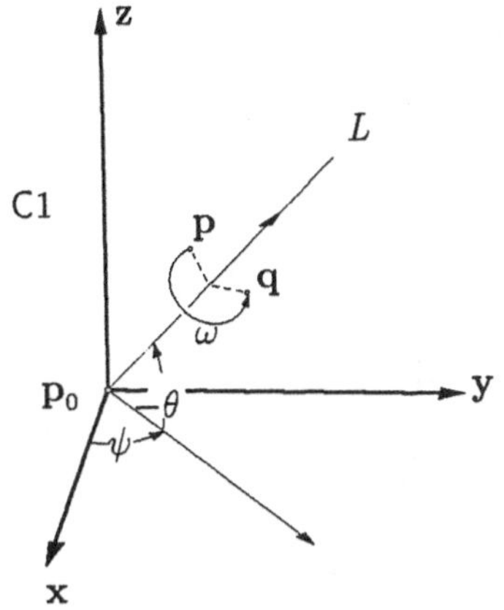

Figure 2.3 Rotation of a body around a line in space

2.6.2 Rotation of a Body Around a Line in Space

When a body B rotates by an angle ω around a line L which passes a fixed point $\mathbf{p}_0$ in the space, we are to determine its final position B by using the coordinate transformation.

Refer to Fig. 2.3. We use coordinate systems C0,C1,C2 and C3 explained below.

C0: the original coordinate system in the space.
C1: The origin is at $\mathbf{p}_0$ and the axes are parallel to those of C0.
C2: the system made by C1 rotating around its z axis by an angle ψ, which is the azimuth angle of the line L.
C3: the system made by C2 rotating around its y axis by an angle θ, which is the elevation angle of L.

In the system C3, the body rotates around the x axis by an angle ω. The final position is seen from C1 and then from C0. Let $\mathbf{p}^{(0)}$ be a point of the body B. In C1 it is $\mathbf{p}^{(1)} = \mathbf{p}^{(0)} - \mathbf{p}_0^{(0)}$. After the rotation, $\mathbf{p}^{(1)}$ moves to $\mathbf{q}^{(1)}$, which is expressed by $\mathbf{p}^{(1)}S^{(1)}$. From eq. (2.30), $S^{(1)}$ is equal to $(T_{13} \cdot S^{(3)} \cdot T_{31})$. So we have

$$\mathbf{q}^{(1)} = \mathbf{p}^{(1)}(T_{13} \cdot S^{(3)} \cdot T_{31}). \tag{2.36}$$

Since $T_{13} = T_{12}T_{23}$, which is given by $T(\psi : z) \cdot T(\theta : y)$, and $S^{(3)}$ is equal to $T(-\omega : x)$ for it is a rotation around the x axis, and $\mathbf{q}^{(1)} = \mathbf{q}^{(0)} - \mathbf{p}_0^{(0)}$, the final position $\mathbf{q}^{(0)}$ is given by

$$\mathbf{q}^{(0)} = \mathbf{p}_0^{(0)} + (\mathbf{p}^{(0)} - \mathbf{p}_0^{(0)})T(\psi : z) \cdot T(\theta : y) \cdot T(-\omega : x) \cdot T(-\theta : y) \cdot T(-\psi : z). \tag{2.37}$$

Exercise. Successive 90-degree rotations of a cube.

The center of a cube of edge size 2 is placed at the origin of the fixed coordinate system with its edges parallel to the coordinate axes. Let rotations of

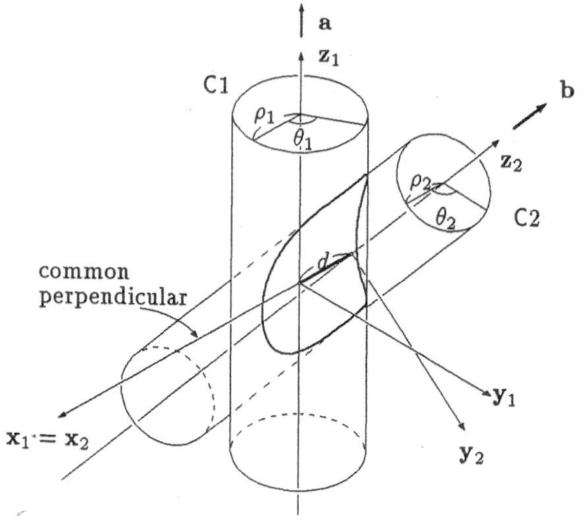

Figure 2.4 Interference of two cylinders

$\pm$ 90 degrees around three axes be written X, X', Y, Y', Z and Z'. Show the following relations by experiments or calculations:

(i) $XX' = I$, $X^4 = Y^4 = Z^4 = I$,
where I is the unit matrix and $X^2 = XX$.
(ii) $XZ = YX = ZY$, $XY = YZ' = Z'X$, $X = YXZ' = ZYZ'$,
$XYZY'ZX' = ZX'$.

2.6.3 Calculation of Geometric Constraints

In calculation of intersection curves of surfaces or motions of mechanisms, selection of the appropriate coordinate systems is necessary for a simple description of the objects concerned. With coordinate transformation matrices, relations among the variables of both the bodies or components are established and the equations to be solved are determined. An example is given below.

The intersection curve(s) of two circular cylinders is (are) to be determined. Let the radii of two cylinders be ρ_1 and ρ_2, unit vectors of the axes of the cylinders be $\mathbf{a}$ and $\mathbf{b}$, and their distance be d. Coordinate systems C1 and C2 are used for the two cylinders.

Referring to Fig. 2.4, the center axes are the z axes of both systems, the common perpendicular of the center axes is the x axis common to C1 and C2, and the y axis is chosen to be orthogonal to both the z and x axes. The parameter θ_i is an angle between the xz plane and a radial plane of a cylinder i. Each cylinder is expressed as

$$\mathbf{r}_i^{(i)} = (\ \rho_i \cos\theta_i \quad \rho_i \sin\theta_i \quad z_i \quad 1\ )^{(i)}, \tag{2.38}$$

where i is 1 or 2. On their intersection curve, a relation

$$\mathbf{r}_1^{(1)} T_{12} = \mathbf{r}_2^{(2)} \tag{2.39}$$

must hold. From this equation, relations between θ_i and z_i are obtained.

To solve this equation, first the transformation matrix T_{12} has to be constructed. Then from the equation $\mathbf{r}_1^{(1)} T_{12} = \mathbf{r}_2^{(2)}$, variables z_1 and θ_1 for the cylinder 1 being eliminated, the relation of z_2 and θ_2 is obtained. The following values are the constants calculated from the given data.

$$(\mathbf{a} \cdot \mathbf{b}) = \cos\phi, \quad |\mathbf{a} \times \mathbf{b}| = \sin\phi.$$

The unit vectors $(\mathbf{i}_1, \mathbf{j}_1, \mathbf{k}_1)$ of the coordinate axes of the system C1 are determined:

$$\mathbf{k}_1 = \mathbf{a}, \tag{2.40}$$
$$\mathbf{i}_1 = -(\mathbf{a} \times \mathbf{b})/\sin\phi, \tag{2.41}$$
$$\mathbf{j}_1 = \mathbf{k}_1 \times \mathbf{i}_1 = -\mathbf{a} \times (\mathbf{a} \times \mathbf{b})/\sin\phi = (\mathbf{b} - \cos\phi\mathbf{a})/\sin\phi. \tag{2.42}$$

The vector triple product appearing in $\mathbf{j}_1$ is changed by using the formulas given in Sect. 1.4.2. The unit vectors $(\mathbf{i}_2, \mathbf{j}_2, \mathbf{k}_2)$ for C2 axes are

$$\mathbf{k}_2 = \mathbf{b}, \tag{2.43}$$
$$\mathbf{i}_2 = (\mathbf{a} \times \mathbf{b})/\sin\phi, \tag{2.44}$$
$$\mathbf{j}_2 = \mathbf{k}_2 \times \mathbf{i}_2 = \mathbf{b} \times (\mathbf{a} \times \mathbf{b})/\sin\phi = (\mathbf{a} - \cos\phi\mathbf{b})/\sin\phi. \tag{2.45}$$

Since the transformation matrix has the form (see eq. (2.22)):

$$T_{12} = (\ \mathbf{i}_1\ \ \mathbf{j}_1\ \ \mathbf{k}_1\ \ \mathbf{o}_1\ )^{(2)T}, \tag{2.46}$$

we obtain

$$T_{12} = \begin{bmatrix} -1 & 0 & 0 & 0 \\ 0 & -\cos\phi & \sin\phi & 0 \\ 0 & \sin\phi & \cos\phi & 0 \\ d & 0 & 0 & 1 \end{bmatrix}. \tag{2.47}$$

Then from the equation $\mathbf{r}_1^{(1)} T_{12} = \mathbf{r}_2^{(2)}$, we get the following three equations:

$$-\rho_1 \cos\theta_1 + d = \rho_2 \cos\theta_2, \tag{2.48}$$
$$-\rho_1 \sin\theta_1 \cos\phi + z_1 \sin\phi = \rho_2 \sin\theta_2, \tag{2.49}$$
$$\rho_1 \sin\theta_1 \sin\phi + z_1 \cos\phi = z_2. \tag{2.50}$$

Squaring the first and the second equations and adding them, we obtain the following quadratic equation for z_1.

$$\sin^2\phi \cdot z_1^2 - 2\rho_1 \sin\theta_1 \cos\phi \sin\phi \cdot z_1 + (\rho_1 \sin\theta_1 \cos\phi)^2 + (-\rho_1 \cos\theta_1 + d)^2 - \rho_2^2 = 0 \tag{2.51}$$

By setting the value of θ_1 and solving the equation, we obtain the value of z_1. Taking the ratio of the second to the first equation, we have

$$\tan\theta_2 = \frac{-\rho_1 \sin\theta_1 \cos\phi + z_1 \sin\phi}{-\rho_1 \cos\theta_1 + d}. \tag{2.52}$$

We can determine the value of θ_2 from eq. (2.52), and z_2 from eq. (2.50). The intersection curve of the two cylinders is thus calculated. According to given values ρ_1, ρ_2, ϕ and d, the shape of intersection curve has two, one and no loop(s) and shows special shapes between their transitions, which has singular points where $dz_1/d\theta_1$ becomes indeterminate.

A more general treatment of the intersection is explained in Chap. 4 and in Chap. 16.

2.7 Expressions of Movement of a Body by Reflection

When a body moves in space, it is difficult for a person to grasp its positions and attitudes without visualizing it, for one cannot imagine correctly a motion of simultaneous translation and rotation. Mathematically, successive displacements of a solid body can be expressed by the successive multiplication of the corresponding 4×4 matrices of displacement. And a discrete displacement of a solid body can be expressed by a translation along an axis and a rotation around it, that is, by a screw motion. For geometric understanding of its screw motion, a method using mirror reflections seems suitable, so it is explained.

Let us consider a mirror M, a body B and its mirror image B' in M. The mirror M is the perpendicular bisection plane of a line segment connecting the corresponding points of B and B'. Since the coordinate system C0 in the space also has its image C0', the description of B by the system C0 is the same as that of B' by the system C0'. However, the description B' by the system C0 is not the same, but has the relation

$$B' = B \cdot M, \tag{2.53}$$

where B and B' are position vectors of bodies B and B', and M is a kind of displacement matrix produced by the mirror M. Hereafter, these definitions of B, B, B', B' and M, M are used rather loosely so long as confusion does not occur. Refer to Fig. 2.5. Let $\mathbf{p}'$ be the image of $\mathbf{p}$ by M, whose surface is defined by its unit normal vector $\mathbf{n}$ and its distance d_0 from the origin of C0, and the position of $\mathbf{p}'$ is given by

$$\mathbf{p}' = \mathbf{p} + 2\mathbf{n}(d_0 - \mathbf{p}\cdot\mathbf{n}). \tag{2.54}$$

To express the above equation in a matrix form, the fourth component 1 has to be added to the three components of a position vector, such as in

$$\mathbf{p} = (\ p_x \ \ p_y \ \ p_z \ \ 1\ ), \tag{2.55}$$

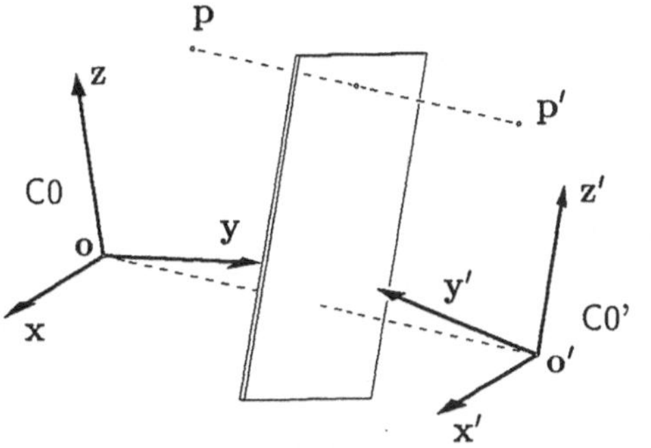

Figure 2.5 Reflection by a mirror

because $\mathbf{p}'$ consists of components proportional to $\mathbf{p}$ and a constant component in eq. (2.54). The matrix M has the form

$$M = \begin{bmatrix} 1-2n_x^2 & -2n_yn_x & -2n_zn_x & 0 \\ -2n_xn_y & 1-2n_y^2 & -2n_zn_y & 0 \\ -2n_xn_z & -2n_yn_z & 1-2n_z^2 & 0 \\ 2n_xd_0 & 2n_yd_0 & 2n_zd_0 & 1 \end{bmatrix}, \tag{2.56}$$

which is derived from eq. (2.54). The value of the determinant of the matrix M is equal to -1. This indicates that a point $\mathbf{p}'$ is an imaginary point seen from the real coordinate system C0.

2.7.1 Translation

Consider two parallel mirrors M_1 and M_2 separated by the distance h. A body B is moved to B' by M_1, and B' to B'' by M_2. Refer to Fig. 2.6(a). These processes are described by

$$B'' = B' \cdot M_2 = B \cdot M_1 \cdot M_2. \tag{2.57}$$

The distance between B and B'' is $2h$.

This is equivalent to the translation of $2h$ in the direction from M_1 to M_2 and does not depend on the absolute position of the mirror pair, that is, the mirror pair (M_1, M_2) can be moved anywhere in the perpendicular direction. If the distance between two mirrors is zero, this is equivalent to no mirrors. Accordingly, translations by even numbers of parallel mirrors are equivalent to those by the two parallel mirrors.

2.7.2 Rotation Around an Axis

Let two mirrors M_1 and M_2 intersect at a line L and their intersection angle be θ. Then a body B rotates around L by an angle 2θ. See Fig. 2.6(b). This rotation does not depend on the absolute position of the mirror pair (M_1, M_2) around the

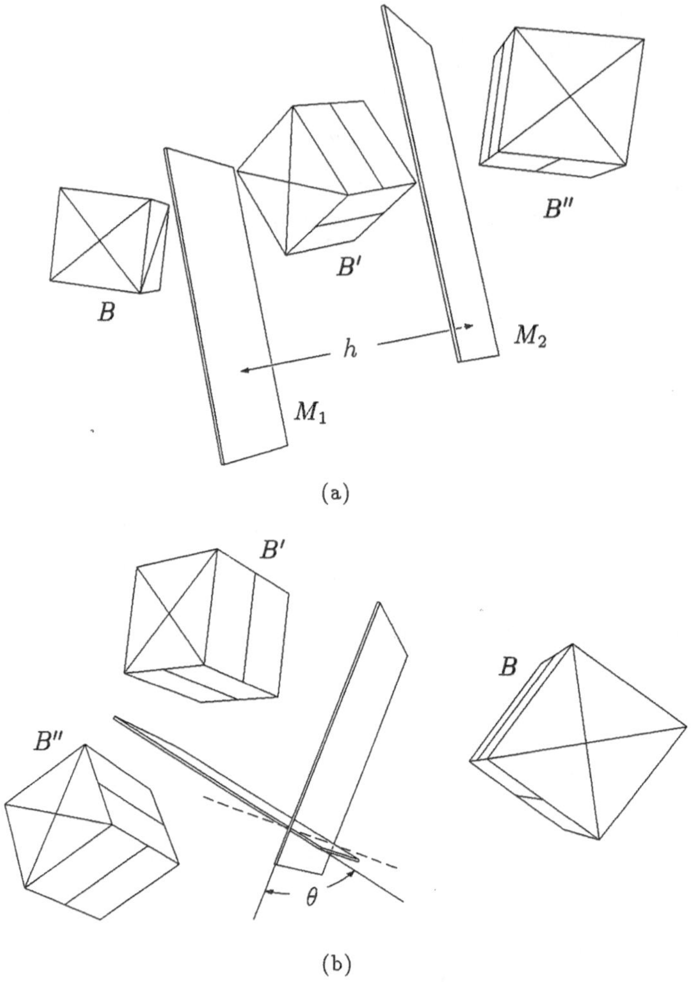

Figure 2.6 Displacement of a body: (a) two parallel mirrors, (b) two intersecting mirrors

axis. The direction of the rotation is determined by the order of application of the mirrors.

If the mirrors are orthogonal, the rotation angle is 180 degrees, so the application order can be interchanged. $M_1 \cdot M_2 = M_2 \cdot M_1$. The rotation by even numbers of mirrors intersecting at the line L is equivalent to rotation by the two mirrors around it. Translation is the case of the intersection line at infinity.

Exercise. Explain the behavior of one's own image in an orthogonal mirror standing at his front right-hand side while he walks.

2.7.3 Movement by Four Mirrors

There are the following three cases.

1. Four mirrors M_1,M_2,M_3 and M_4 are all parallel or intersect in one line, then they are equivalent to two mirrors.
2. If the intersection line of M_1 and M_2 intersects with that of M_3 and M_4, then M_2 and M_3 can be eliminated by making them coincide with suitable rotations of the pair (M_1, M_2) and the pair (M_3, M_4) around the respective axes. Another interpretation is that successive rotations of a body around lines through a fixed point are equivalent to a rotation around one axis through that fixed point.
3. If the intersection line of M_1 and M_2 does not intersect with that of M_3 and M_4, the displacement is described finally by two parallel mirrors and two intersecting mirrors which are orthogonal to the parallel mirrors. The displacement consists of a rotation and a translation, around and along the axis determined by the two intersecting mirrors.

This displacement is represented by a screw motion along and around the screw axis.

Proof of 3. Refer to Fig. 2.7(a). Let a line L_1 be the intersection line of M_1 and M_2, and a line L_2 be that of M_3 and M_4, and the common perpendicular of L_1 and L_2 be L_3.

If M_1 and M_2 are parallel, then there is no intersection line L_1, and rotating the pair (M_3, M_4) around L_2 until M_3 becomes orthogonal to M_2, then interchanging the order of application of M_2 and M_3, we get two pairs (M_1, M_3) and (M_2, M_4), which we consider the initial states of M_1, M_2, M_3 and M_4 by renaming.

Then by the following procedures the rotation axis and the rotation angle are determined.

1. Rotate the pair (M_1, M_2) around L_1 until M_2 includes L_3 which is the common perpendicular of L_1 and L_2. Similarly rotate the pair (M_3, M_4) around L_2 until M_3 contains L_3. The two planes M_2 and M_3 intersect along L_3. See Fig. 2.7(b).
2. Then rotate the pair (M_2, M_3) around L_3 by 90 degrees without moving the mirrors M_1 and M_4. We have two orthogonal pairs (M_1, M_2) and (M_3, M_4), whose axes are L_4 and L_5. See Fig. 2.7(c).
3. Rotate each of the orthogonal pairs around the respective axes L_4 and L_5 to make M_1 and M_3 parallel. The intersection of M_2 and M_4 becomes orthogonal to M_1 and M_3.
4. Change the application order from $M_2 \cdot M_3$ to $M_3 \cdot M_2$. This does not affect the result, because they are orthogonal. Then we have a parallel pair ($M_1 \cdot M_3$) and an intersecting pair ($M_2 \cdot M_4$) orthogonal to the former. The screw axis and the rotation angle and the translation distance are determined from these pairs. See Fig. 2.7(d).

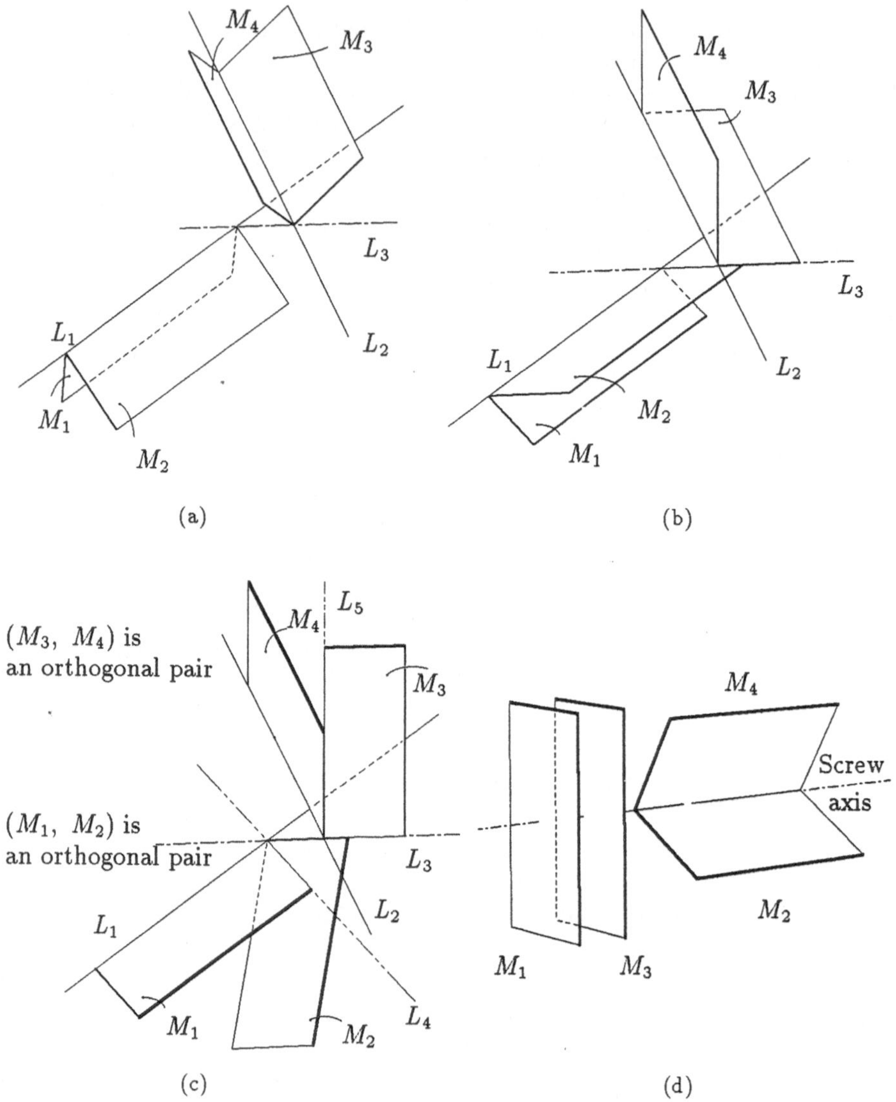

Figure 2.7 Conversion of two pairs of mirrors to a parallel pair and an intersecting pair orthogonal to the former: (a) initial locations of two pairs, (b) rotation of each pair to include the common perpendicular L_3, (c) rotation of (M_2, M_3) around L_3, (d) final state of two pairs

2.7.4 Determination of Screw Axis, Rotation Angle and Translation Distance

Suppose a body in the initial position B moves to the final position B", then its screw axis, rotation angle and translation distance can be determined by four mirrors which give the same effect.

First, we give a simple theorem concerning two mutually mirror-imaged bodies B' and B", which have a common point $\mathbf{p}'$.

Theorem. By a rotation around an axis through $\mathbf{p}'$ and a reflection by a mirror through $\mathbf{p}'$ and orthogonal to the axis, two bodies B' and B" can be made to coincide.

This mirror plane M_4 is determined by three points in the space: $\mathbf{p}'$ and two mid-points of any two line segments joining corresponding points of the two bodies B' and B". The axis of rotation which is represented by the mirror pair (M_2, M_3) passes through the point $\mathbf{p}'$ and is orthogonal to the mirror M_4. The rotation angle by the mirror pair (M_2, M_3) is determined by an angle between two planes which include the axis of the pair and each of the corresponding points of the bodies.

Using the above theorem, we can determine the screw axis and the rotation angle between the initial position B and the moved position B". The body B is moved to B' by a mirror M_1 which is the perpendicular bisection plane of the corresponding points of B and B". Since B' and B" are mutually mirror-imaged bodies with a common point, the displacement of B and B" are represented by the four mirrors M_1, M_2, M_3 and M_4. By rotating the pair (M_2, M_3) around their intersection line to make M_2 orthogonal to M_1, we have two orthogonal pairs (M_1, M_2) and (M_3, M_4), from which we can determine the screw axis, the rotation angle and the displacement by the method of Sect. 2.7.3.

2.7.5 Displacement Matrix S and Mirror Matrix M

The displacement matrix or screw matrix S, which is introduced in Sect. 2.5, can be expressed by two parallel mirrors and two intersecting mirrors orthogonal to the parallel mirrors. The product of two displacement matrices S_1 and S_2 can be synthesized by their corresponding mirror matrices.

Let $S_1 = M_1 \cdot M_2 \cdot M_3 \cdot M_4$ where M_1 and M_2 are parallel and their screw axis is L_1, and let $S_2 = M_5 \cdot M_6 \cdot M_7 \cdot M_8$ where M_5 and M_6 are parallel and their screw axis is L_2, and let the common perpendicular of L_1 and L_2 be L_3.

Perform the following procedures:

1. Move the pairs (M_1, M_2) and (M_5, M_6) along the respective screw axes to make M_2 and M_5 include their common perpendicular L_3.

2. Rotate the pair (M_2, M_5) around L_3 by 90 degrees, then M_2 includes L_1 and M_5 includes L_2.
3. Rotate (M_3, M_4) around L_1 to eliminate M_2 and M_3 and rotate (M_7, M_8) around L_2 to eliminate M_5 and M_7. Then two orthogonal pairs (M_1, M_4) and (M_6, M_8) remain.

Then, as in Sect. 2.7.3, the new screw axis and the four mirrors are determined.

3 Lines, Planes and Polyhedra

3.1 Introduction

In advanced CAD, its software has to process information on the models of 3D objects as though a person treats real ones. In construction and processing of the models, their information must be sufficient and have no contradiction, and their processing methods must be robust, efficient and flexible, because the result is directly connected to real processes of physical objects. From this point of view, a polyhedron is a fundamental object to be modelled in the computer. Its information structure and processing methods are clear, and numerical approximation enters least compared with those of other 3D objects. Polygons and polyhedra or wire-frames are also useful in describing objects with free-form surfaces, because free-form curves and surfaces are defined by their control points (refer to Chap. 9). So we explain in this chapter their basic equations as well as their geometric and topological properties and their interference.

3.2 Equations of Straight Line and Intersection of Line Segments

A line which passes a point $\mathbf{p}$ and has a direction of a unit vector $\hat{\mathbf{a}}$ is expressed with a parameter t by

$$\mathbf{r}(t) = \mathbf{p} + \hat{\mathbf{a}} \cdot t \, , \qquad -\infty < t < \infty. \tag{3.1}$$

This is a line of infinite length. An equation of a line segment joining two points $\mathbf{p}_0$ and $\mathbf{p}_1$ is

$$\mathbf{r}(t) = \mathbf{p}_0 \cdot (1-t) + \mathbf{p}_1 \cdot t \, , \qquad 0 \le t \le 1. \tag{3.2}$$

Next, we are to determine an intersection point of two line segments L_1 and L_2 both of which are on a plane. Equations of these line segments joining $\mathbf{p}_0$ and $\mathbf{p}_1$, and joining $\mathbf{p}_2$ and $\mathbf{p}_3$ are given by

$$L_1 \; : \quad \mathbf{r}_1(t) = \mathbf{p}_0 \cdot (1-t) + \mathbf{p}_1 \cdot t, \tag{3.3}$$
$$L_2 \; : \quad \mathbf{r}_2(t) = \mathbf{p}_2 \cdot (1-t) + \mathbf{p}_3 \cdot t. \tag{3.4}$$

Since these two lines are on a plane, three vectors made from the end points of the line segments must be coplanar, that is

$$[\mathbf{p}_1 - \mathbf{p}_0, \mathbf{p}_2 - \mathbf{p}_0, \mathbf{p}_3 - \mathbf{p}_0] = 0. \tag{3.5}$$

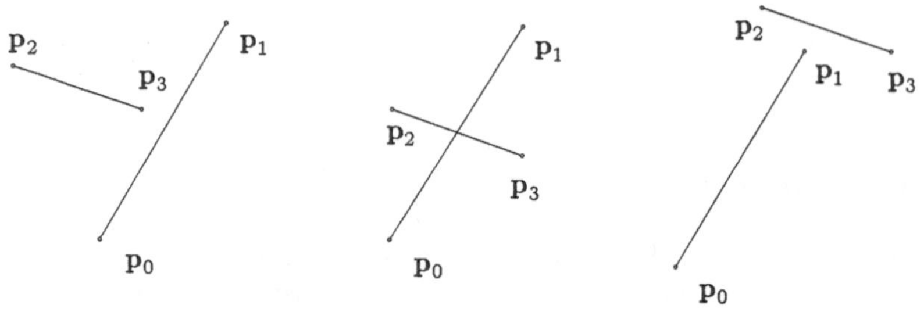

Figure 3.1 Intersection of two line segments

The unit normal vector of the plane is given by

$$\mathbf{n} = \frac{(\mathbf{p}_1 - \mathbf{p}_0) \times (\mathbf{p}_3 - \mathbf{p}_2)}{|(\mathbf{p}_1 - \mathbf{p}_0) \times (\mathbf{p}_3 - \mathbf{p}_2)|}. \tag{3.6}$$

For the two line segments L_1 and L_2 to intersect, the two end points of L_1 must be on both sides of L_2 and those of L_2 must be on both sides of L_1. These conditions are formulated in the following way: referring to Fig. 3.1, if signs of two scalar triple products

$$[(\mathbf{p}_2 - \mathbf{p}_0),\ (\mathbf{p}_1 - \mathbf{p}_0),\ \mathbf{n}] \text{ and } [(\mathbf{p}_3 - \mathbf{p}_0),\ (\mathbf{p}_1 - \mathbf{p}_0),\ \mathbf{n}]$$

are different, and also signs of

$$[(\mathbf{p}_0 - \mathbf{p}_2),\ (\mathbf{p}_3 - \mathbf{p}_2),\ \mathbf{n}] \text{ and } [(\mathbf{p}_1 - \mathbf{p}_2),\ (\mathbf{p}_3 - \mathbf{p}_2),\ \mathbf{n}]$$

are different, the line segments L_1 and L_2 intersect.

The intersection point is determined by the equation $\mathbf{r}_1(t_1) = \mathbf{r}_2(t_2)$, which is

$$(\mathbf{p}_1 - \mathbf{p}_0)t_1 - (\mathbf{p}_3 - \mathbf{p}_2)t_2 = \mathbf{p}_2 - \mathbf{p}_0. \tag{3.7}$$

Multiplying $(\mathbf{p}_1 - \mathbf{p}_0) \times \mathbf{n}$ with the above equation to eliminate t_1, we obtain

$$t_2 = \frac{[\ \mathbf{p}_0 - \mathbf{p}_2,\ \mathbf{p}_1 - \mathbf{p}_0,\ \mathbf{n}\ ]}{[\ \mathbf{p}_3 - \mathbf{p}_2,\ \mathbf{p}_1 - \mathbf{p}_0,\ \mathbf{n}\ ]}. \tag{3.8}$$

Similarly, multiplying $(\mathbf{p}_3 - \mathbf{p}_2) \times \mathbf{n}$, we obtain

$$t_1 = \frac{[\ \mathbf{p}_0 - \mathbf{p}_2,\ \mathbf{p}_3 - \mathbf{p}_2,\ \mathbf{n}\ ]}{[\ \mathbf{p}_3 - \mathbf{p}_2,\ \mathbf{p}_1 - \mathbf{p}_0,\ \mathbf{n}\ ]}. \tag{3.9}$$

With these parameter values, the intersecting point of two line segments L_1 and L_2 is given by $\mathbf{r}_1(t_1)$ or $\mathbf{r}_2(t_2)$ of eq. (3.3) and eq. (3.4).

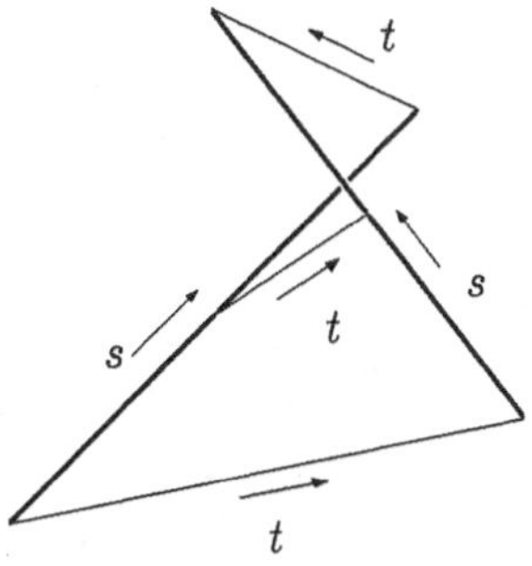

Figure 3.2 Generation of a surface patch of degree one

3.3 Control Polygons and Menelaus' Theorem

In eq. (3.2), if the two end points $\mathbf{p}_0$ and $\mathbf{p}_1$ are not static points, but move on further line segments L_a and L_b as linear functions of t, then $\mathbf{r}(t)$ becomes the equation of a curve of degree two of t. If the end points of the line segments L_a and L_b move also as linear functions of t on yet further line segments, a curve equation of degree three can be obtained.

When the points $\mathbf{p}_0$ and $\mathbf{p}_1$ move on lines as linear functions of another parameter s independently of t, $\mathbf{r}(t)$ becomes also a function of s, that is, $\mathbf{r}(t,s)$ represents an equation of a surface patch of degree one. See Fig. 3.2. If the corner points of a surface patch $\mathbf{r}(t,s)$ move as functions of t and s on other patches of degree one, then the surface patch $\mathbf{r}(t,s)$ becomes of degree two. In these kinds of curves or surfaces, their end points or corner points determine their shapes. These points are closely related to the control points, which have important functions for analysis and synthesis of curves and surfaces. They are treated in detail in Chap. 9.

Two useful theorems in the primary geometry for treating the control points are explained in the following.

(I) Refer to Fig. 3.3.

- L_0 and L_1 are adjacent edges of a polyline made from three points $\mathbf{p}_0$, $\mathbf{p}_1$ and $\mathbf{p}_2$. Let a point $\mathbf{q}_0$ on L_0 and a point $\mathbf{q}_1$ on L_1 divide each edge in the ratio $t : 1-t$, where $0 \leq t \leq 1$, and let the line segment joining $\mathbf{q}_0$ and $\mathbf{q}_1$ be L_2.
- Let a points $\mathbf{q}'_0$ on L_0 and $\mathbf{q}'_1$ on L_1 also divide the edges L_0 and L_1 in the ratio $t' : 1-t'$, where $0 \leq t' \leq 1$, and let the line segment connecting $\mathbf{q}'_0$ and $\mathbf{q}'_1$ be L'_2.
- Then L_2 divides L'_2 in the ratio $t : 1-t$ and L'_2 divides L_2 in the ratio $t' : 1-t'$.

The inverse theorem says that a point $\mathbf{q}_2$ which divides L_2 in the ratio $t' : 1-t'$ is collinear with $\mathbf{q}'_0$ and $\mathbf{q}'_1$.

Since these theorems are closely related to *Menelaus' theorem* which is explained below, in this book we call such divisions of edges a *Menelaus division* of

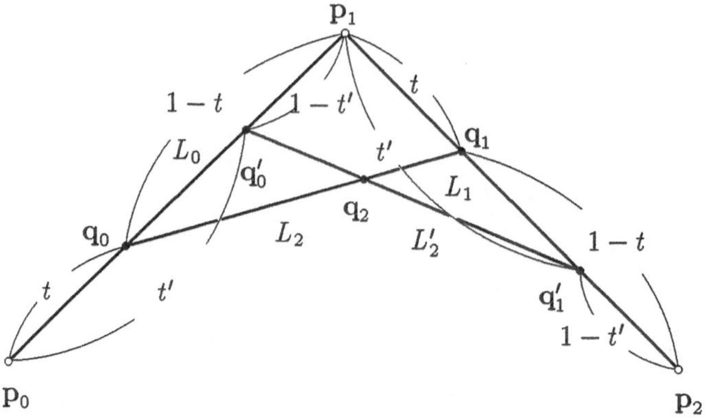

Figure 3.3 Menelaus division of edges

edges. For the application of the Menelaus division, the following procedure (refer to Chap. 10) is used in the connection of Bézier polygons or the construction of a spline polygon:

- Let three consecutive vertices of a polygon be $\mathbf{p}_0$, $\mathbf{p}_1$ and $\mathbf{p}_2$, and one edge $(\mathbf{p}_0\mathbf{p}_1)$ be divided into n sub-edges in the dividing ratio $(\lambda_1 : \lambda_2 : \cdots : \lambda_n)$ at the dividing points $\mathbf{p}_{0,i}$ for $i = 1 \ldots n-1$.
- The other edge $(\mathbf{p}_1\mathbf{p}_2)$ is divided into n sub-edges in the ratio $(\lambda_2 : \lambda_3 : \cdots : \lambda_{n+1})$ at the dividing points $\mathbf{p}_{1,i}$ for $i = 1..n-1$.
- We make an opposite edge of the vertex $\mathbf{p}_1$ by connecting $\mathbf{p}_{0,i}$, and $\mathbf{p}_{1,i-1}$, and divide this line in the ratio $(\lambda_2 : \lambda_3 : \cdots : \lambda_n)$ at the dividing points $\mathbf{p}_{0,i,j}$, for $j = 1..n-2$.
- Then $\mathbf{p}_{0,j}$, $\mathbf{p}_{0,i,j-1}$, and $\mathbf{p}_{1,j-1}$ are collinear, or $\mathbf{p}_{0,j,i-1}$ and $\mathbf{p}_{0,i,j-1}$ are coincident, where $\mathbf{p}_{0,j,i}$ is the i-th dividing point of a line joining $\mathbf{p}_{0,j}$ and $\mathbf{p}_{1,j-1}$ divided in the ratio $(\lambda_2 : \lambda_3 : \cdots : \lambda_n)$.

(II) Refer to Fig. 3.4. Let $\mathbf{p}_0\mathbf{p}_1\mathbf{p}_2$ be a triangle, whose edges are $\mathbf{a}_0$, $\mathbf{a}_1$ and $\mathbf{a}_2$, and points $\mathbf{q}_0$, $\mathbf{q}_1$ and $\mathbf{q}_2$ divide $\mathbf{a}_0$, $\mathbf{a}_1$ and $\mathbf{a}_2$ in the ratio $(t_0 : 1-t_0)$, $(t_1 : 1-t_1)$ and $(t_2 : 1-t_2)$ as shown in the figure, where $0 \leq t_0, t_1, t_2 \leq 1$. If

$$\frac{t_0}{(1-t_0)}\frac{t_1}{(1-t_1)}\frac{t_2}{(1-t_2)} = 1,$$

then line segments $\overline{\mathbf{p}_0\mathbf{q}_0}$, $\overline{\mathbf{p}_1\mathbf{q}_1}$ and $\overline{\mathbf{p}_2\mathbf{q}_2}$ are coincident. This is *Ceva's theorem.* The inverse of the *Ceva's theorem* is as follows: let an intersection point of line segments $\overline{\mathbf{p}_0\mathbf{q}_0}$, $\overline{\mathbf{p}_1\mathbf{q}_1}$ be $\mathbf{o}$, then the three points $\mathbf{p}_2$, $\mathbf{o}$ and $\mathbf{q}_2$ are collinear. For the application of this theorem, see Sect. 13.4.1.

The above theorems can be proved easily by applying *Menelaus' Theorem.* Referring to Fig. 3.5, it states that for a triangle abc and any line intersecting

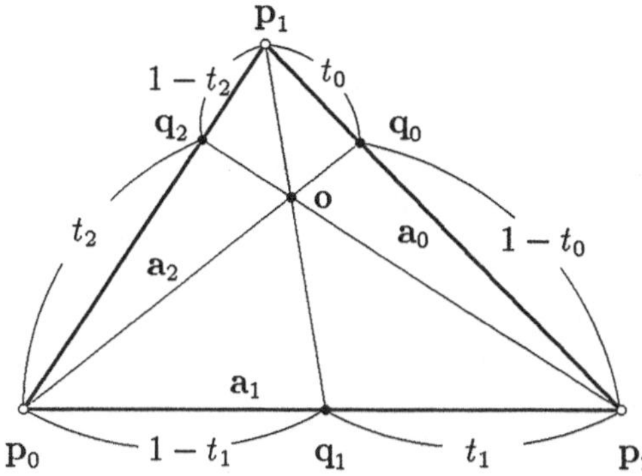

Figure 3.4 *Ceva's theorem*

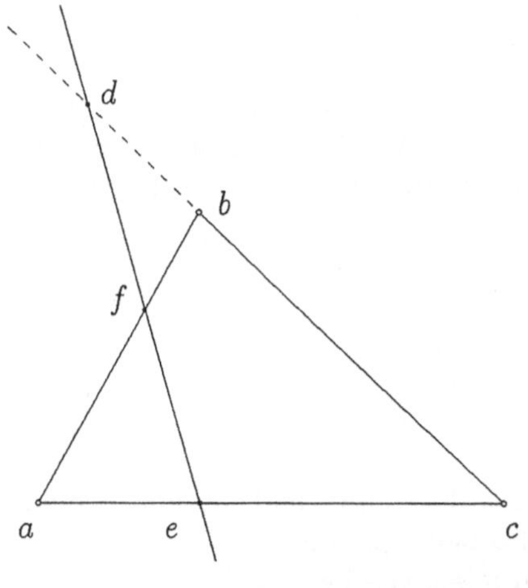

Figure 3.5 *Menelaus' theorem*

with three edges bc, ca and ab or their extensions at d, e, and f, where the distance between points b and d is denoted by bd or db, the relation

$$\frac{bd}{dc}\frac{ce}{ea}\frac{af}{fb} = 1 \tag{3.10}$$

holds. Inversely, three points d, e and f (or d, c and b) are collinear if the above relation holds in the triangle abc (or the triangle afe).

3.4 Equations of Plane and Intersection of Line and Plane

To distinguish the front side from the rear side of a plane, the direction of the unit normal vector $\mathbf{n}$ of the plane is defined to be from its rear side to its front side. The direction of a boundary line or curve which encloses a part of a front

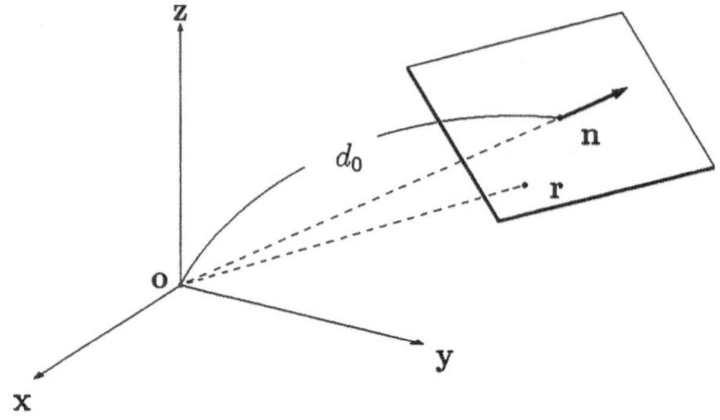

Figure 3.6 Symbols for a plane

plane is defined to be positive when the enclosed part of the plane is on the left-hand side of the boundary.

Refer to Fig. 3.6. Let $\mathbf{r}$ be the position vector of a point on a plane, and let the distance from the origin to the plane be d_0 (d_0 is defined positive, if the origin looks at the rear side of the plane, so the distance from the plane to the origin is $-d_0$), then the relation

$$\mathbf{r} \cdot \mathbf{n} = d_0 \tag{3.11}$$

is an equation of the plane. Another expression

$$\mathbf{r}(u, v) = \mathbf{p}_0 + u\mathbf{e}_1 + v\mathbf{e}_2 \tag{3.12}$$

is sometimes used for a plane which passes through $\mathbf{p}_0$ and contains two different non-zero basis vectors $\mathbf{e}_1$ and $\mathbf{e}_2$, u and v being parameters.

A line through $\mathbf{p}_0$ with its direction $\hat{\mathbf{a}}$ is given by

$$\mathbf{r}(t) = \mathbf{p}_0 + \hat{\mathbf{a}} \cdot t. \tag{3.13}$$

By substituting eq. (3.13) into eq. (3.11), we obtain the parameter value of the intersection point of the line and the plane :

$$t = \frac{(d_0 - \mathbf{p}_0 \cdot \mathbf{n})}{(\hat{\mathbf{a}} \cdot \mathbf{n})}. \tag{3.14}$$

The numerator of the above equation expresses the direct distance

$$d = (d_0 - \mathbf{p}_0 \cdot \mathbf{n}) \tag{3.15}$$

from the point $\mathbf{p}_0$ to the plane, and the denominator the cosine of the angle between the line and the normal of the plane. It follows that t gives the distance along the line from the point $\mathbf{p}_0$ to the plane.

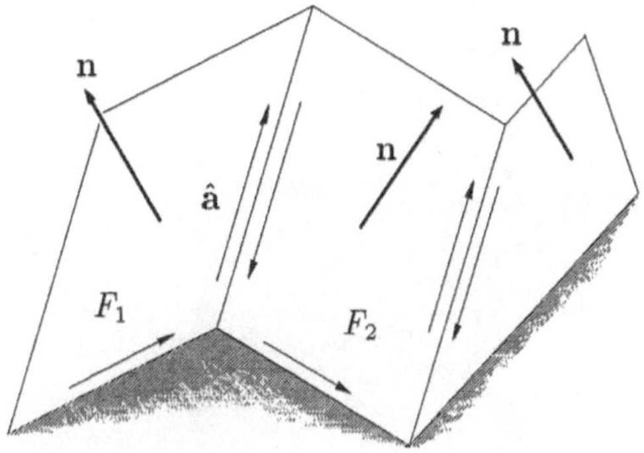

Figure 3.7 Convex and concave edges

If the distance d is positive, $\mathbf{p}_0$ looks at the rear side of the plane, if negative, it looks at its front side and if zero, it is on the plane.
Exercise. Determine the common perpendicular of two lines in space. (Hint: the direction of the common perpendicular is that of the vector product of the directions of the two lines. Determine a plane containing one of the lines and parallel to the common perpendicular. The intersection of the plane and the other line is on the common perpendicular.)

3.5 Polyhedron and Its Geometric Properties 1

A polyhedron is a body enclosed by plane segments called faces. If all the dihedral angles between adjacent faces of a polyhedron, which are measured between the front sides, are greater than 180 degrees, it is called a *convex polyhedron.* Any points which do not look at the front sides of any faces of a convex polyhedron are inside the polyhedron. If the plane equations of all the faces of a convex polyhedron are given, the polyhedron is uniquely defined.

But if a general polyhedron is to be uniquely defined, not only the equations of its face planes but also the relation of adjacency of its faces have to be given. From these data, the topological relations between its edges, vertices and its faces as well as their geometric values are obtained.

Let the equation of a plane π_1 including a face F_1 be $\mathbf{r}_1 \cdot \mathbf{n}_1 = d_1$ and that of its adjacent face F_2 be $\mathbf{r}_2 \cdot \mathbf{n}_2 = d_2$. The direction of the intersection line of the two adjacent faces F_1 and F_2 (refer to Fig. 3.7) is given by

$$\hat{\mathbf{a}} = \frac{\mathbf{n}_1 \times \mathbf{n}_2}{|\mathbf{n}_1 \times \mathbf{n}_2|}. \tag{3.16}$$

If the dihedral angle is greater than 180 degrees, the edge between them is named a convex edge, if smaller than 180 degrees, it is a concave edge. Whether

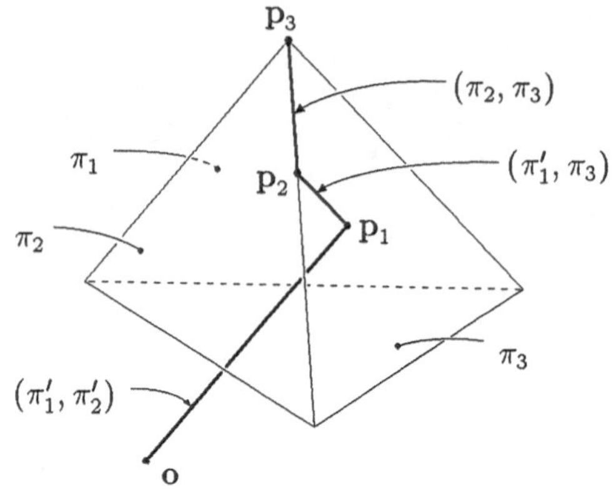

Figure 3.8 Intersection of three planes

an edge is convex or concave is determined not by the vector $\hat{\mathbf{a}}$ only, but by it together with the direction of the edge which is a boundary of a face F_1. If the directions of the intersection line and the edge are the same, the edge is convex, otherwise concave. For the adjacent face F_2, as the edge direction is reversed, its boundary line vector $\hat{\mathbf{a}}$ should be written

$$\hat{\mathbf{a}} = \frac{\mathbf{n}_2 \times \mathbf{n}_1}{|\mathbf{n}_2 \times \mathbf{n}_1|} \tag{3.17}$$

instead of eq. (3.16).

Next, when the intersection line L_{12} of the planes π_1 and π_2 is given, let this line L_{12} intersect with the third plane π_3: $\mathbf{r}_3 \cdot \mathbf{n}_3 = d_3$. The intersecting point is determined as follows. The value of the parameter t from a point $\mathbf{p}_0$ on L_{12} to the intersection point with the plane π_3 is given by eq. (3.14). Accordingly, the intersection point of the three planes π_1 ,π_2 and π_3 is determined by eq. (3.14) and eq. (3.16) being substituted into eq. (3.13) as

$$\mathbf{r} = \mathbf{p}_0 + \frac{(d_3 - \mathbf{p}_0 \cdot \mathbf{n}_3)(\mathbf{n}_1 \times \mathbf{n}_2)}{[\mathbf{n}_1, \mathbf{n}_2, \mathbf{n}_3]}. \tag{3.18}$$

From this point, two intersection lines of the planes (π_1,π_3) and (π_2,π_3) start. Their directions are given by eq. (3.16) or (3.17) with $\mathbf{n}_3$ substituting $\mathbf{n}_1$ or $\mathbf{n}_2$.

Example. Refer to Fig. 3.8. Edge tracing and an intersection point of three planes π_1,π_2 and π_3.

Let a plane through the origin and parallel to the plane π_1 be π_1', and a similar plane parallel to the plane π_2 be π_2'. Starting from the origin along the intersection line of π_1' and π_2', which is denoted by (π_1',π_2'), one can reach the intersecting point $\mathbf{p}_1$ with π_3, which is denoted by (π_1',π_2',π_3). $\mathbf{p}_1$ is given by eq. (3.18) with $\mathbf{p}_0 = 0$:

$$\mathbf{p}_1 = \frac{d_3(\mathbf{n}_1 \times \mathbf{n}_2)}{[\mathbf{n}_1, \mathbf{n}_2, \mathbf{n}_3]}. \tag{3.19}$$

Starting from $\mathbf{p}_1$ along the intersection line (π_3, π_1') of the planes π_3 and π_1', we get to the intersecting point $\mathbf{p}_2$ with π_2, which is (π_1',π_2,π_3) and given by

$$\mathbf{p}_2 = \mathbf{p}_1 + \frac{(d_2 - \mathbf{p}_1 \cdot \mathbf{n}_2)(\mathbf{n}_3 \times \mathbf{n}_1)}{[\mathbf{n}_1, \mathbf{n}_2, \mathbf{n}_3]}.$$

From $\mathbf{p}_2$ along the intersection line (π_2, π_3) of π_3 and π_2 , we reach the intersection point $\mathbf{p}_3$ with π_1 ,which is (π_1,π_2,π_3) and given by

$$\mathbf{p}_3 = \mathbf{p}_2 + \frac{(d_1 - \mathbf{p}_2 \cdot \mathbf{n}_1)(\mathbf{n}_2 \times \mathbf{n}_3)}{[\mathbf{n}_1, \mathbf{n}_2, \mathbf{n}_3]}. \tag{3.20}$$

In the above equations, $(\mathbf{p}_1 \cdot \mathbf{n}_2)$, $(\mathbf{p}_1 \cdot \mathbf{n}_1)$ and $(\mathbf{p}_2 \cdot \mathbf{n}_1)$ are zero because they include factors $[\mathbf{n}_1,\mathbf{n}_2,\mathbf{n}_2]$ and $[\mathbf{n}_1,\mathbf{n}_1,\mathbf{n}_3]$. So the intersection point $\mathbf{p}_3$ of the three planes π_1,π_2 and π_3 is given by

$$\mathbf{p}_3 = \frac{d_3(\mathbf{n}_1 \times \mathbf{n}_2) + d_2(\mathbf{n}_3 \times \mathbf{n}_1) + d_1(\mathbf{n}_2 \times \mathbf{n}_3)}{[\mathbf{n}_1, \mathbf{n}_2, \mathbf{n}_3]}. \tag{3.21}$$

This is the solution of linear simultaneous equations of three variables, or the intersection point of three planes.

3.6 Polyhedron and Its Geometric Properties 2

Each face of the polyhedron is separated from its adjacent faces by the edges. Both ends of an edge are vertices. For a closed polyhedron, if the numbers of its faces, its vertices and its edges are F, V and E, the following Eulerian condition holds:

$$F + V - E = 2. \tag{3.22}$$

In a general polyhedron, its edges are convex or concave. Refer to Fig. 3.9. Its vertices, to which the edges gather, are convex, concave or mixed, according to the characteristics of the gathered edges. Its faces which are bounded by the edges are also convex, concave or mixed according to the properties of the bounding edges. A polyhedron of all convex vertices can be converted to a concave polyhedron by inverting the direction of its face normals. Generally, convex vertices of a polyhedron do not connect directly to its concave vertices, but do so through the mixed vertices. The same is true for its faces; there are mixed domains between convex and concave regions.

A polyhedron with a mixed domain can be divided by a plane, which coincides with a face, into two parts. If the dividing plane is large enough, the divided parts may be more than two, but the dividing process is stopped when one loop of the dividing lines is generated. With this process, concave edges contained in the dividing plane are eliminated. When the polyhedron becomes hollow, its openings must be closed by putting in face planes. With this repeated division of the polyhedron, the original one is resolved into a set of convex polyhedra.

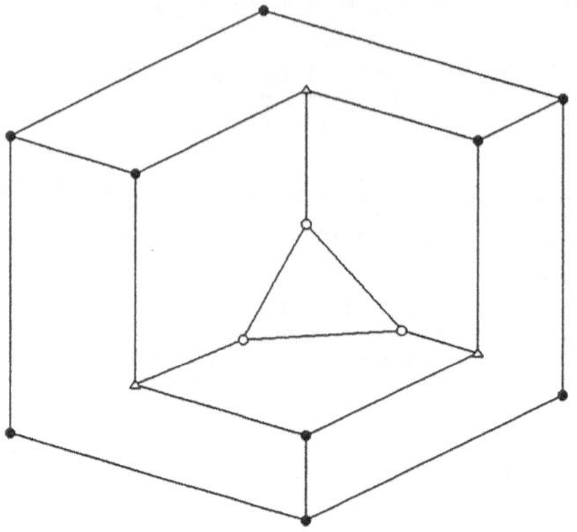

Figure 3.9 Edges of a polyhedron

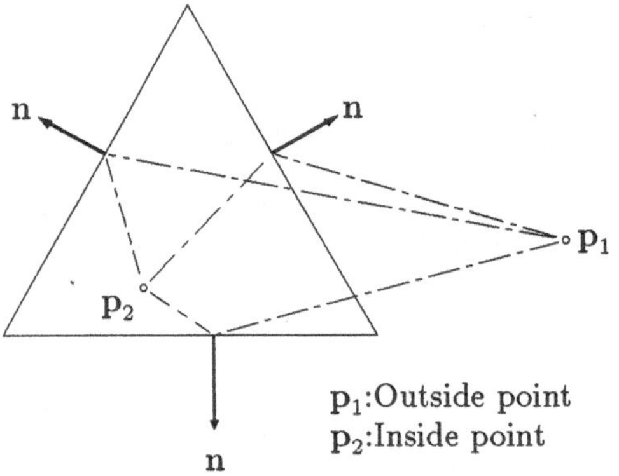

Figure 3.10 Points and a convex polyhedron

The geometric characteristics of a convex polyhedron are simple. See Fig. 3.10, where some examples are shown.

- A point is inside of the polyhedron if it looks at rear sides of all the faces ("looks at a rear side" means that the sign of inner product of the normal vector of a face and the vector from the point to the face is positive). A point is outside of the polygon if it looks at both the rear side and the front side faces, and it is on a face if it looks at all the rear side faces except the one on which the point exists.
- A line segment joining two points outside a convex polyhedron does not intersect with it, if the two end points look at a common face of the polyhedron or if any point P on the line segment and one of its end points look at a common face of the polyhedron, and the same point P and the other end point look at

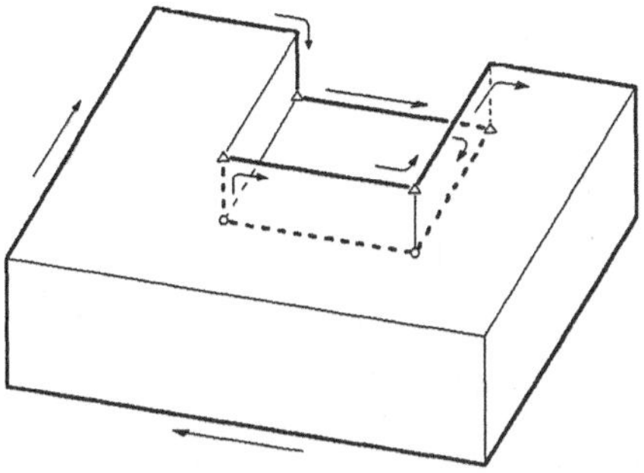

Figure 3.11 Silhouette edge loop

another common face. In this case each of the line segments is outside of the convex polyhedron.
Looking from a point outside of a polyhedron, one can classify its faces into two groups; those which the point looks and those which it does not. The boundaries of the two groups necessarily make a loop, otherwise its ends go to infinity. This loop is a silhouette line of the polyhedron looking from the point. For a convex polyhedron this loop is simple. In a general polyhedron, there may be more than one boundary loop which interfere with each other or with themselves on the projected plane. The whole loops or parts of the loops become the silhouette lines of the polyhedron. See Fig. 3.11.

3.7 Interference of Polyhedra

An edge of a polyhedron is an intersecting line of neighboring face planes π_1 and π_2, and its end point, which is a vertex, is an intersecting point of the edge with the their common adjacent face plane π_3. Edge traversing of a polyhedron is intersection line traversing of face planes of the polyhedron. The process of traversing intersection lines of two polyhedra is similar to that of tracing on its own edges, but the intersection lines are those generated by face planes π_1 and π_2 which belong to different polyhedra B_1 and B_2.

Suppose that a pair of faces π_1 and π_2 of the interfering polyhedra has been determined and names of adjacent faces of each face are given. The end point of the intersection line of the pair is an intersecting point with one of the candidate faces, which are the adjacent faces of π_1 or π_2. See Fig. 3.12. By applying eq. (3.14) to each of the candidate faces, we select the nearest true intersecting point with one of the adjacent faces. Then one of the current faces π_1 or π_2 is replaced by the new one and the new pair of interfering planes generates an intersection line which joins the previous intersection line segment. In the figure, pairs of intersecting faces are $(\pi_1\pi_2)$, $(\pi_2\pi_3)$, $(\pi_3\pi_{10})$, $(\pi_{10}\pi_5)$, $(\pi_5\pi_8)$, $\cdots$.

In the usual case one of the face planes of the pair is replaced by a new face

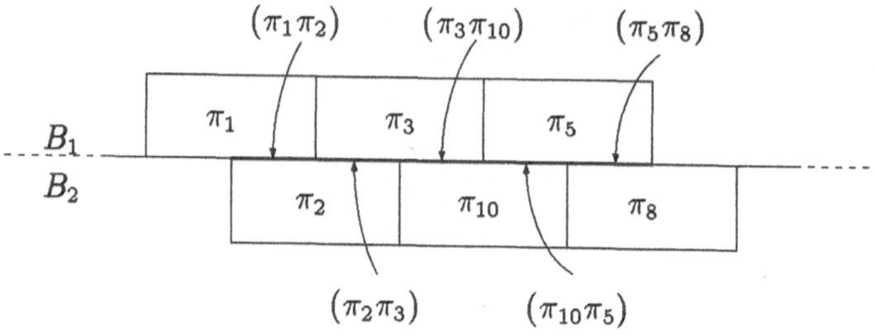

Figure 3.12 Intersection of faces of polyhedra

plane to make a new pair, but in the special case when the previous intersection line passes through a vertex of either polyhedron or when the vertices of the two polyhedra coincide, the previous face pair ends there and a new face pair starts, which is selected from the faces around the vertex or vertices concerned. Continuing the above process of determining the intersection until we reach the starting intersection line, we complete an intersection loop.

Though the procedure stated above can apply to general polyhedra, if we divide the original polyhedra into sets of convex polyhedra, we can determine their intersection loops easier than by attacking the problem without dividing.

3.8 Local Operations for Deformation of Polyhedron

The shape of a polyhedron can be modified by so-called local operations such as sweeping a face, moving an edge and its vertices, generating a new vertex and new edges or eliminating some of them from the original polyhedron so as not to violate the Eulerian relation. This process does not require interference calculations, which usually take a long time. Since the Eulerian relation is derived only from a topological net of faces, edges and vertices, and does not concern with realizability of a polyhedron, it is sometimes necessary to check the realizability or interference of the generated parts of the polyhedron after the modification of shape by the local operations.

Figure 3.13 shows a simple example of creation of a polyhedron by surface sweeping, which is one of the local operations.

(i) Sweep a square, which is on the xy plane and of size a, in the z direction by length a. This makes a cube and increases the number of faces, vertices and edges by 4, 4 and 8 respectively,

(ii) Make a small square of size b at the corner of the front face A of the cube. This creates one face, three vertices and four edges.

(iii) Sweep the new small square on the face A in the direction x by length b. This creates two faces, three vertices and five edges.

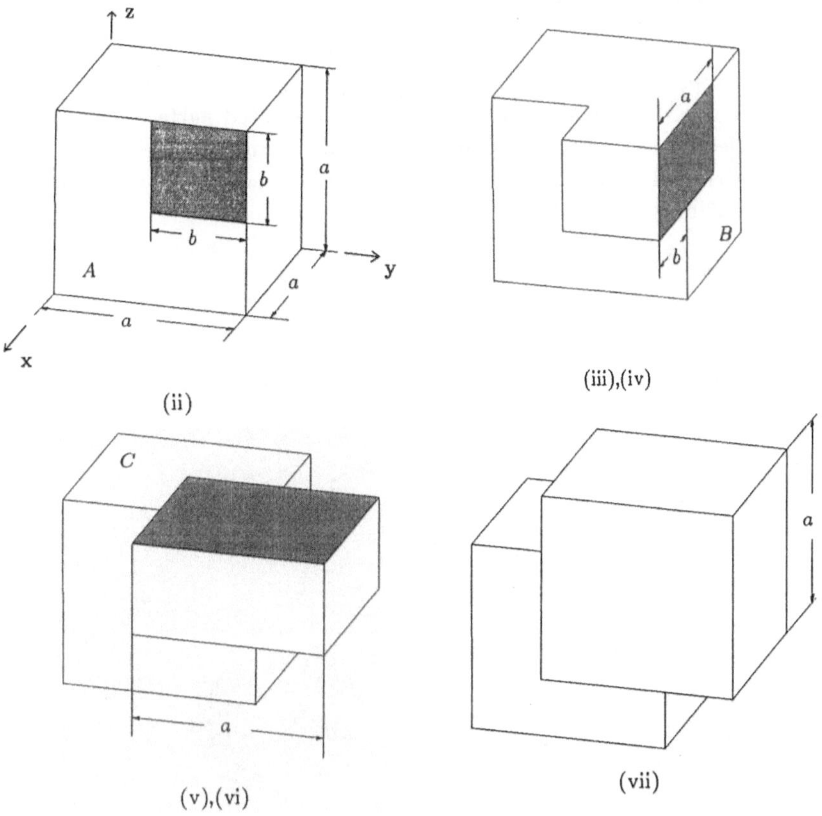

Figure 3.13 Example of Euler operations

(iv) Make a $b \times a$ rectangle at the corner on the side face B of the body. This creates one face, two vertices and three edges.

(v) Sweep the new face in the y direction by length $(a - b)$. This creates one face, two vertices and three edges.

(vi) Make an $a \times a$ square on the upper face C . This creates one face, one vertex and two edges.

(vii) Sweep the $a \times a$ square on the face C in the direction z by length $(a - b)$. This creates two faces, one vertex and three edge.

The final body created is the same as that of one cube of size a interfering with another cube of the same size displaced by $(a - b)$ in the x, y and z directions. In each process described above, the increase in the number of faces and vertices is equal to the increase in the number of edges. So, the processes do not violate the Eulerian relation of eq. (3.22).

A general form of eq. (3.22) is given by

$$F + V - E = 2(S - H),$$

where S and H are the numbers of shells (connected surfaces) and holes of the solid object. An advanced explanation and discussion of the topology and its use for construction of data structure of geometric models are to be treated in the book by Kimura.

4 Conics and Quadrics

4.1 Introduction

Conics and quadrics are frequently used in various parts of shapes in engineering products. Though they are only one degree higher than lines and planes, their expressive capability is far greater than the latter. Since their geometric and algebraic properties have been fully investigated theoretically, we can use them with confidence. Knowledge of them is necessary not only for their appropriate use, but also for estimating characteristics of the shape around a point on a free-form surface and understanding various techniques of analyzing surface problems.

Moreover, solid bodies can interfere in practice and this is a rather complicated problem, but using quadrics of special types their interference analysis is not difficult. So we treat such cases in this chapter and leave general ones to Chap. 16. Conics are also be expressed as the rational Bézier curves of degree two, which are treated in Chap. 12.

4.2 Conics

4.2.1 Equation of Conics

A general implicit expression of a conic is given by

$$ax^2 + 2hxy + by^2 + 2gx + 2fy + d = 0, \tag{4.1}$$

where a,b,f,g and h are constants, and $a \geq 0$ can be assumed. Its matrix form is

$$(x\ y\ 1) \cdot \begin{bmatrix} a & h & g \\ h & b & f \\ g & f & d \end{bmatrix} \cdot (x\ y\ 1)^T = 0. \tag{4.2}$$

Using a symbol C for the above central matrix, which we call the characteristic matrix of a conic, we write the above equation:

$$(x\ y\ 1) \cdot C \cdot (x\ y\ 1)^T = 0. \tag{4.3}$$

4.2.2 Transformation of Equation

We transform this general equation to the special forms in which it is easy to grasp their characteristics. At first we translate the origin of this coordinate system C1 to C2, whose origin is at (x_0, y_0) of C1. Its transformation matrix and its inverse are given by

$$T_{12} = \begin{bmatrix} 1 & 0 & 0 \\ 0 & 1 & 0 \\ -x_0 & -y_0 & 1 \end{bmatrix}, \quad T_{21} = \begin{bmatrix} 1 & 0 & 0 \\ 0 & 1 & 0 \\ x_0 & y_0 & 1 \end{bmatrix}.$$

Applying the following relation

$$(x\ y\ 1)^{(1)} = (x\ y\ 1)^{(2)} T_{2,1}$$

to eq. (4.2), we obtain

$$(x\ y\ 1)^{(2)} \cdot T_{21} \cdot C \cdot T_{21}^T \cdot (x\ y\ 1)^{(2)\mathrm{T}} = 0.$$

Given to no risk of confusion, we omit the superscript indicating the coordinate system C2. The matrix part $T_{21} \cdot C \cdot T_{21}^T$ of eq. (4.2) is changed to

$$\begin{bmatrix} a & h & g' \\ h & b & f' \\ g' & f' & d' \end{bmatrix}, \tag{4.4}$$

where

$$g' = ax_0 + hy_0 + g, \quad h' = hx_0 + by_0 + f, \quad d' = gx_0 + fy_0 + d. \tag{4.5}$$

From the first two equations of (4.5), we can determine x_0 and y_0 so as to make g' and h' vanish if the value of the determinant taken from the coefficients of x_0 and y_0 is not zero. Denoting the left-upper 2×2 sub-matrix of C as D and assuming its determinant is not zero:

$$det(D) = \begin{vmatrix} a & h \\ h & b \end{vmatrix} \neq 0, \tag{4.6}$$

we obtain

$$x_0 = \begin{vmatrix} -g & h \\ -f & b \end{vmatrix} \div det(D), \qquad y_0 = \begin{vmatrix} a & -g \\ h & -f \end{vmatrix} \div det(D), \tag{4.7}$$

and substituting the above values into eq. (4.5), we have

$$d' = \frac{det(C)}{det(D)}. \tag{4.8}$$

Then, eq. (4.3) becomes

$$(x\ y\ 1) \cdot \begin{bmatrix} a & h & 0 \\ h & b & 0 \\ 0 & 0 & d' \end{bmatrix} \cdot (x\ y\ 1)^T = 0. \tag{4.9}$$

By rotating the coordinate system around its origin, we can transform the matrix (4.9) into a diagonal form. We explained in Sect. 2.6 the eigenvalue method for obtaining a diagonal matrix. Using the eigenvalues λ_1, λ_2 of the upper-left 2×2 submatrix in eq. (4.9), the equation (4.9) is transformed to the following form:

$$(x\ y\ 1) \cdot \begin{bmatrix} \lambda_1 & 0 & 0 \\ 0 & \lambda_2 & 0 \\ 0 & 0 & d' \end{bmatrix} \cdot (x\ y\ 1)^T = 0, \tag{4.10}$$

where

$$\lambda_1, \lambda_2 = \{a + b \pm \sqrt{(a-b)^2 + 4h^2}\}/2. \tag{4.11}$$

Corresponding to these λ_1,λ_2, the directions θ and $\theta + \pi/2$ of the eigenvectors are given by

$$\tan\theta = \frac{(\lambda_1 - a)}{h} = \frac{h}{(\lambda_1 - b)}. \tag{4.12}$$

Exercise. Show the same results are obtained by rotation θ of the conic around its origin.
Hint. Since the rotation matrix S (refer to Sect. 2.6.1) is given by

$$S = T(-\theta : z) = \begin{bmatrix} \cos\theta & \sin\theta \\ -\sin\theta & cos\theta \end{bmatrix}$$

the rotational part of eq. (4.8) is

$$S \cdot \begin{bmatrix} a & h \\ h & b \end{bmatrix} \cdot S^T.$$

4.2.3 Classification of Conics

Equation (4.10) is

$$a'x^2 + b'y^2 + d' = 0, \tag{4.13}$$

where $a' = \lambda_1$ and $b' = \lambda_2$.
For $ab - h^2 \neq 0$, conics have a center given by eq. (4.5). We can classify these conics:

- an imaginary ellipse if a', b' and d' have all the same sign,
- an ellipse if $a' > 0, b' > 0$ and $d' < 0$,
- a hyperbola if $a' > 0, b' < 0$ and $d' < 0$,
- two straight lines if $d' = 0$, which is equivalent to $det(C) = 0$ from eq. (4.8).

For $det(D) = ab - h^2 = 0$, eq. (4.1) is transformed into

$$(\sqrt{a}x + \sqrt{b}y)^2 + 2gx + 2fy + d = 0. \tag{4.14}$$

Using the coordinate system $(u,\ v)$ instead of $(x,\ y)$:

$$\begin{aligned} u &= \sqrt{a}x + \sqrt{b}y + (g/\sqrt{a} + f/\sqrt{b})/2, \\ v &= \sqrt{a}x - \sqrt{b}y + (g/\sqrt{a} + f/\sqrt{b})/4 - d/(g/\sqrt{a} + f/\sqrt{b}), \end{aligned}$$

we transform eq. (4.14) into

$$u^2 + (g/\sqrt{a} - f/\sqrt{b})v = 0.$$

This equation is classified as

- a parabola if $f/g \neq \sqrt{b/a}$,
- a real line if $(g/\sqrt{a} - f/\sqrt{b}) = 0$,
- two real or imaginary lines if $f = g = 0$.

The conics are expressed in parametric forms in various ways. A rational Bézier expression of degree two covers all kinds of conics. This is treated in Chap. 12.
Exercise. Prove that eq. (4.1) is resolved into a product of two linear functions of x and y by the condition $det(C) = 0$. In this case a conic becomes two straight lines.

4.2.4 Intersection of Conics

Intersection points of two conics are generally given by roots of a quartic equation which is produced by eliminating one of the variables from the given two conics. The number of intersecting points is at most four. Since the number of intersection points of a line and a conic is two, if we can reduce the problem of the intersection of two conics to that of the intersection of two lines and one conic, we need not solve a quartic equation.

Let characteristic matrices of two conics be C_1 and C_2. For brevity, we call these conics also C_1 and C_2. Now we consider a conic C_3, whose characteristic matrix is a linear combination of C_1 and C_2, for instance $C_3 = C_2 - \mu C_1$ where μ is a constant. The conic C_3 intersects with the conics C_1 or C_2 at the same points as those of intersection of the conics C_1 and C_2, because the intersecting points satisfy each equation of the conics C_1, C_2 and C_3. The conic C_3 is called a *pencil* of C_1 and C_2.

If we can make an appropriate pencil which becomes straight lines, our objective is attained. The condition that a conic equation is resolved into a product of two straight lines is that the value of the determinant of its characteristic matrix is zero. This is proved from equations (4.8) and (4.9). The characteristic matrix C_3 of the pencil is equal to that of $C_2 - \mu C_1$, where μ is a constant to be determined. The condition that the pencils are straight lines is

$$det(C_2 - \mu C_1) = 0. \tag{4.15}$$

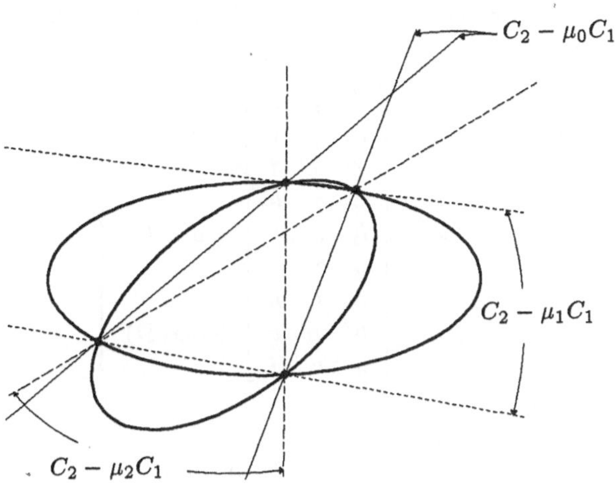

Figure 4.1 Three set of pencils of two ellipses

Since the degree of the above equation for μ is three, there is at least one real root, which corresponds to the two lines. Intersections of these lines and one of the conics give desired points. The number of intersecting points is from zero to four. When there are three real roots, we have three set of two straight lines which represent the pencils. Their intersecting points are those of the conics C_1 and C_2. An example is shown in Fig. 4.1.

4.3 Quadrics

4.3.1 Coordinate Transformation

A quadric is expressed in a matrix form as

$$(x, y, z, 1) \cdot Q \cdot (x, y, z, 1)^T = 0, \tag{4.16}$$

where

$$Q = \begin{bmatrix} a & h & g & l \\ h & b & f & m \\ g & f & c & n \\ l & m & n & d \end{bmatrix}. \tag{4.17}$$

Let the left-upper 3×3 sub-matrix of the matrix (4.17) be

$$D = \begin{bmatrix} a & h & g \\ h & b & f \\ g & f & c \end{bmatrix}. \tag{4.18}$$

If det(D) is not zero, we can translate the origin of the coordinates to (x_0, y_0, z_0), which are given by

$$\left.\begin{array}{l} x_0 = \begin{bmatrix} l & g & h \\ m & f & b \\ n & c & f \end{bmatrix} \div det(D) \\ y_0 = \begin{bmatrix} a & g & l \\ h & f & m \\ g & c & n \end{bmatrix} \div det(D) \\ z_0 = \begin{bmatrix} a & l & h \\ h & m & b \\ g & n & f \end{bmatrix} \div det(D) \end{array}\right\} . \tag{4.19}$$

The matrix Q is changed to

$$Q' = \begin{bmatrix} a & h & g & 0 \\ h & b & f & 0 \\ g & f & c & 0 \\ 0 & 0 & 0 & d' \end{bmatrix}, \tag{4.20}$$

where

$$d' = lx_0 + my_0 + nz_0 + d = det(Q)/det(D). \tag{4.21}$$

To transform the matrix (4.18) into a diagonal form, we have to determine eigenvalues of the matrix D, whose characteristic equation is (refer to Sect. 2.6):

$$\begin{bmatrix} a-\lambda & h & g \\ h & b-\lambda & f \\ g & f & c-\lambda \end{bmatrix} = 0. \tag{4.22}$$

This is a cubic equation in λ:

$$\lambda^3 - (a+b+c)\lambda^2 + (ab+bc+ca-f^2-g^2-h^2)\lambda - det(D) = 0. \tag{4.23}$$

It has three real roots λ_1, λ_2 and λ_3, which are the eigenvalues. The coefficients of λ^2 ,λ^1 and λ^0 of eq. (4.23),

$$(a+b+c), \quad (ab+bc+ca-f^2-g^2-h^2), \quad det(D), \tag{4.24}$$

are used for determining the eigenvalues, which are independent of the coordinate systems used, and consequently they are invariant and are equal to

$$(\lambda_1+\lambda_2+\lambda_3), \quad (\lambda_1\lambda_2+\lambda_2\lambda_3+\lambda_3\lambda_1), \quad (\lambda_1\lambda_2\lambda_3). \tag{4.25}$$

Let the principal directions corresponding to λ_i be $(\alpha_i, \beta_i, \gamma_i)$; from eq. (4.21), their ratio is given by

$$\alpha_i : \beta_i : \gamma_i = \begin{vmatrix} (b-\lambda_i) & f \\ f & (c-\lambda_i) \end{vmatrix} : -\begin{vmatrix} h & f \\ g & (c-\lambda_i) \end{vmatrix} : \begin{vmatrix} h & (b-\lambda_i) \\ g & f \end{vmatrix} . \tag{4.26}$$

Since the three eigenvectors (principal directions) are mutually orthogonal, they make a coordinate system. Then the matrix (4.20) becomes diagonal, elements of which are λ_1, λ_2, λ_3 and d'. Accordingly, the equation of a quadric is given by

$$\lambda_1 x^2 + \lambda_2 y^2 + \lambda_3 z^2 = -d'. \tag{4.27}$$

4.3.2 Classification

The shapes of quadrics are classified as follows:

1. λ_1,λ_2,λ_3 and d' have the same sign: an imaginary ellipsoid.
2. λ_1,λ_2,λ_3 are of the same sign and d' has the opposite sign: a real ellipsoid.
3. one or two of λ's are negative and d' is not zero: a hyperboloid, which is classified into two types according to the sign of the λ's and d':
   (i) $\lambda_1 > 0$, $\lambda_2 < 0$, $\lambda_3 < 0$ and $d' < 0$: eq. (4.27) becomes a hyperboloid of two sheets.
   (ii) λ_1, $\lambda_2 > 0$, $\lambda_3 < 0$ and $d' < 0$: eq. (4.27) represents a hyperboloid of one sheet and can be factored into
   $$\{\sqrt{\lambda_1}x + \sqrt{-\lambda_3}z\}\{\sqrt{\lambda_1}x - \sqrt{-\lambda_3}z\} = \{\sqrt{-d'} + \sqrt{\lambda_2}y\}\{\sqrt{-d'} - \sqrt{\lambda_2}y\}. \tag{4.28}$$
   Accordingly, there are two sets of generator lines which satisfy the above equation,
   $$\left.\begin{aligned}\{\sqrt{\lambda_1}x + \sqrt{-\lambda_3}z\} &= \gamma\{\sqrt{-d'} + \sqrt{\lambda_2}y\}\\ \{\sqrt{\lambda_1}x - \sqrt{-\lambda_3}z\} &= (1/\gamma)\{\sqrt{-d'} - \sqrt{\lambda_2}y\}\end{aligned}\right\}, \tag{4.29}$$
   $$\left.\begin{aligned}\{\sqrt{\lambda_1}x + \sqrt{-\lambda_3}z\} &= \gamma\{\sqrt{-d'} - \sqrt{\lambda_2}y\}\\ \{\sqrt{\lambda_1}x - \sqrt{-\lambda_3}z\} &= (1/\gamma)\{\sqrt{-d'} + \sqrt{\lambda_2}y\}\end{aligned}\right\}, \tag{4.30}$$
   where γ is an arbitrary constant.
4. one or two of the λ's are negative and d' is zero: a cone. A cone also has a generator. Those quadric surfaces which have generators are ruled surfaces (refer to Sect. 8.9).

In the above cases, since the point (x_0, y_0, z_0) is determined from eq. (4.19), these quadrics are called *central quadrics*.

When $\det(D) = 0$, the center cannot be determined, but eq. (24) becomes

$$\lambda^3 - (a+b+c)\lambda^2 + (ca + ab + bc - f^2 - g^2 - h^2)\lambda = 0. \tag{4.31}$$

One of its roots λ_3 is zero; let the others be λ_1, λ_2. Taking the corresponding principal directions $(\alpha_i, \beta_i, \gamma_i)$ as the coordinate axes, we transform eq. (4.17) into the following form:

$$(x\ y\ z\ 1) \cdot \begin{bmatrix} \lambda_1 & 0 & 0 & l' \\ 0 & \lambda_2 & 0 & m' \\ 0 & 0 & 0 & n' \\ l' & m' & n' & d \end{bmatrix} \cdot (x\ y\ z\ 1)^T = 0, \tag{4.32}$$

where (l', m', n') are given by

$$(l, m, n)\begin{bmatrix} \alpha_1 & \alpha_2 & \alpha_3 \\ \beta_1 & \beta_2 & \beta_3 \\ \gamma_1 & \gamma_2 & \gamma_3 \end{bmatrix}. \tag{4.33}$$

Equation (4.32) is

$$\lambda_1 x^2 + \lambda_2 y^2 + 2l'x + 2m'y + 2n'z + d = 0. \tag{4.34}$$

Translating the coordinate origin of eq. (4.34) to eliminate its terms including x and y, we have one of the following forms:

- $\lambda_1 x^2 + \lambda_2 y^2 = -2n'z$: elliptical or hyperbolic paraboloid.
- $\lambda_1 x^2 + \lambda_2 y^2 = d'$: elliptical or hyperbolic cylinder.

When two eigenvalues are zero, eq. (4.32) for $\lambda_2 = 0$ can be transformed into the following forms by a translation parallel to the x axis and a rotation around the x axis:

- $\lambda_1 x^2 = 2m'y$: parabolic cylinder.
- $\lambda_1 x^2 = d'$: two parallel planes.

They are *non-central quadrics.* A hyperbolic paraboloid has the generators:

$$\left.\begin{aligned} \{\sqrt{\lambda_1}x + \sqrt{-\lambda_2}y\} &= \gamma(-2n')z \\ \{\sqrt{\lambda_1}x - \sqrt{-\lambda_3}z\} &= (1/\gamma) \end{aligned}\right\}, \tag{4.35}$$

$$\left.\begin{aligned} \{\sqrt{\lambda_1}x + \sqrt{-\lambda_2}y\} &= \gamma \\ \{\sqrt{\lambda_1}x - \sqrt{-\lambda_3}z\} &= (1/\gamma)(-2n')z \end{aligned}\right\}. \tag{4.36}$$

All cylinders have the generators which are parallel to the z axis. General quadrics can be transformed into the canonical forms of central or non-central quadrics. Some of them have generator lines.

The quadrics mostly used in CAD are those of revolution, which have two equal eigenvalues. Their equations and names are

sphere	$x^2 + y^2 + z^2 = r^2$	
circular cylinder	$x^2 + y^2 = r^2$	
circular cone	$x^2 + y^2 = c^2 z^2$	
circular paraboloid	$x^2 + y^2 = c^2 z$	
circular prolate ellipsoid	$x^2 + y^2 = r^2 - c^2 z^2$	$c^2 < 1$
circular oblate ellipsoid	$x^2 + y^2 = r^2 - c^2 z^2$	$c^2 > 1$
circular hyperboloid of one sheet	$x^2 + y^2 = r^2 + c^2 z^2$	
circular hyperboloid of two sheets	$x^2 + y^2 = c^2 z^2 - r^2$	

(4.37)

Since the above bodies of revolution are parameterized, calculation of their interference becomes a little easier.

4.4 Intersection of Two Quadrics

The intersection of two planes can easily be obtained, because both of their equations are linear. If the value of one of three variables $(x,\ y,\ z)$ of the intersection line is given, the other two are determined by its linear functions. When one of the intersecting surfaces is a quadric, two variables of the intersection curve are determined as roots of a quadratic equation of the other one variable. When both of the surfaces are quadrics, generally an equation to determine their intersection curve becomes quartic. So, it cannot be solved easily. However, as we explained in the previous section some of the quadrics have generators which are expressed by two linear equations, that is, a straight line, so if one of the surfaces belongs to such a class, solution of their intersection curve is obtained from the roots of the quadratic equations.

Even when both of the intersecting quadrics lack generators, we can produce a quadric with generators, which intersects with the same intersection curve as that of the original quadrics [4][5]. The method of producing such a quadric is similar to that of producing a pencil of conics from two intersecting conics explained in Sect. 4.2.4. In this case, the pencil must be a non-central quadric, excluding an elliptical paraboloid which does not have generators. For this purpose the method requires numerical manipulations of the equations such as affine transformations together with solving a characteristic equation of a matrix of degree four. After making generators of a desired pencil, intersecting points of a modified quadric with the generators have to be calculated and then transformed to the original coordinate system. These operations are apt to introduce numerical errors, so this method of calculating intersection curves [7] is not always the best one.

Alternatively, the following method is easier in the calculation [8]. If we cut the intersecting quadrics by a plane parallel to one of the coordinate planes, we get two intersecting conics, whose intersecting points can be obtained by the pencil method. The intersecting points of a conic and one or two lines are points on the intersection curve in 3D. By moving the cutting plane in its normal direction, we can trace the intersection points with the cutting planes.

Let Q_1 and Q_2 be two intersecting quadrics. Let the equations of two conics generated by the cut of Q_1 and Q_2 with a plane $z = z_0$ be C_1 and C_2, which are obtained from eq. (4.16).

$$C_i \ : \ \begin{bmatrix} a_i & h_i & g_i'(z_0) \\ h_i & b_i & f_i'(z_0) \\ g_i'(z_0) & f_i'(z_0) & d_i'(z_0) \end{bmatrix}, \quad i = 1, 2, \tag{4.38}$$

where $g_i'(z_0)$, $f_i'(z_0)$ and $d_i'(z_0)$ are functions of z_0. Let the characteristic equation of their pencil be $D(\mu, z_0) = 0$, which is a cubic function of μ and has at least one real root μ_0. For the root, the pencil become a straight line, whose intersection with one of the conics gives intersection points of the two conics. For $z = z_0 + \Delta z$, we obtain $\mu = \mu_0 + \Delta\mu$ by using the Newton-Raphson method

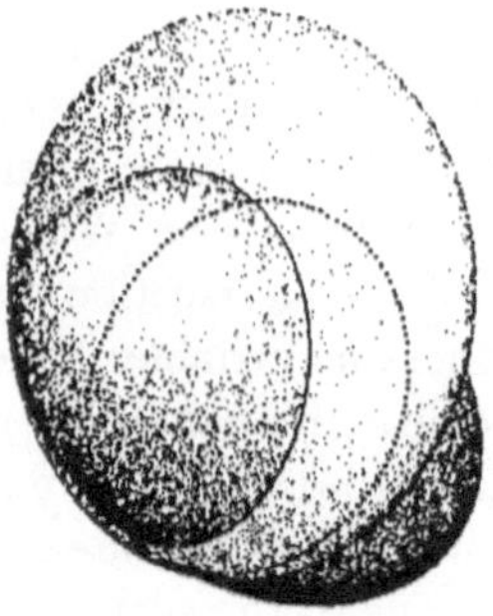

Figure 4.2 Interference of two ellipsoids, intersection curves are shown

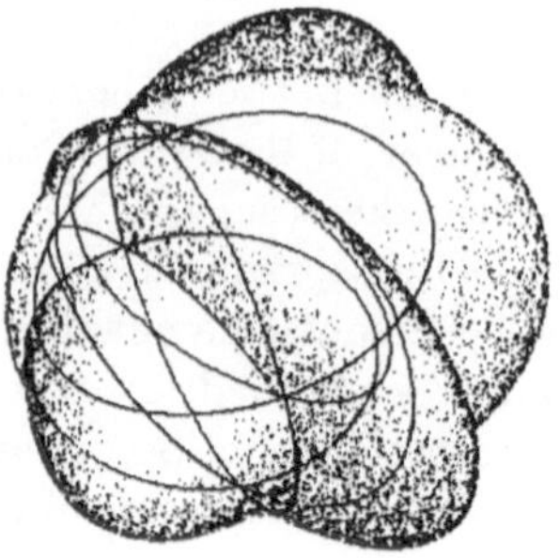

Figure 4.3 Interference of three ellipsoids, intersection curves are also shown

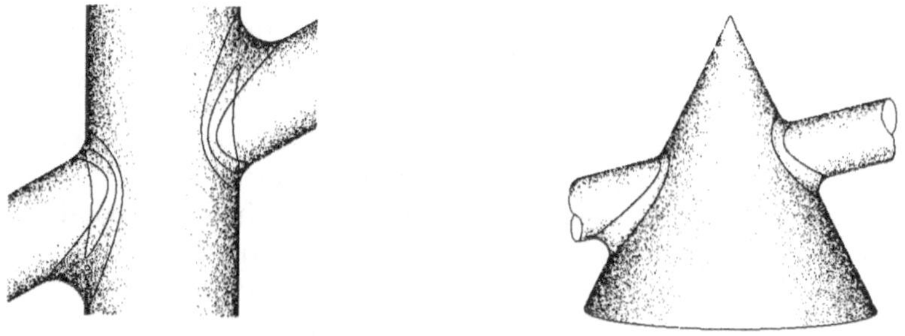

(a) (b)

Figure 4.4 Fillet generation by rolling ball method: (a) interference of two cylinders, (b) interference of a cylinder and a cone, ball-contact curves are also shown

starting from z_0 and μ_0. For a new μ, we obtain new intersecting points. Examples of intersection of two ellipsoids obtained by this method are shown in Fig. 4.2.

This method is extended easily to a case of intersection of multiple quadrics. Figure 4.3 shows an example of the intersection of three ellipsoids. When the intersecting quadrics are circular cones or cylinders, their offset surfaces are also the same type. Since the intersection curves of their offset surfaces are easily obtained, they are used as loci of the center of a rolling ball to make fillet surfaces along the intersecting corners. Figure 4.4 shows these examples.

Another method of intersection tracing of two surfaces represented by two implicit equations is to use the numerical methods of solving differential equations. Details of this method are explained in Sect. 16.3. It is efficient and robust and critical cases can also be treated.

5 Theory of Curves

5.1 Introduction

Curves are not only the fundamental geometric elements, but also they give us various information on shapes. In CAD/CAM, free-form curves have to be mathematically defined or extracted so that their properties can be controlled and evaluated by the computer. Accordingly, knowledge of the fundamentals of the differential geometry of curves is required to apply geometry to practical problems and it also gives the concepts and methods needed to understand the theory of surfaces which is explained in the next two chapters [1]. The curves are usually expressed in parametric forms, and arc length of the curve is used for the parameter in theoretical treatments because of its simplicity of expression, but for practical uses the parameter is changed from arc length s to a more manageable variable parameter t which monotonically increases with arc length. When treating interference problems of curves or surfaces, the curves expressed in implicit forms are sometimes convenient in numerical calculations or theoretical analysis. Therefore, we introduce a method of converting from a parametric form to an implicit one in the last section.

5.2 Tangent and Curvature of Curve

When a position vector $\mathbf{r}$ is a function of one variable t, $\mathbf{r}$ represents a curve in space:

$$\mathbf{r}(t) = \{x(t), y(t), z(t)\}. \tag{5.1}$$

The independent variable is any parameter which is convenient to calculate, but when the geometric properties of a curve are to be described concisely, an arc length s measured along the curve is used as the parameter. So, initially we use s and express the curve as $\mathbf{r}(s)$, and afterwards convert it to $\mathbf{r}(t)$ for practical uses.

We introduce a unit tangent vector $\mathbf{t}$, a unit normal vector $\mathbf{n}$ and a curvature κ to describe the behavior of a curve $\mathbf{r}(s)$ in the vicinity of a point on the curve. A unit tangent vector $\mathbf{t}$ of a curve is a unit vector from $\mathbf{r}(s)$ to $\mathbf{r}(s+ds)$, which is given by

$$\mathbf{t} = \frac{d\mathbf{r}}{|d\mathbf{r}|} = \lim_{ds \to 0} \frac{\mathbf{r}(s+ds) - \mathbf{r}(s)}{|\mathbf{r}(s+ds) - \mathbf{r}(s)|} = \mathbf{r}' \frac{ds}{|d\mathbf{r}|} \tag{5.2}$$

Here, a prime mark indicates differentiation with respect to the arc s. When ds becomes infinitesimal, $|d\mathbf{r}|$ is equal to ds,

$$|d\mathbf{r}| = ds, \tag{5.3}$$

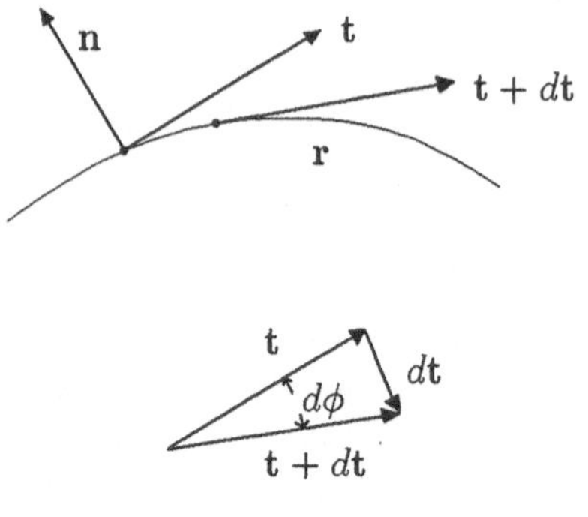

Figure 5.1 Tangent vectors and their difference and normal vector

and we obtain from eq. (5.2)

$$\mathbf{t} = \mathbf{r}'. \tag{5.4}$$

A unit tangent vector of a curve is equal to the first derivative with respect to s.

Let $\mathbf{n}$ be the unit vector orthogonal to $\mathbf{t}$ in the plane determined by $\mathbf{t}$ and $\mathbf{t} + d\mathbf{t}$ and let its direction be 90 degrees anti-clockwise to that of $\mathbf{t}$.

The *curvature* κ at a point on a curve is the ratio of the anti-clockwise rotation angle of the unit tangent vector $\mathbf{t}$ to the infinitesimal increment ds of arc length. Let $\mathbf{t}$ and $\mathbf{t} + d\mathbf{t}$ be the unit tangent vectors on the curve separated by ds, and $d\phi$ an angle between these vectors, then the curvature κ is defined by

$$\kappa = \frac{d\phi}{ds}. \tag{5.5}$$

As shown in Fig. 5.1, $|\delta\phi| \approx (|\delta\mathbf{t}|)/|\mathbf{t}|$, and $|\mathbf{t}| = 1$ by definition, so the relation $|\delta\phi| \approx |\delta\mathbf{t}|$ holds. It follows that we get $\delta\mathbf{t} \approx \mathbf{n}\delta\phi$, and from eq. (5.5) we obtain the relation between the unit tangent and the unit normal vectors :

$$\frac{d\mathbf{t}}{ds} = \kappa\mathbf{n}. \tag{5.6}$$

The curvature κ is positive if $d\mathbf{t}$ and $\mathbf{n}$ are in the same direction. When a curve proceeds anti-clockwise, its curvature is positive. From equations (5.4) and (5.6), we obtain

$$\mathbf{r}'' = \kappa\mathbf{n}. \tag{5.7}$$

The second derivative of a curve by an arc s is equal to the normal vector multiplied by the value of its curvature.

The plane determined by $\mathbf{t}$ and $\mathbf{t} + d\mathbf{t}$ is considered to be the plane determined by three points $\mathbf{r}(s)$, $\mathbf{r}(s + ds)$ and $\mathbf{r}(s + 2ds)$ in their limit and is called the *osculating plane* at the point on the curve. The normal vector $\mathbf{n}$ is in the osculating plane and orthogonal to $\mathbf{t}$ in the anti-clockwise direction. The vector $\mathbf{n}$ is called the *principal normal* vector. Let the radius of a circle passing through these three points determining the osculating plane be ρ. Since there is a relation

$$\rho d\phi = ds, \tag{5.8}$$

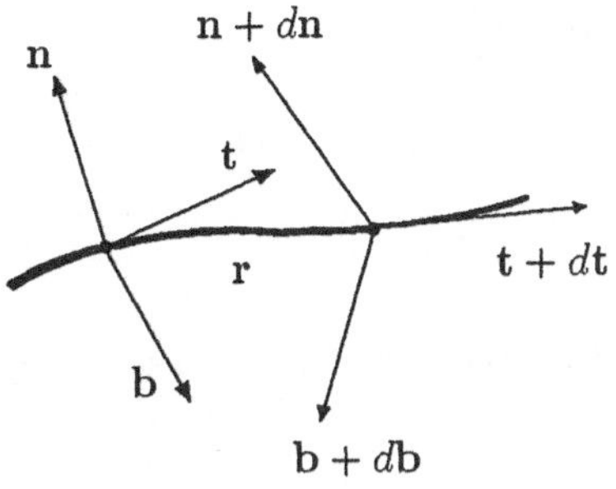

Figure 5.2 Binormal and torsion

we obtain from equations (5.5) and (5.8)

$$\rho = \frac{1}{\kappa}. \tag{5.9}$$

The radius ρ is called the *radius of curvature*, and the center of the circle is called the *center of curvature*, whose location is

$$\mathbf{r}_c = \mathbf{r} + \rho\mathbf{n}.$$

5.3 Binormal and Torsion of Curve

There is a unit vector $\mathbf{b}$ orthogonal to $\mathbf{t}$ and $\mathbf{n}$, called the **binormal vector**, which is the unit normal vector of the osculating plane and has the following relation from equations (5.4), (5.7) and (5.9):

$$\mathbf{b} = \mathbf{t} \times \mathbf{n} = \rho(\mathbf{r}' \times \mathbf{r}''). \tag{5.10}$$

By differentiating the relations $\mathbf{b} \cdot \mathbf{b} = 1$ and $\mathbf{t} \cdot \mathbf{b} = 0$ with respect to s, we obtain

$$\mathbf{b} \cdot \mathbf{b}' = 0, \quad \mathbf{t}' \cdot \mathbf{b} + \mathbf{t} \cdot \mathbf{b}' = 0.$$

From eq. (5.6) and from eq. (5.10), $(\mathbf{t}' \cdot \mathbf{b})$ becomes $\kappa[\mathbf{n}, \mathbf{t}, \mathbf{n}]$, which equals zero, and from the above relation, we get $(\mathbf{t} \cdot \mathbf{b}') = 0$. This means $\mathbf{b}'$ is orthogonal to $\mathbf{t}$ as well as to $\mathbf{b}$, so it has a component of $\mathbf{n}$ only. We can express it using the proportional constant τ as

$$\frac{d\mathbf{b}}{ds} = -\tau\mathbf{n}. \tag{5.11}$$

This is interpreted as follows: when advancing along the curve, the rate of rotation of the osculating plane is τ and the rotation is around the tangent in the right screw direction if τ is positive. See Fig. 5.2. The parameter τ is called the *torsion*.

Differentiating $\mathbf{n} = \mathbf{b} \times \mathbf{t}$, which is obtained from eq. (5.10), we get

$$\frac{d\mathbf{n}}{ds} = \frac{d\mathbf{b}}{ds} \times \mathbf{t} + \mathbf{b} \times \frac{d\mathbf{t}}{ds}.$$

Substituting equations (5.6) and (5.11) into the above, we obtain

$$\frac{d\mathbf{n}}{ds} = -\tau\mathbf{n} \times \mathbf{t} + \kappa\mathbf{b} \times \mathbf{n} = \tau\mathbf{b} - \kappa\mathbf{t}. \tag{5.12}$$

Equations (5.6), (5.11) and (5.12) are called *Serret-Frenet's formulas.* They are written collectively:

$$\left.\begin{array}{rcl} \mathbf{t}' & = & \kappa\mathbf{n} \\ \mathbf{n}' & = & \tau\mathbf{b} - \kappa\mathbf{t} \\ \mathbf{b}' & = & -\tau\mathbf{n} \end{array}\right\}. \tag{5.13}$$

Like the expression of curvature in eq. (5.7), we make an expression for torsion. Differentiating $\mathbf{r}(s)$ repeatedly, we obtain

$$\mathbf{r}' = \mathbf{t}, \quad \mathbf{r}'' = \kappa\mathbf{n}, \quad \mathbf{r}''' = \kappa'\mathbf{n} + \kappa(\tau\mathbf{b} - \kappa\mathbf{t}). \tag{5.14}$$

The following triple product

$$[\mathbf{r}',\ \mathbf{r}'',\ \mathbf{r}'''] = (\mathbf{r}' \times \mathbf{r}'') \cdot \mathbf{r}'''$$

is transformed to

$$\kappa\mathbf{b} \cdot \kappa'\mathbf{n} + \kappa^2\mathbf{b} \cdot (\tau\mathbf{b} - \kappa\mathbf{t}) = \kappa^2\tau.$$

From this we have

$$\tau = \left(\frac{1}{\kappa^2}\right)[\mathbf{r}',\ \mathbf{r}'',\ \mathbf{r}''']. \tag{5.15}$$

Next, we determine an approximate shape of a curve near a point on it. By expanding $\mathbf{r}(s)$, with $\mathbf{r}(0) = 0$ as the origin, we get using equations (5.14),

$$\begin{aligned} \mathbf{r}(s) &= \mathbf{r}'s + (1/2)\mathbf{r}''s^2 + (1/6)\mathbf{r}'''s^3 + \cdots \\ &= (s - \kappa^2\frac{s^3}{6})\mathbf{t} + (\kappa\frac{s^2}{2} + \kappa'\frac{s^3}{6} + \cdots)\mathbf{n} + (\kappa\tau\frac{s^3}{6} + \cdots)\mathbf{b}. \end{aligned} \tag{5.16}$$

To express the shape of a curve near the origin with $\mathbf{t}$, $\mathbf{n}$ and $\mathbf{b}$ as the coordinate axes, we take the first term of each coefficient of $\mathbf{t}$, $\mathbf{n}$ and $\mathbf{b}$ in eq. (5.16) and replace $\mathbf{t}$, $\mathbf{n}$ and $\mathbf{b}$ by x, y and z, then we have

$$x = s, \ y = \kappa\frac{s^2}{2}, \ z = \kappa\tau\frac{s^3}{6}. \tag{5.17}$$

Eliminating s in eq. (5.17), we obtain

$$y = \frac{1}{2}\kappa x^2, \ y^3 = \frac{9}{2}\frac{\kappa}{\tau^2}z^2, \ z = \frac{1}{6}\kappa\tau x^3. \tag{5.18}$$

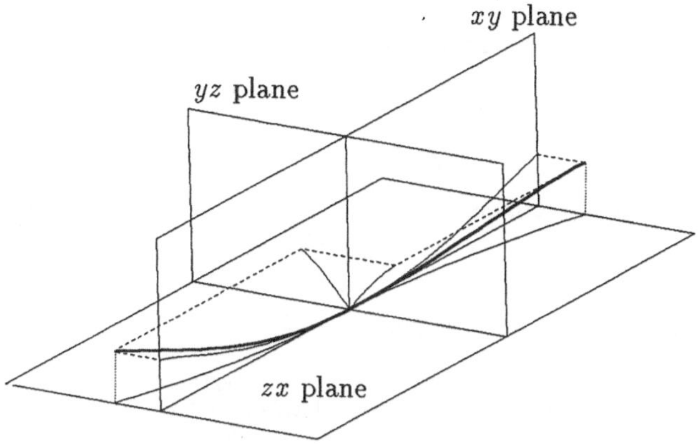

Figure 5.3 Shape of a curve near its origin. Its projections on three coordinate planes are shown

Referring to Fig. 5.3, from the above equations, we can understand the local shape of a curve. The projection of the curve on the osculating plane(the xy plane) is a symmetrical second-degree curve with a tangent to the x axis at the origin. The projection on the normal plane (the yz plane) is symmetrical with respect to the z axis and has a cusp at the origin. The projection on the tangential plane (the zx plane) is anti-symmetrical with respect to the origin and has a tangent to the x axis. If the curvature does not take an extremum value there, the shape of the curve in the $\mathbf{n}$ direction distorts anti-symmetrically, because of the term $\kappa' s^3/6$ in eq. (5.16).

5.4 Expressions with Parameter t

So far, we have treated $\mathbf{r}$ as a function of the arc s, but in practice it is difficult to describe curves as functions of s, because we do not know the function of a curve before designing a shape which determines the independent variable s. Accordingly, instead of s, for convenience, we use a parameter t which monotonically increases with s.

With the use of parameter t, calculation of position on the curve is not difficult, but derivation of the geometric quantities of the curve is not as simple as doing so in terms of s. So, in the following, the derivation of $\mathbf{r}'$,$\mathbf{r}''$ and $\mathbf{r}'''$ as functions of the parameter t is described first. For brevity, we denote $d\mathbf{r}/dt$, $d^2\mathbf{r}/dt^2$ and $d^3\mathbf{r}/dt^3$ by $\dot{\mathbf{r}}$, $\ddot{\mathbf{r}}$ and $\mathbf{r}^{(3)}$. Changing the independent variable from s to t, we have

$$\dot{\mathbf{r}} = \frac{d\mathbf{r}}{ds} \cdot \frac{ds}{dt} = \mathbf{t} \cdot \frac{ds}{dt}. \tag{5.19}$$

Hence, a vector $\dot{\mathbf{r}}$ has the same direction as $\mathbf{t}$. By squaring both sides of the above equation, we get

$$\dot{\mathbf{r}}^2 = \mathbf{t}^2 \cdot (\frac{ds}{dt})^2 = (\frac{ds}{dt})^2, \tag{5.20}$$

so we have

$$\frac{ds}{dt} = \sqrt{\dot{\mathbf{r}} \cdot \dot{\mathbf{r}}} = |\dot{\mathbf{r}}|, \tag{5.21}$$

and the unit tangent vector is

$$\mathbf{t} = \frac{\dot{\mathbf{r}}}{|\dot{\mathbf{r}}|}. \tag{5.22}$$

Differentiating eq. (5.19) by t, we obtain

$$\ddot{\mathbf{r}} = \frac{d^2\mathbf{r}}{ds^2} \cdot (\frac{ds}{dt})^2 + \frac{d\mathbf{r}}{ds} \cdot \frac{d^2 s}{dt^2} = \kappa\mathbf{n} \cdot (\frac{ds}{dt})^2 + \mathbf{t} \cdot \frac{d^2 s}{dt^2}. \tag{5.23}$$

Multiplying $\mathbf{n}$ with both sides of the above equation, we have an expression for the curvature by using eq. (5.21),

$$\kappa = \frac{(\mathbf{n} \cdot \ddot{\mathbf{r}})}{\dot{\mathbf{r}}^2}. \tag{5.24}$$

And differentiating eq. (5.21) and using eq. (5.22) we have

$$\frac{d^2 s}{dt^2} = \frac{(\ddot{\mathbf{r}} \cdot \dot{\mathbf{r}})}{(\sqrt{\dot{\mathbf{r}} \cdot \dot{\mathbf{r}}})} = (\ddot{\mathbf{r}} \cdot \mathbf{t}). \tag{5.25}$$

From eq. (5.23) and eq. (5.25), the vector $\kappa\mathbf{n}$ becomes

$$\kappa\mathbf{n} = \frac{\{\ddot{\mathbf{r}} - \mathbf{t} \cdot (\ddot{\mathbf{r}} \cdot \mathbf{t})\}}{\dot{\mathbf{r}}^2}. \tag{5.26}$$

Next, the torsion is treated. By differentiating eq. (5.23) by t,

$$\frac{d^3\mathbf{r}}{dt^3} = \mathbf{n} \cdot (\frac{d}{dt})(\kappa\dot{\mathbf{r}}^2) + \kappa\mathbf{n}' \cdot (\frac{ds}{dt})^3 + \mathbf{t}' \cdot (\frac{ds}{dt}) \cdot \frac{d^2 s}{dt^2} + \mathbf{t} \cdot \frac{d^3 s}{dt^3},$$

then eliminating $\mathbf{t}'$ and $\mathbf{n}'$ by using *Serret-Frenet's formulas* (5.13), and multiplying $\mathbf{b}$ with its result, we obtain

$$\mathbf{b} \cdot \frac{d^3\mathbf{r}}{dt^3} = \kappa\tau|\dot{\mathbf{r}}|^3.$$

From this, the torsion is given by

$$\tau = (\mathbf{b} \cdot \frac{d^3\mathbf{r}}{dt^3})/(\kappa|\dot{\mathbf{r}}|^3). \tag{5.27}$$

For a curve of degree two, there is no torsion because $d^3\mathbf{r}/dt^3$ vanishes.

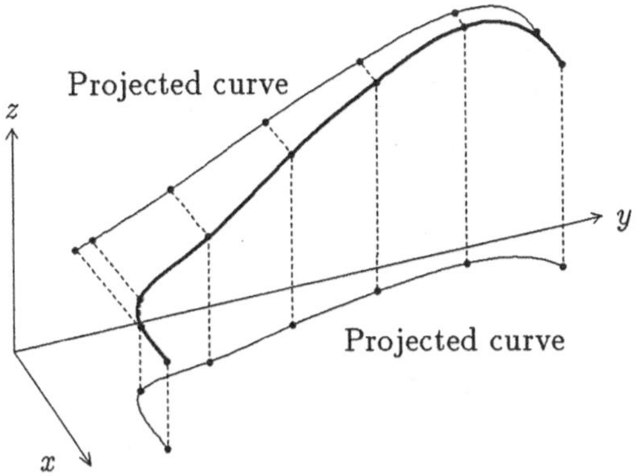

Figure 5.4 Space curve determined by its two orthogonal projections

5.5 Curvature of Space Curve and Its Projection

This section is an example of the application of the previous sections. When measurement data of a space curve in a real three-dimensional model are not available, designers have to define a space curve they intended to make with drawings of its projection curves. See Fig. 5.4. Consequently, sometimes it is required from these orthogonally projected curves to construct a space curve expression for use in the computer. The expression must have the same geometric properties as those the designers expressed on the drawings.

For the projection curves, we assume their expressions have already been determined independently. They are denoted by $\mathbf{r}_{xy}$ and $\mathbf{r}_{yz}$. And their respective curvatures at the ends κ_{xy} and κ_{yz} are assumed known. From the tangent directions and the curvatures of the corresponding end points of the projected curves, we are to obtain the tangent vector $\mathbf{t}$, the normal and binormal vectors $\mathbf{n}$ and $\mathbf{b}$ and the curvature κ of the space curve segment.

Since a space curve near the origin is given by eq. (5.16), its x, y, z components are, higher degrees than s^2 being neglected,

$$\begin{aligned} &x = st_x + (1/2)\kappa s^2 n_x, \quad y = st_y + (1/2)\kappa s^2 n_y, \\ &z = st_z + (1/2)\kappa s^2 n_z, \\ &\text{where} \quad \mathbf{t} = (t_x, t_y, t_z), \quad \mathbf{n} = (n_x, n_y, n_z). \end{aligned} \tag{5.28}$$

The components of $\mathbf{t}$ are obtained from the tangent directions of the projected curves, because of the relation $(dx : dy : dz) = (t_x : t_y : t_z)$ at the end point.

The curvatures of the curves on the projected (x, y) and (y, z) planes are expressed by

$$\kappa_{xy} = \frac{y''x' - x''y'}{(x'^2 + y'^2)^{3/2}}, \quad \kappa_{yz} = \frac{z''y' - y''z'}{(y'^2 + z'^2)^{3/2}}. \tag{5.29}$$

Substituting the equations in (5.28) into the above equations, we obtain

$$\kappa_{xy} = \frac{\kappa(n_y t_x - n_x t_y)}{(t_x^2 + t_y^2)^{3/2}}, \quad \kappa_{yz} = \frac{\kappa(n_z t_y - n_y t_z)}{(t_y^2 + t_z^2)^{3/2}}. \tag{5.30}$$

By setting

$$A = (t_x^2 + t_y^2)^{3/2}\kappa_{xy}, \quad B = (t_y^2 + t_z^2)^{3/2}\kappa_{yz}, \tag{5.31}$$

eq. (5.30) becomes

$$A = \kappa(n_y t_x - n_x t_y), \quad B = \kappa(n_z t_y - n_y t_z). \tag{5.32}$$

Taking the ratio of the two equations of (5.32), we obtain

$$Bt_y n_x - (Bt_x + At_z)n_y + At_y n_z = 0. \tag{5.33}$$

Since $\mathbf{t}$ and $\mathbf{n}$ are orthogonal, there is a relation:

$$t_x n_x + t_y n_y + t_z n_z = 0. \tag{5.34}$$

From (5.33) and (5.34) the components of $\mathbf{n}$ can be written in the forms:

$$\left.\begin{array}{rcl} n_x &=& C\{-(Bt_x + At_z)t_z - At_y^2\} \\ n_y &=& C\{-Bt_y t_z + At_x t_y\} \\ n_z &=& C\{Bt_y^2 + (Bt_x + At_z)t_x\} \end{array}\right\}. \tag{5.35}$$

The constant C is determined from the normalizing condition: $n_x^2 + n_y^2 + n_z^2 = 1$, as

$$C = \{(t_y^2 + t_x^2)B^2 + (t_y^2 + t_z^2)A^2\}^{-1/2}. \tag{5.36}$$

We can determine $\mathbf{n}$ by equations (5.35), (5.31) and (5.36). The curvature κ is given from equations (5.32) and (5.35) by

$$\kappa = 1/(Ct_y). \tag{5.37}$$

The direction of the tangent vector $\mathbf{t}$ of the space curve is easily determined from the tangent vectors of the projected curves, and the normal vector $\mathbf{n}$ is calculated from equations (5.31), (5.36) and (5.35). Accordingly, the binormal vector $\mathbf{b}$ at the stating point is obtained from $\mathbf{n} \times \mathbf{t}$. The curvature of the space curve is given by eq. (5.37) and the shape of the curve near its starting point in space is given by eq. (5.28.). Since these values are determined at both ends of the space curve segment, it can be expressed by a Bézier curve of degree five or higher.

5.6 Implicit Expression of a Parametric Curve

In parametric expressions of plane curves $\mathbf{r}_1(t)$ and $\mathbf{r}_2(t)$, there is no relation between parameters t though the same symbol is used. When considering the intersection of the two curves, the two t's have to be changed to t_1 and t_2 to avoid confusion. The intersection points are determined from the solution of the following vector equation

$$\mathbf{r}_1(t_1) = \mathbf{r}_2(t_2).$$

Generally this represents a set of simultaneous non-linear equations, from which t_1 and t_2 are obtained as values of the intersection points. Except for straight line segments, their solution is hard to obtain.

In a case involving plane curves, if either of their expressions can be converted into an implicit form $f(x, y) = 0$, then we substitute the expressions for $x(t)$ and $y(t)$ of the other curve into $f(x, y) = 0$ to obtain an algebraic equation $f\{x(t), y(t)\} = 0$ in a single variable t. So long as its degree in t is not very high, we can use well established solving methods for algebraic equations.

Then the problem is implicitization of a parametric expression. We present a method of conversion [5][6]. From the x and y components of $\mathbf{r}$(t) we define two polynomial equations $X(t) = 0$ and $Y(t) = 0$:

$$X(t) = a_n t^n + a_{n-1}t^{n-1} + \cdots + a_0 = 0, \tag{5.38}$$
$$Y(t) = b_n t^n + b_{n-1}t^{n-1} + \cdots + b_0 = 0. \tag{5.39}$$

When $\mathbf{r}(t)$ is expressed by a polynomial of t, the variable x is contained in a_0 and the variable y in b_0 only. When $\mathbf{r}$(t) is a rational expression which is treated in Chap. 12, x is contained in all a_i and y in all b_i, but a coefficient of x in each a_i and that of y in each b_i are the same.

If a value of t satisfies both equations (5.38) and (5.39), a pair (x, y) is on the curve $\mathbf{r}(t)$. This value of t is a common root of equations (5.38) and (5.39). Even if we make an equations of degree $n-1$ from them by elimination of t^n, the result contains the common roots. There are n ways of elimination, as shown below.

An operation $b_n \times X(t) - a_n \times Y(t)$ gives

$$c_{0,n-1}t^{n-1} + c_{0,n-2}t^{n-2} + \cdots + c_{0,0} = 0, \tag{5.40}$$

where the coefficients $c_{0,n-1}, c_{0,n-2}, \cdots, c_{0,0}$ of powers of t contain only linear functions of x, y and *const* and are easily determined.

By parenthesizing the first two terms of eq. (5.38) and eq. (5.39) with common t^{n-1} outside we change them into

$$X(t) = (a_n t + a_{n-1})t^{n-1} + \cdots + a_0 = 0, \tag{5.41}$$
$$Y(t) = (b_n t + b_{n-1})t^{n-1} + \cdots + b_0 = 0. \tag{5.42}$$

An operation $\{(b_n t + b_{n-1}) \times (5.41) - (a_n t + a_{n-1}) \times (5.42)\}$ eliminates the first terms and gives

$$c_{1,n-1}t^{n-1} + c_{1,n-2}t^{n-2} + \cdots + c_{1,0} = 0, \tag{5.43}$$

where $c_{i,n-i}$ are constants containing only linear functions of x, y and *const*. Generally, by parenthesizing the first $i+1$ terms of eq. (5.38) and eq. (5.39), we get

$$\begin{aligned} X(t) &= (a_n t^i + a_{n-1}t^{i-1} + \cdots + a_{n-i})t^{n-i} + \cdots + a_0 = 0, & (5.44)\\ Y(t) &= (b_n t^i + b_{n-1}t^{i-1} + \cdots + b_{n-i})t^{n-i} + \cdots + b_0 = 0. & (5.45) \end{aligned}$$

By a similar elimination process of coefficients for t^{n-i} we obtain an equation of degree $n-1$ of the form:

$$c_{i,n-1}t^{n-1} + c_{i,n-2}t^{n-2} + \cdots + c_{i,0} = 0. \tag{5.46}$$

Continuing these parenthesizing and eliminating processes, finally we obtain n equations of degree $n-1$ in t, which have a common root:

$$\begin{bmatrix} c_{0,n-1} & c_{0,n-2} & \cdots & c_{0,0} \\ c_{1,n-1} & c_{1,n-2} & \cdots & c_{1,0} \\ \vdots & \vdots & \ddots & \vdots \\ c_{n-1,n-1} & \cdots & \cdots & c_{n-1,0} \end{bmatrix} \times \begin{bmatrix} t^{n-1} \\ t^{n-2} \\ \vdots \\ t^0 \end{bmatrix} = 0. \tag{5.47}$$

We interpret these n equations as n homogeneous linear equations for n variables $(t^{n-1}, t^{n-2}, \cdots, t, 1)$. And if these linear simultaneous equations have solutions other than zero, the determinant of the coefficient matrix must be zero.

An element of this determinant is a linear function with respect to x and y. The development of this determinant gives an algebraic equation of degree n in x and y, which is the implicit form of $\mathbf{r}(t)$.

Example. Implicitization of a rational Bézier expression of degree two, which is treated in Sect. 12.6.

Let control points be $\mathbf{p}_0 = (x_0, y_0)$, $\mathbf{p}_1 = (x_1, y_1)$, $\mathbf{p}_2 = (x_2, y_2)$, and the parameter t be $0 \le t \le 1$.

A rational Bézier curve is given by

$$\mathbf{r}(t) = \frac{(1-t)^2\mathbf{p}_0 + 2t(t-1)w\mathbf{p}_1 + 3t^2\mathbf{p}_2}{(1-t)^2 + 2t(1-t)w + t^2}. \tag{5.48}$$

For simplicity, the parameter t is transformed to u by a relation $t = u/(1+u)$, and eq. (5.48) becomes

$$\mathbf{r}(t) = \frac{\mathbf{p}_0 + 2uw\mathbf{p}_1 + u^2\mathbf{p}_2}{1 + 2uw + u^2}. \tag{5.49}$$

Since $\mathbf{r}$ is (x, y), the above equation is transformed to

$$\begin{aligned} X(u) &= u^2(x - x_2) + 2uw(x - x_1) + (x - x_0) = 0, \\ Y(u) &= u^2(y - y_2) + 2uw(y - y_1) + (y - y_0) = 0. \end{aligned}$$

Applying the elimination method described above, we obtain the following two equations of u:

$$\begin{aligned}
& (y-y_2)X(u)-(x-x_2)Y(u) \\
= \;& 2w\{(x-x_1)(y-y_2)-(x-x_2)(y-y_1)\}u+ \\
& \{(x-x_0)(y-y_2)-(x-x_2)(y-y_0)\} \\
= \;& 0,
\end{aligned}$$

$$\begin{aligned}
& \{(y-y_2)u+2w(y-y_1)\}X(u)-\{(x-x_2)u+2w(x-x_1)\}Y(u) \\
= \;& \{(x-x_0)(y-y_2)-(x-x_2)(y-y_0)\}u \\
& +2w\{(x-x_0)(y-y_1)-(x-x_1)(y-y_0)\} \\
= \;& 0.
\end{aligned}$$

Eliminating u from the above two equations, we obtain

$$\begin{aligned}
& 4w^2\{(y_1-y_2)x+(x_2-x_1)y+x_1y_2-x_2y_1\} \\
& \quad \{(y_0-y_1)x+(x_1-x_0)y+x_0y_1-x_1y_0\} \\
& -\{(y_0-y_2)x+(x_2-x_0)y+x_0y_2-x_2y_0\}^2=0.
\end{aligned}$$

Now, for simplicity, we put $y_0=y_2=0$, $y_1=h$, $-x_2=x_0=a$, $x_1=b$, and the above equation becomes

$$h^2x^2-2bhxy+\{c^2/w^2-(a^2-b^2)\}y^2+2a^2hy-a^2h^2=0. \tag{5.50}$$

This is a conic equation, which is treated in Chap. 5.

6 Basic Theory of Surfaces

6.1 Introduction

Quadric surfaces, especially bodies of revolution such as spheres, cylinders, cones, or also ruled surfaces have been widely used in engineering products. This is because their geometric properties are easily specified by their designers and the shapes are easily manufactured by machines. On the other hand, free-form surfaces have only been used in special cases and their design and manufacture have required special talent and skill. But in recent years their uses have increased because of strong demands for high performance and aesthetic quality in engineering products. And computers have been used for their design and manufacture. Accordingly, various mathematical expressions of free-forms for engineering use have been developed.

With these advanced requirements, appropriate mathematical expressions for free-form surfaces and methods of design and evaluation for their shapes have been studied. Accordingly, knowledge of the fundamental geometric theory of surfaces is indispensable for engineers in this field. Since the theory of the differential geometry of surfaces is not new, but has been treated from a mathematical point of view and not from engineering one, practical engineers have not wanted to approach it. In textbooks on differential geometry, tensor notation is widely used, because it is concise in analysis and description of geometry in n-dimensional space. But since uses of the surfaces in CAD is limited to three dimensions (mathematically two dimensions), adoption of the tensor notation has little merit for practical engineers unless it is used advantageously in other domains of their tasks. Moreover, since we emphasize geometrical intuition and understanding rather than elegance of the mathematical formalism, we use the conventional notations in this book.

First, we introduce the basic framework for description of the geometric characteristics of the surfaces, then explain the features which determine the local shape of surfaces such as the tangential plane and the principal curvatures and the lines of curvature.

Since the behavior of normal vectors of the surface gives important information for analysis and understanding of the shape features of the surface, we explain the formulas for the differential of a unit normal vector, and the local shape around a point on the surface. In this chapter we treat the fundamental parts of the theory of surfaces. Its advance topics, such as various curves which characterize the surfaces and the offset surfaces, are left to the next chapter.

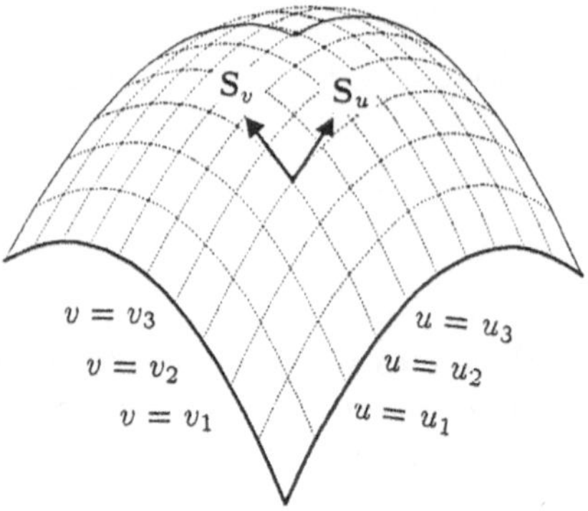

Figure 6.1 Surface patch with u-v parameter net

6.2 The Basic Vectors and the Fundamental Magnitudes

A surface segment or a surface patch is denoted in a parametric form by

$$\mathbf{s}(u,v) = \{x(u,v), y(u,v), z(u,v)\} \quad 0 \le u, v \le 1. \tag{6.1}$$

If there is any functional relation between u and v, $\mathbf{s}(u,v)$ represents a curve on the surface. Values of $\mathbf{s}(u,v)$ of $u = const$ make a family of curves on the surface and those of $\mathbf{s}(u,v)$ of $v = const$ generate another family of curves. They constitute a net of curves on the surface. Usually, any point on the surface is determined by the intersection of two curves, each of which belongs to one of the families. See Fig. 6.1. A partial derivative $\partial\mathbf{s}/\partial u$ represents a tangent vector of the curves $v = const$ and $\partial\mathbf{s}/\partial v$ is a tangent vector of the curves $u = const$. They are denoted by $\mathbf{s}_u$ and $\mathbf{s}_v$ for brevity and are called *the basic vectors* of the surface.

They span the tangent plane of the surface. The vector $d\mathbf{s}$ which connects two points on the surface from $\mathbf{s}(u,v)$ to $\mathbf{s}(u+du, v+dv)$ is given by

$$d\mathbf{s} = \mathbf{s}_u du + \mathbf{s}_v dv. \tag{6.2}$$

The square of the absolute value of $d\mathbf{s}$ is

$$(d\mathbf{s})^2 = d\mathbf{s} \cdot d\mathbf{s} = \mathbf{s}_u^2 (du)^2 + 2\mathbf{s}_u \cdot \mathbf{s}_v du dv + \mathbf{s}_v^2 (dv)^2. \tag{6.3}$$

From the basic vectors of the surface, the following quantities are defined.

$$E = \mathbf{s}_u^2, \;\; F = \mathbf{s}_u \cdot \mathbf{s}_v, \;\; G = \mathbf{s}_v^2. \tag{6.4}$$

They are called the *fundamental magnitudes of the first order*. Then eq. (6.3) becomes

$$ds^2 = E(du)^2 + 2F du dv + G(dv)^2. \tag{6.5}$$

Let an angle between the basic vectors $\mathbf{s}_u$ and $\mathbf{s}_v$ be ω, then their inner product F and the absolute value H of vector product of the basic vectors can be expressed by the fundamental magnitudes of the first order:

$$\begin{aligned} F &= |\mathbf{s}_u| \cdot |\mathbf{s}_v| \cos\omega = (\sqrt{EG}) \cos\omega, & (6.6) \\ H &= |\mathbf{s}_u \times \mathbf{s}_v| = |\mathbf{s}_u| \cdot |\mathbf{s}_v| \sin\omega \\ &= \sqrt{EG(1-\cos^2\omega)} = \sqrt{EG - F^2}. & (6.7) \end{aligned}$$

Since usually the angle ω is not equal to 90 degrees, the parameter net on the surface constitutes an oblique coordinate system with its basic vectors. In analysis and description in the oblique coordinate systems, the tensor notation is convenient, but in this case, as the oblique coordinate system is only two-dimensional, its merit is lost to those who are not familiar with the tensor notation and its operations.

The third coordinate axis of the local coordinate system is taken in the direction orthogonal to $\mathbf{s}_u$ and $\mathbf{s}_v$, that is, along the normal vector of the surface. The *unit normal vector* of a point on the surface is given by

$$\mathbf{n} = \frac{(\mathbf{s}_u \times \mathbf{s}_v)}{H}. \qquad (6.8)$$

6.3 Normal Section and Normal Curvature

A pencil of tangent vectors at P on the surface is in the tangent plane there. One of the *unit tangent vector* $\mathbf{t}$ is expressed from eq. (6.2) as

$$\mathbf{t} = \frac{d\mathbf{s}}{ds} = \mathbf{s}_u(\frac{du}{ds}) + \mathbf{s}_v(\frac{dv}{ds}), \qquad (6.9)$$

where $ds = |d\mathbf{s}|$, and a plane determined by $\mathbf{t}$ and $\mathbf{n}$ is called a *normal plane.* The intersection of the surface by a normal plane is called a *normal section.* It is a plane curve. See Fig. 6.2.

The curvature κ at P of the normal section is called the *normal curvature* at P, which is obtained as follows. Differentiating $\mathbf{t}$ with respect to the arc s of the normal section, we obtain

$$\begin{aligned} \frac{d\mathbf{t}}{ds} &= \mathbf{s}_u \frac{d^2u}{ds^2} + \mathbf{s}_v \frac{d^2v}{ds^2} + \\ & + \mathbf{s}_{uu}(\frac{du}{ds})^2 + 2\mathbf{s}_{uv}(\frac{du}{ds})(\frac{dv}{ds}) + \mathbf{s}_{vv}(\frac{dv}{ds})^2. \qquad (6.10) \end{aligned}$$

This is equal to $\kappa\mathbf{n}$ from eq. (5.6) of Chap. 5, when we consider the sign of curvature. Multiplying the normal vector $\mathbf{n}$ with the above equation (6.10), we can eliminate the first and the second terms of the right-hand side, because they are

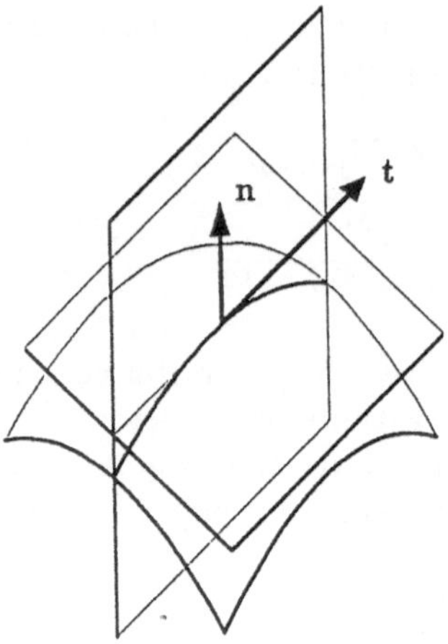

Figure 6.2 Normal section and tangential plane

in the tangential plane. Introducing the following symbols called the *fundamental magnitudes of the second order*,

$$L = \mathbf{n} \cdot \mathbf{s}_{uu}, \quad M = \mathbf{n} \cdot \mathbf{s}_{uv}, \quad N = \mathbf{n} \cdot \mathbf{s}_{vv}, \tag{6.11}$$

we obtain the relation:

$$(\mathbf{n} \cdot \mathbf{n})\kappa = L(\frac{du}{ds})^2 + 2M(\frac{du}{ds})(\frac{dv}{ds}) + N(\frac{dv}{ds})^2. \tag{6.12}$$

Using eq. (6.5), eq. (6.12) becomes

$$\kappa = \frac{L(du)^2 + 2M\,dudv + N(dv)^2}{E(du)^2 + 2F\,dudv + G(dv)^2}. \tag{6.13}$$

Values of the fundamental magnitudes of the first and the second order E, G, F, L, M and N are fixed on a point of the surface and do not depend on the direction of the normal plane through the point P. But the direction depends on the ratio $du : dv$. So the normal curvature κ has different values in different directions of the normal section.

When a cutting plane passing through a point on the surface is not orthogonal to the tangential plane at the point, that is, the section is not a normal section, the curvature κ_c of this section at the point is given by

$$\kappa_c = \frac{\kappa}{\cos\theta}, \tag{6.14}$$

where κ is the curvature of the normal section and θ is the angle between the normal plane and the oblique cutting plane. See Fig. 6.3. This relation is called *Meunier's theorem.*

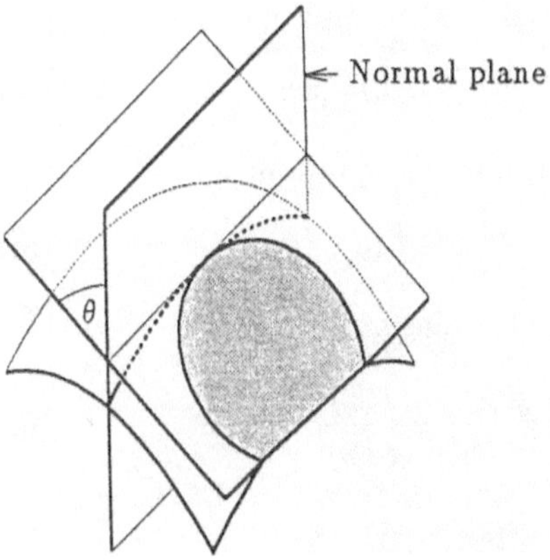

Figure 6.3 Non-normal section and Meunier's theorem

Its proof is given as follows. Let $\mathbf{n}'$ be the unit normal of the cutting section at the point, $\mathbf{t}$ be the unit tangent there and κ_c be the curvature of the section. There is a relation $\kappa_c \mathbf{n}' = d\mathbf{t}/ds$, whose right-hand side is given by eq. (6.10). Multiplying both sides of the equation by $\mathbf{n}$, we obtain $(\kappa_c \mathbf{n} \cdot \mathbf{n}') = (\mathbf{n} \cdot d\mathbf{t}/ds)$, which is $\kappa_c \cos\theta = \kappa$. So we have (6.14).

Exercise. Prove that the curvature κ_c of an intersection curve of two surfaces $\mathbf{s}_1$ and $\mathbf{s}_2$ is given by

$$\kappa_c = \sqrt{(\kappa_1^2 + \kappa_2^2 - 2\kappa_1\kappa_2 \cos\phi)}(\sin^{-1}\phi),$$

where κ_1 and κ_2 are curvatures of the normal sections of $\mathbf{s}_1$ and $\mathbf{s}_2$, which include the tangent of the intersection curve, and ϕ is the angle between the normals of $\mathbf{s}_1$ and $\mathbf{s}_2$ at the point.

6.4 Principal Curvatures

Next, we consider the extremum values of the normal curvatures. When the normal plane rotates around the normal, the shape of the normal section changes with its tangent direction, and so does the normal curvature. The shape returns to the initial state after half a rotation of the normal plane. Now, setting

$$\gamma = \frac{dv}{du} \tag{6.15}$$

and rewriting κ as $\kappa(\gamma)$, a function of γ, we get the following equation from eq. (6.13),

$$\{L - \kappa(\gamma) \cdot E\} + 2\{M - \kappa(\gamma) \cdot F\}\gamma + \{N - \kappa(\gamma) \cdot G\}\gamma^2 = 0. \tag{6.16}$$

This equation gives the relation between κ and γ. With the rotation of the normal plane, $\kappa(\gamma)$ takes the extremum values, which are determined by $d\kappa(\gamma)/d\gamma = 0$.

Differentiating eq. (6.16) with respect to γ together with the above extremum condition, and then writing κ and γ as $\tilde{\kappa}$ and $\tilde{\gamma}$, we obtain an equation:

$$(M - \tilde{\kappa}F) + (N - \tilde{\kappa}G)\tilde{\gamma} = 0, \tag{6.17}$$

and substituting this into eq. (6.15), we get another equation:

$$(L - \tilde{\kappa}E) + (M - \tilde{\kappa})\tilde{\gamma} = 0. \tag{6.18}$$

Then from the above equations, we obtain the relations:

$$\tilde{\gamma} = \frac{-(M - \tilde{\kappa}F)}{(N - \tilde{\kappa}G)} = \frac{-(L - \tilde{\kappa}E)}{(M - \tilde{\kappa}F)}, \tag{6.19}$$

$$\tilde{\kappa} = \frac{(M + \tilde{\gamma}N)}{(F + \tilde{\gamma}G)} = \frac{(L + \tilde{\gamma}M)}{(E + \tilde{\gamma}F)}. \tag{6.20}$$

From eq. (6.19) we have the equation for $\tilde{\kappa}$,

$$(EG - F^2)\tilde{\kappa}^2 - (EN + LG - 2MF)\tilde{\kappa} + LM - M^2 = 0. \tag{6.21}$$

The coefficient of $\tilde{\kappa}^2$ of the above equation is positive because of eq. (6.7), and this equation is the second degree for $\tilde{\kappa}$. Let its roots be κ_1 and κ_2, which are called the *principal curvatures of surface*, then we get the following expressions from the relation between roots and coefficients of a quadratic equation,

$$K_m = \frac{1}{2}(\kappa_1 + \kappa_2) = \frac{1}{2}\frac{(EN + LG - 2MF)}{(EG - F^2)}, \tag{6.22}$$

$$K_g = \kappa_1\kappa_2 = \frac{(LN - M^2)}{(EG - F^2)}. \tag{6.23}$$

The expression K_m is called the *mean curvature* and the expression K_g is the *Gaussian curvature* of the surface. The local shape of a surface and the sign of its Gaussian curvature have the following relation:

$K_g > 0$. The normal curvature has the same sign in all directions, so the tangent plane touches the surface at one point. The usual convex or concave surfaces correspond to this.

$K_g < 0$. The normal curvature becomes zero twice during the half rotation of the normal plane around the normal. The tangent plane intersects with the surface in these directions of zero curvature.
Since the denominator $EG - F^2$ of K_m and K_g is positive from eq. (6.7), the sign of the Gaussian curvature is determined from the sign of the numerator $LN - M^2$. So, when L or N or both are zero, or L and N have different sign, the Gaussian curvature is surely negative. A surface patch $\mathbf{s}(u, v)$ of degree one, the inner ring surface of a torus and a hyperboloid of one sheet are examples.

$K_g = 0$. When the Gaussian curvature is zero everywhere, the surface is developable, because at least one of the principal curvature is zero, hence the line of curvature on the surface is straight (refer to the next section). Such surfaces include general cones, cylinders and the surface swept out by a tangent to any twisted space curve.

From equations (6.22) and (6.23), we have

$$\kappa_1 = K_m + \sqrt{(K_m^2 - K_g)}, \qquad \kappa_2 = K_m - \sqrt{(K_m^2 - K_g)}, \tag{6.24}$$

for the two principal curvatures.

6.5 Principal Directions and Lines of Curvature

Eliminating $\tilde{\kappa}$ from eq. (6.20), we obtain the equation of $\tilde{\gamma}$ (hereafter we omit the tilde mark for simplicity):

$$(MG - NF)\tilde{\gamma}^2 + (GL - NE)\tilde{\gamma} + FL - ME = 0,$$

$$\text{or} \quad (MG - NF)dv^2 + (GL - NE)dudv + (FL - ME)du^2 = 0. \tag{6.25}$$

This is the *equation of lines of curvature* which is explained in this section. Since the equation is of degree two, it has two roots γ_1 and γ_2 and has the relation:

$$\gamma_1 + \gamma_2 = \frac{-(GL - NE)}{(MG - NF)}, \quad \gamma_1\gamma_2 = \frac{(FL - ME)}{(MG - NF)}. \tag{6.26}$$

At a point on the surface, in the directions determined by γ_1 and γ_2, the normal curvatures take extremum values. Since a tangent vector on the surface is $(\mathbf{s}_u du + \mathbf{s}_v dv)$, the inner product of the two tangent vectors corresponding to γ_1 and γ_2 is given by

$$(d\mathbf{s})_1 \cdot (d\mathbf{s})_2 = \{(\mathbf{s}_u + \mathbf{s}_v\gamma_1) \cdot (\mathbf{s}_u + \mathbf{s}_v\gamma_2)\}(du)_1(dv)_2. \tag{6.27}$$

With equations (6.4) and (6.26), we convert the inside of { } of the above expression to

$$\{E(MG - NF) - F(GL - NE) + G(FL - ME)\}/(MG - NF),$$

and this is equal to zero.

It follows that the two tangent directions of the normal sections of the principal curvature are orthogonal. These directions are called the *principal directions.* Except for singular points on the surface, any point has principal directions. If the tangents of curves on a surface coincide with the principal directions there, the curves are called the *lines of curvature.* Examples are shown in Fig. 7.1 and Fig. 7.5.

The lines of curvature make an orthogonal net. The pattern of lines of curvature of a surface is an inherent attribute independent of the coordinate systems used. They are discussed in more detail in the next chapter.

6.6 Derivatives of a Unit Normal and Rodrigues' Formula

For designing and manufacturing surfaces in engineering products or for evaluating the surfaces from technical and aesthetic viewpoints, analysis of the behavior of their normal vector becomes important. The surface normal vector $\mathbf{n}$ is a function of the parameters u and v, and since its derivative vectors $\mathbf{n}_u$ and $\mathbf{n}_v$ are orthogonal to $\mathbf{n}$, that is, in the tangential plane at that point, they can be expressed by the basic vectors of the surface.

$$\mathbf{n}_u = \alpha\mathbf{s}_u + \beta\mathbf{s}_v, \quad \mathbf{n}_v = \alpha'\mathbf{s}_u + \beta'\mathbf{s}_v. \tag{6.28}$$

The constants α,β, α' and β' are determined in the following way. Multiplying $\mathbf{s}_u$ and $\mathbf{s}_v$ to the first equation of (6.28) and considering $L = \mathbf{n}\cdot\mathbf{s}_{uu} = -\mathbf{n}_u\cdot\mathbf{s}_u$ and $M = \mathbf{n}\cdot\mathbf{s}_{uv} = -\mathbf{n}_v\cdot\mathbf{s}_u$, which are derived from differentiation of the relation $\mathbf{n}\cdot\mathbf{s}_u = 0$ with respect to u or v, we obtain the equations:

$$-L = \alpha E + \beta F, \qquad -M = \alpha F + \beta G.$$

By solving these simultaneous equations for α and β, we obtain :

$$\alpha = \frac{(MF - LG)}{H^2}, \qquad \beta = \frac{(LF - ME)}{H^2}, \tag{6.29}$$

where $H^2 = EG - F^2$. In the same way, using the relation $\mathbf{n}\cdot\mathbf{s}_v = 0$ we obtain from the second equation of (6.28),

$$\alpha' = \frac{(NF - MG)}{H^2}, \qquad \beta' = \frac{(MF - NE)}{H^2}. \tag{6.30}$$

The equations (6.28) with (6.29) and (6.30) are called *Weingarten's equations*, which express the derivatives of the normal $\mathbf{n}$ in terms of the basic vectors.

We can deduce the following relations using equations (6.28)–(6.30) and the expressions of mean curvature K_m and Gaussian curvature K_g:

$$\mathbf{n}_u^2 = 2K_mL - K_gE, \qquad \mathbf{n}_v^2 = 2K_mN - K_gG, \tag{6.31}$$

$$\mathbf{n}_u\cdot\mathbf{n}_v = 2K_mM - K_gF, \tag{6.32}$$

$$\mathbf{n}_u\times\mathbf{n}_v = K_gH\mathbf{n}. \tag{6.33}$$

Multiplying $\mathbf{n}$ by both sides of eq. (6.33), we obtain another expression of the Gaussian curvature

$$K_g = \frac{[\mathbf{n}, \mathbf{n}_u, \mathbf{n}_v]}{H} \tag{6.34}$$

When the differential $d\mathbf{n}$ of the normal vector is given at a point on a surface $\mathbf{s}(u, v)$, we are required to determine the location $\mathbf{s}(u + du, v + dv)$ which has the specified normal vector $\mathbf{n} + d\mathbf{n}$. Since there is a relation:

$$d\mathbf{n} = \mathbf{n}_u du + \mathbf{n}_v dv, \tag{6.35}$$

we can determine du and dv by multiplying $\mathbf{n} \times \mathbf{n}_v$ or $\mathbf{n} \times \mathbf{n}_u$ and using eq. (6.34):

$$du = -\frac{[\mathbf{n}, \mathbf{n}_v, d\mathbf{n}]}{HK_g}, \quad dv = \frac{[\mathbf{n}, \mathbf{n}_u, d\mathbf{n}]}{HK_g}. \tag{6.36}$$

Accordingly, if an absolute values of the Gaussian curvature is small, values of du and dv become large, which may violate the assumption of the Weingarten's equation.

Next we treat the differential $d\mathbf{n}$ of a unit normal vector in the principal direction in terms of a principal curvature $\tilde{\kappa}$. The relation is given by *Rodrigues' formula*:

$$d\mathbf{n} + \tilde{\kappa} d\mathbf{s} = 0. \tag{6.37}$$

This is proved as follows. Substituting the Weingarten's equations into eq. (6.35), we obtain

$$d\mathbf{n} = (\alpha + \alpha'\gamma)\mathbf{s}_u du + (\beta/\gamma + \beta')\mathbf{s}_v dv, \tag{6.38}$$

where $\gamma = dv/du$. On the lines of curvature γ becomes $\tilde{\gamma}$, which is given by eq. (6.19). Using this and equations (6.29) and (6.30), the right-hand side of eq. (6.38) is transformed to $-\tilde{\kappa}\mathbf{s}_u du - \tilde{\kappa}\mathbf{s}_v dv$, which is equal to $-\tilde{\kappa} d\mathbf{s}$. Hence we have eq. (6.37).

In a general space curve, Frenet-Serret's formula (5.12) gives the relation

$$d\mathbf{n}/ds + \kappa\mathbf{t} = \tau\mathbf{b}, \qquad (\mathbf{t} = d\mathbf{s}/ds),$$

where $\mathbf{n}$ is the principal normal of the curve, so $d\mathbf{n}$ and $d\mathbf{s}$ are not always parallel. But along a line of curvature Rodrigues's formula (6.37) assures that $d\mathbf{n}$, which is variation of the unit normal of the surface, is parallel to $d\mathbf{s}$: $d\mathbf{n}$ is in the same or opposite direction to $d\mathbf{s}$ according to negative or positive principal curvature. In the region of negative Gaussian curvature, there is two directions in which $d\mathbf{n}$ is zero.

If the tangent $\mathbf{t}$ of a curve on a surface is parallel to $d\mathbf{n}$, then the curve is a line of curvature of the surface. This is the inverse of Rodrigues' formula and proved as follows: for an arbitrary constant λ, if $d\mathbf{n} = \lambda d\mathbf{s}$, then using eq. (6.38), we get

$$\{\alpha + \alpha'\gamma - \lambda\}\mathbf{s}_u du + \{\beta/\gamma + \beta' - \lambda\}\mathbf{s}_v dv = 0.$$

For the above equation to hold everywhere, the insides of { } must be zero, that is, the following relations must hold.

$$\lambda = \alpha + \alpha'\gamma = \beta/\gamma + \beta'.$$

Substituting equations (6.29), (6.30) into the above equation, we obtain equation (6.25), which indicates the lines of curvature.

6.7 Local Shape of Surface

We can express an approximate shape of a surface around a point taken as the origin by expanding $\mathbf{s}$:

$$ds = \mathbf{s}_u du + \mathbf{s}_v dv + (1/2)(\mathbf{s}_{uu} du^2 + 2\mathbf{s}_{uv} du dv + \mathbf{s}_{vv} dv^2). \tag{6.39}$$

Multiplying the unit normal $\mathbf{n}$ with this equation and using the fundamental magnitudes, we obtain an equation of shape near the origin, z being the distance from the tangential plane at the origin,

$$z = (1/2)(L du^2 + 2M du dv + N dv^2). \tag{6.40}$$

We transform this into an orthogonal coordinate system, whose axes coincide with the principal direction of the surface, since the unit vectors $\mathbf{i}$ and $\mathbf{j}$ of the principal directions are given by

$$\mathbf{i} = (\mathbf{s}_u + \mathbf{s}_v \gamma_1)/A, \qquad \mathbf{j} = (\mathbf{s}_u + \mathbf{s}_v \gamma_2)/B, \tag{6.41}$$

where

$$A^2 = E + 2F\gamma_1 + G{\gamma_1}^2, \qquad B^2 = E + 2F\gamma_2 + G{\gamma_2}^2, \tag{6.42}$$

and γ_1 and γ_2 are determined from eq. (6.25). Let a point (du, dv) on the tangential plane expressed by the basic vectors $\mathbf{s}_u$ and $\mathbf{s}_v$ be a point (x, y) expressed by $\mathbf{i}$ and $\mathbf{j}$. There is a relation:

$$\mathbf{s}_u du + \mathbf{s}_v dv = x\mathbf{i} + y\mathbf{j}.$$

Multiplying $\mathbf{n} \times \mathbf{s}_u$ or $\mathbf{n} \times \mathbf{s}_v$ with both sides of the above equation and considering eq. (6.41), we obtain

$$du = x/A + y/B, \qquad dv = x\gamma_1/A + y\gamma_2/B. \tag{6.43}$$

Then eq. (6.40) is transformed to

$$\begin{aligned} z &= \frac{1}{2}(L du^2 + 2M du dv + N dv^2) \\ &= \frac{1}{2}\{(L + 2M\gamma_1 + N\gamma_1^2)(\frac{x^2}{A^2}) \\ &\quad +2(L + M(\gamma_1 + \gamma_2) + N\gamma_1\gamma_2)(\frac{xy}{AB}) \\ &\quad + (L + 2M\gamma_2 + N\gamma_2^2)(\frac{y^2}{B^2})\}. \end{aligned} \tag{6.44}$$

From equations (6.26) we know the coefficient of xy in eq. (6.44) is zero. So, eq. (6.44) takes the following form:

$$z = \frac{1}{2}(K_1 x^2 + K_2 y^2), \tag{6.45}$$

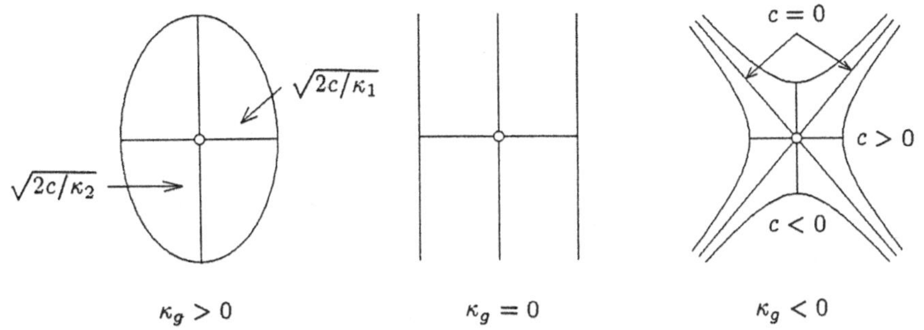

Figure 6.4 Section patterns of a surface by its offset tangential planes

where

$$K_i = \frac{(L + 2M\gamma_i + N\gamma_i^2)}{(E + 2F\gamma_i + G\gamma_i^2)} \quad \text{for} \quad i = 1, 2. \tag{6.46}$$

Considering that eq. (6.46) corresponds to eq. (6.13) and the γ_i's to the principal directions, we have $K_1 = \kappa_1$, $K_2 = \kappa_2$ and eq. (6.45) becomes

$$z = \frac{1}{2}(\kappa_1 x^2 + \kappa_2 y^2). \tag{6.47}$$

Since the curvature of the normal section is given by eq. (6.13),

$$\kappa = \frac{(Ldu^2 + 2Mdudv + Ndv^2)}{(Edu^2 + 2Fdudv + Gdv^2)}, \tag{6.48}$$

and using eq. (6.43), this is also transformed to

$$\kappa = \kappa_1 \frac{x^2}{(x^2 + y^2)} + \kappa_2 \frac{y^2}{(x^2 + y^2)} = \kappa_1 \cos^2\theta + \kappa_2 \sin^2\theta, \tag{6.49}$$

where θ is an angle between the x axis and the tangent of the normal section at the origin. Equation (6.49) is called *Euler's formula*.

Equation (6.47) represents the shape of the surface in the vicinity of the origin. The intersection of the surface with a plane $z = c$ which is parallel to the xy plane has three kinds of form:

- $K_g\{= \kappa_1\kappa_2\} > 0$: the intersection is an ellipse for $c > 0$ or nothing for $c < 0$. Its long and short radii are $\sqrt{(2c/\kappa_1)}$ and $\sqrt{(2c/\kappa_2)}$.
- $K_g < 0$: The intersection curve is a hyperbola, which intersects with the x axis or the y axis according to the sign of c. If c equals zero, the intersection is two straight lines through the origin, which become its asymptotes.
- $K_g = 0$: the intersection is two straight lines equidistant from one of the axes.

See Fig. 6.4; this diagram is called Dupin's indicatrix. Equation (6.47) is used also for approximate calculation of the intersection curve of surfaces.

7 Advanced Applications of Theory of Surfaces

7.1 Introduction

Special topics on the geometric properties of surfaces are treated in this chapter. They are developed from the fundamental theory of surfaces which has been explained in the previous chapter. Knowledge of these topics is useful for treating free-form surfaces in advanced problems. First we discuss the umbilics and lines of curvature. On a free-form surface, there are points and regions which have the special characters inherent to its shape, and whose locations do not depend on the coordinate system adopted. The lines of curvature make orthogonal nets on a surface, and the pattern they form exhibits inherent features of the surface. The umbilics are singular points, or curves or regions seeing from the lines of curvature. On the free-form surface the umbilics appear more frequently than our expectation. There are other curves on the surface which depend not only on its inherent features, but also on its orientation with respect to its observers or its environments or its surface physical properties. These curves are useful for describing or evaluating the objects from engineering and aesthetic criteria.

After treating these problems, we explain the application of the surface theory to two special surfaces: an offset surface and a ruled surface. Determining the offset surfaces of an original surface is important in engineering, but the solution in closed analytical form is very difficult to obtain. Since in determining the intersection curves of a surface with an offset surface, we need the fundamental magnitudes of the first and the second order of the latter and not necessarily its whole shape, we can deduce them from the properties of the original surface. In the last section, an equation of a ruled surface is introduced and its geometric properties are treated.

7.2 Umbilics

An *umbilic* is a point at which the principal curvatures are equal. It follows that all the normal curvatures there are equal: it represents an infinitesimally small region of a spherical or planar surface. The condition of equal principal curvatures is given from the condition of equal roots of eq. (6.21) of the previous chapter, which determines the principal curvatures of a surface. From the zero discriminant value of the quadratic equation, we have

$$(EN + GL - 2FM)^2 - 4(LN - M^2)(EG - F^2) = 0. \qquad (7.1)$$

This is transformed to

$$(EN - GL)^2 - 4(EM - FL)(FN - GM) = 0,$$

which is the same to the condition of equal roots of the equation of the lines of curvature which is given by eq. (6.25). This equation gives differential form $dv/du = -(GL - EN)/2(GM - FN)$, but the double roots mean the same directions of tangents of the orthogonal curves. This is a contradiction. So the roots must be indefinite. A relation

$$E : F : G = L : M : N \tag{7.2}$$

is sufficient for the above condition.

A line element on a surface is given from eq. (6.5) by

$$ds^2 = Edu^2 + 2Fdudv + Gdv^2. \tag{7.3}$$

If we set $du : dv$ in the principal directions, using eq. (6.19), ds^2 is transformed to

$$(EG - F^2)(M - \kappa_i F)(\kappa_1 + \kappa_2 - 2\kappa_i), \quad \text{where } i = 1 \text{ or } 2. \tag{7.4}$$

At the umbilic this becomes zero. So we cannot trace the line of curvature there by the differential equation solvers.

Examples of the umbilics which constitutes a curve are shown. If in a small region of a surface, its shape is smooth and mirror-symmetric (reflection), the intersection curve of the surface and the mirror plane is the line of curvature, because the differential of the normal $d\mathbf{n}$ along the intersection is in the plane of symmetry and the inverse of Rodrigues' formula assures us that the intersection is the line of curvature. If another line of curvature of the surface intersects with this intersection curve obliquely, the intersection point is an umbilic, because there are more than two lines of curvature through the point. In this case all the points of the symmetrical intersection curve are umbilics.

On a surface of revolution, when the curvature at a point on its meridian curve is equal to that of the parallel of latitude through that point, all the points on the parallel are umbilics. Surfaces on which all the points are umbilics are a sphere or a plane. On these surfaces any orthogonal net can give the lines of curvature.

On free-form surfaces the umbilics appears frequently. For example, in the patch shown in Fig. 7.1 there seem to be four umbilics, but we cannot recognize them from its contour curves shown in Fig. 7.2.

If we take the origin at an umbilic of a surface and the z axis in the direction of the normal, the shape of the surface around the umbilic is given approximately by

$$\begin{aligned} z &= \mathbf{n} \cdot \mathbf{s} = S(du, dv) = (1/2)\kappa ds^2 + \\ &\quad + (1/6)(S_{uuu}du^3 + 3S_{uuv}du^2dv + 3S_{uvv}dudv^2 + S_{vvv}dv^3), \end{aligned} \tag{7.5}$$

where κ is a constant normal curvature.

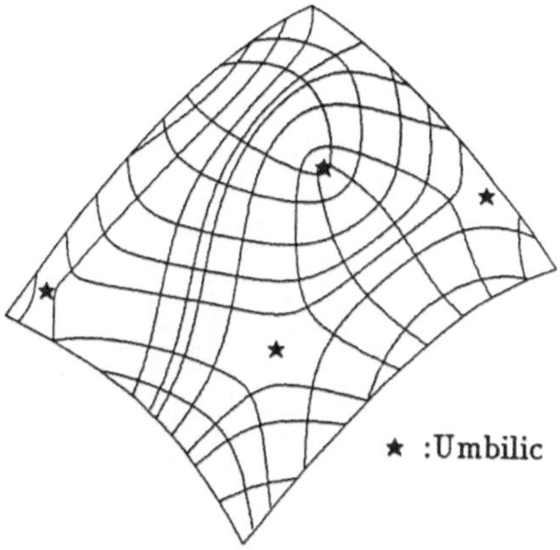

Figure 7.1 Lines of curvature and locations of umbilics of the patch

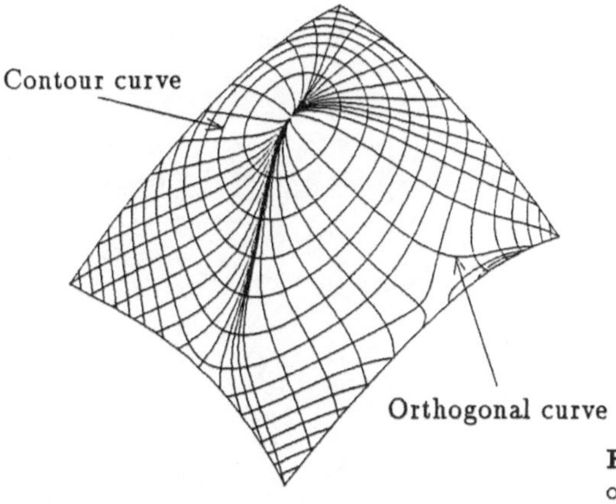

Figure 7.2 Contour curves and their orthogonal curves

Examples of shapes given by eq. (7.5) are shown in Fig. 7.3 and Fig. 7.4 together with their patterns of the lines of curvature. In Fig. 7.3 (a) and (b), the umbilics are in region of positive Gaussian curvature and the pattern (a) appears in a region of elliptical shape. The pattern (b) appears also in a region of nearly elliptical shape with disturbing third-degree terms. When we observe shapes of the surfaces by their contour curve patterns, we cannot imagine that the patterns of their lines of curvature are so different. Clearly we can see an invisible singular curve on which tracing the lines of curvature by a differential equation solver fails. This is an umbilic of the curve type. In Fig. 7.4 an umbilic is an isolated point of zero curvature surrounded by a region of negative Gaussian curvature. A pattern like (c) appears also in a region of positive Gaussian curvature.

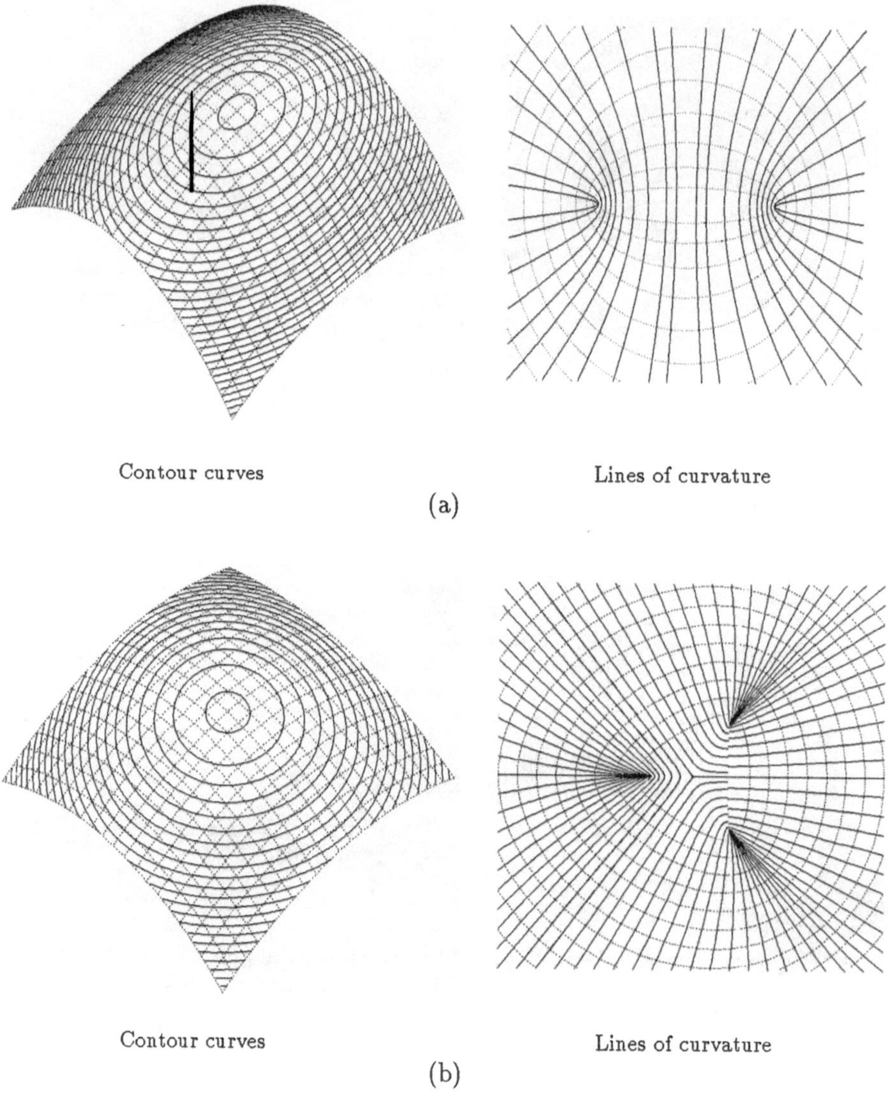

Figure 7.3 (a), (b) Contour curves of two patches and their lines of curvature in a region of positive Gaussian curvature

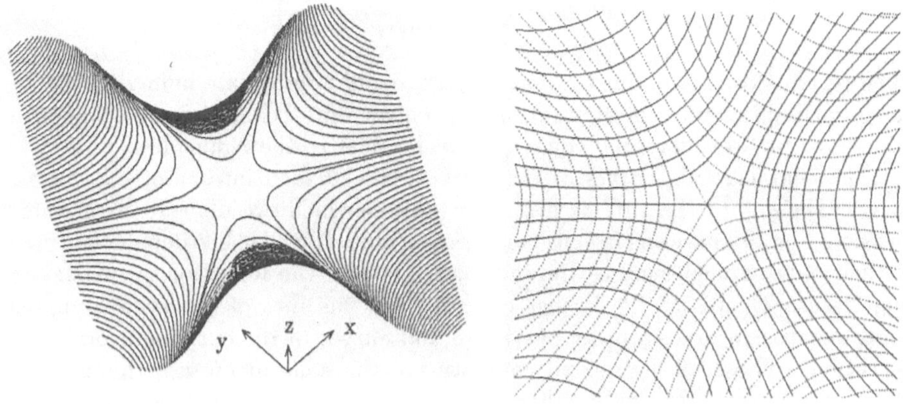

Contour curves Lines of curvature

Figure 7.4 Contour curves of a patch and its lines of curvature in region of negative Gaussian curvature

7.3 Characteristic Curves on a Surface 1

7.3.1 General Remarks

There are curves on a surface which characterize its features. Some of them are useful for understanding the shape of the surface or evaluating its surface characteristics. We introduce their equations and examples in the following subsections [18]. Generally, a curve on a surface $\mathbf{s}(u, v)$ is determined by a relation between the parameters u and v. The relation is mostly given by a differential equation of a form:

$$f(u,v)du + g(u,v)du = 0. \tag{7.6}$$

Otherwise, a relation $F(u,v) = 0$ may be given. If solving $F(u,v)$ for successive values of given u or v is not easy, it is converted into the form of eq. (7.6) from its differential. To trace the curve, after solving the above equation numerically with the initial condition, we calculate $\mathbf{s}(u,v)$ step by step for the obtained set of u and v.

If the shape of a curve is simple to trace, we can integrate eq. (7.6) with u or v as the independent variable. Otherwise, the above equation is transformed into the following simultaneous differential equations with the independent variable of arc length s.

$$\frac{du}{ds} = \psi(u,v)g(u,v), \qquad \frac{dv}{ds} = -\psi(u,v)f(u,v), \tag{7.7}$$

where

$$\psi(u,v) = \pm\frac{1}{\sqrt{(Eg^2 - 2Ffg + Gf^2)}}.$$

The last equation is obtained from eq. (7.3). Using appropriate numerical methods of solving the differential equations with given initial conditions, we can trace the curve. As for differential equation solvers, refer to the Appendix.

The characteristic curves of the surfaces are divided into three categories: curves inherent to the surface geometry, curves which are dependent on their positions and orientations in space and certain curves, which also depend features of their environment such as the incident light and the reflection properties of their faces. Curves of the former category are the lines of curvatures, curves of zero Gaussian or zero mean curvature and curves of the extremum principal curvature. Those of the second category are the contour curves, the contour orthogonal curves, the equi-gradient curves, the extremum gradient curves, the extremum search curves and so on. Some of these curves represent features of the surface.

For example, the extremum gradient curves are loci of the maximum or minimum gradient point on each of the contour curves. These curves are the analog of ridge or valley lines in a topographical map. The third category includes silhouette patterns, equi-brightness curves, equi-highlight curves and reflection patterns on surfaces, which relate to the incident light, the view points, the reflection characteristics of the surface, and the environmental as well as the geometric conditions of the surfaces.

The points on the surface which are determined by the conditions $f(u,v) = 0$, $g(u,v) = 0$ of eq. (7.6) are singular points of the family of the curves described by this differential equation. These points have certain special properties.

7.3.2 Lines of Curvature

The equation of lines of curvature is given by eq. (6.25):

$$(GM - FN)dv^2 + (GL - EN)dudv + (FL - EM)du^2 = 0. \tag{7.8}$$

Since this equation gives an orthogonal net which is inherent to a surface, features of the surface seems to be extracted from its pattern. But we have not yet succeeded in their categorization. Equation (7.8) is resolved into a product of two equations of type (7.6), each of which corresponds to one family of lines of curvature.

Since each point of a surface determines its principal directions, when tracing a line of curvature by numerical integration we must be careful not to be deviated from the curve by numerical errors. If we introduce an arc length s as the independent variable for the integration of the equation, the amount of calculation increases but control of the steps of the integration procedure becomes easier than the integration without it. In small regions around the umbilics,

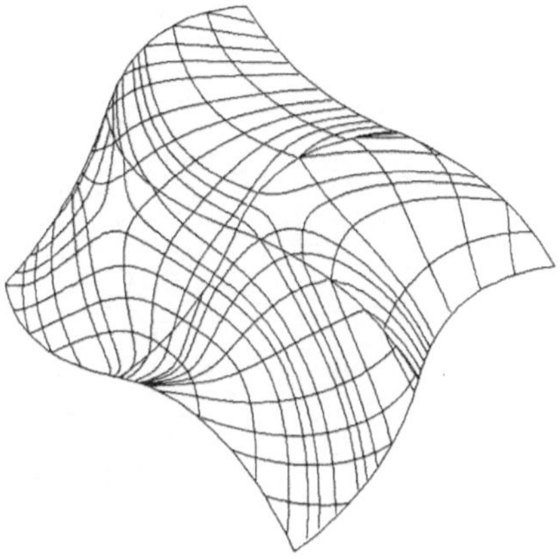

Figure 7.5 Lines of curvature of patches connected in $C^{(1)}$

the calculation becomes unstable. Beyond that point the sign of ds has to be reversed.

It should be noticed that the umbilics make the lines of curvature form special patterns. Figure 7.1 is an example of the lines of curvature of a surface whose contour curves are shown in Fig. 7.2. This is a Bézier patch of degree 3×3. We can notice a few special features indicated by arrows in Fig.7.1. These are local patterns of the lines of curvature showing the existence of umbilics, but in Fig. 7.2 we cannot recognize any special features at the same locations. Figure 7.5 shows the pattern of the lines of curvature of four patches connected in $C^{(1)}$. The pattern shows clearly the connecting boundary curves of the patches and also the differences of their surface characters. In Fig. 7.6(a) and (b), examples of variations of patterns of the contour curves and the lines of curvature for the same displacement $\Delta z = 6\%$ of the control point $\mathbf{p}_{22}$ of the Bézier patch used in Fig. 7.1, which is of degree 3×3. Variation of the curve patterns is more sensitive in the line of curvature than in the contour curve.

Uses of the lines of curvature are currently rather limited, because of their cumbersome calculation and lack of experience for interpreting their patterns. Since the lines of curvature of an offset surface are easily determined from those of the original surface, they can be used in the analysis of offset surfaces. This is treated in Sect. 7.5.

7.3.3 Extremum Search Curves

We frequently need to find locations of the maximum or the minimum height or the saddle points or paths from the lowest to the highest points of a surface to use these data for other purposes. Hence, we need to obtain this information as simply as possible.

Let $\mathbf{n}_f$ be the unit vector directing upwards and let a component of $\mathbf{s}(u, v)$

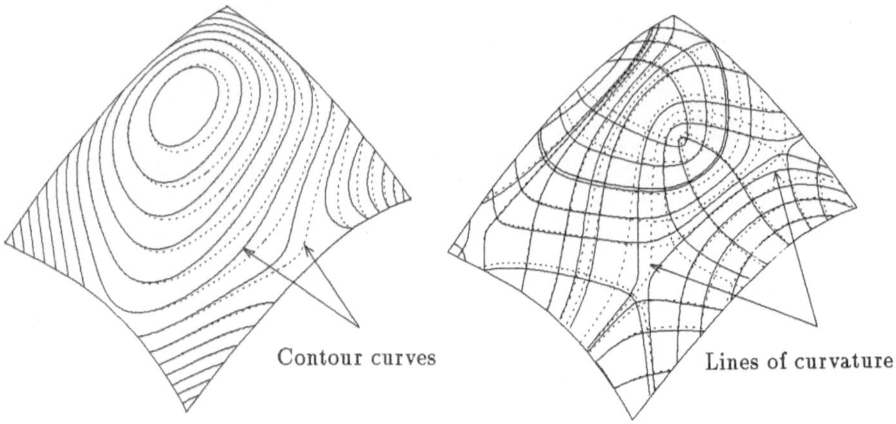

Figure 7.6 Variation of pattern of contour curves and lines of curvature with a displacement $\Delta z = 6\%$ of a control point $\mathbf{p}_{2,2}$

in that direction be $\phi(u, v)$.

$$\phi(u, v) = \mathbf{n}_f \cdot \mathbf{s}(u, v). \tag{7.9}$$

The extremum values of $\mathbf{s}(u, v)$ are obtained by solutions of the simultaneous equations:

$$\phi_u(u, v) = 0, \quad \phi_v(u, v) = 0, \tag{7.10}$$

Since usually it is difficult to solve these equations directly, we trace the curves represented by them individually by transforming them into the differential equations with appropriate initial conditions:

$$\phi_{uu}du + \phi_{uv}dv = 0, \quad \phi_{uv}du + \phi_{vv}dv = 0. \tag{7.11}$$

These curves are the loci of the extremum points on the curves of constant parameter u or v of $\mathbf{s}(u, v)$. The intersecting points of these curves indicate the locations of the extremum points on the surface. Their initial conditions are given on the boundary curves or on constant parameter curves inside the boundaries.

These curves not only give the extremum points at their intersection, but also roughly depict the shape of the surface together with its contour curves. On these curves one can determine the starting points of the contour curves which do not reach the boundary of the surface. Figure 7.7 shows examples of the extremum search curves of a complicated surface, and Fig. 7.8 shows their overlay with the contour curves of the surface. Eight cross points of the curves correspond to the lowest and the highest points and the saddle points of the surface shape. When drawing the contour curves, these extremum search curves were very helpful, otherwise some extremum points would be missed.

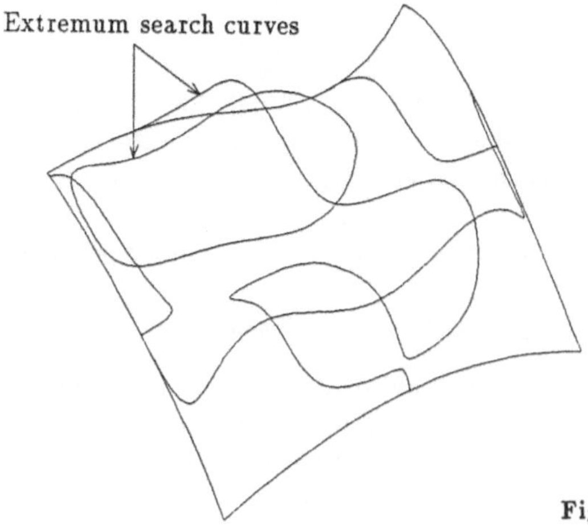

Figure 7.7 Extremum search curves

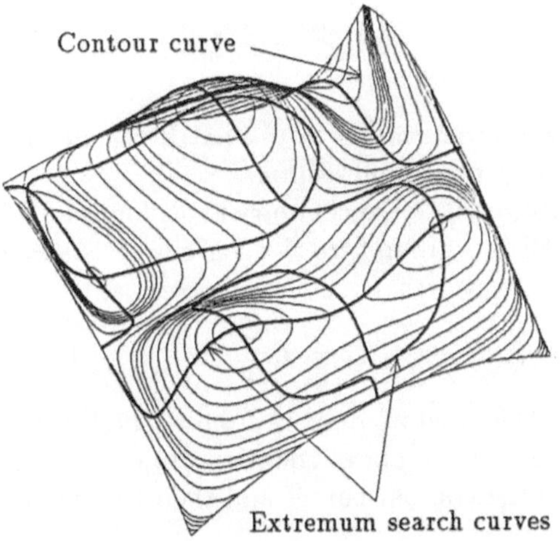

Figure 7.8 Contour curves of a patch with the extremum search curves

To trace the extremum curves given by equations (7.11), we transform the equation into the form (7.7) introducing the arc length s as the independent variable. The initial conditions of the above equations are given usually on the boundaries of the patch using equations (7.10).

Other than equations (7.10), if we trace the curves given by the equations

$$\phi_{uu}(u,v)=0, \quad \phi_{vv}(u,v)=0 \tag{7.12}$$

from the boundaries, we can find the intersections with internal loops made by (7.11), because curves specified by equations (7.12) are the loci of the inflection points of the constant parameter curves. Locus curves of the derivatives of higher degree show simple behavior because of the lower degree of the powers of their polynomials and because they intersect with the loops of the locus of the derivative of one degree less.

The extremum search curves given by equations (7.10) divides the surface into regions with a monotonic gradient and each point determined by the intersection of the two curves given by (7.12) is the maximum gradient point in each one of the regions.

The idea of extremum search curves can be applied for similar problems.

7.3.4 Contour Curves and Their Orthogonal Curves

The contour curves of a surface are represented by the intersection of the surface with equi-distance parallel planes. They have widely been used to represent the shapes of free-form surfaces in two dimensions. Let $\mathbf{n}_f$ be the unit normal vector of the cutting planes. The equation of the contour curves is given by

$$z = \mathbf{n}_f \cdot \mathbf{s}(u, v) \equiv \phi(u, v) = const \equiv h \tag{7.13}$$

Differentiating eq. (7.13), we have

$$\phi_u du + \phi_v dv = 0. \tag{7.14}$$

This is the differential equation of the contour curves of the patch $\mathbf{s}(u, v)$. Its extremum search curves are given in the previous subsection.

We introduce a group of curves orthogonal to the contour curves for detailed description of the surface shape. We call this group here the contour orthogonal curves, whose equation is given by

$$(\phi_v E - \phi_u F)du + (\phi_v F - \phi_u G)dv = 0. \tag{7.15}$$

Since the inner product of the tangent vectors derived from eq. (7.14) and eq. (7.15) becomes zero, the curves given by the above equation are orthogonal to the contour curves. In Fig. 7.2, both the contour curves and their orthogonal curves are shown.

7.3.5 Equi-gradient Curves

Let σ be an angle between the unit normal $\mathbf{n}$ of the patch $\mathbf{s}$ and the vertical direction $\mathbf{n}_f$. An equation of points of constant σ is given by

$$\cos \sigma = \mathbf{n}_f \cdot \mathbf{n} = const. \tag{7.16}$$

Sets of these points make the equi-gradient curves. If $\mathbf{n}_f$ represents a direction of parallel rays and the diffused reflection of the surface is assumed, these curves are the same as equi-brightness curves. When σ is 90 degrees, the curve includes

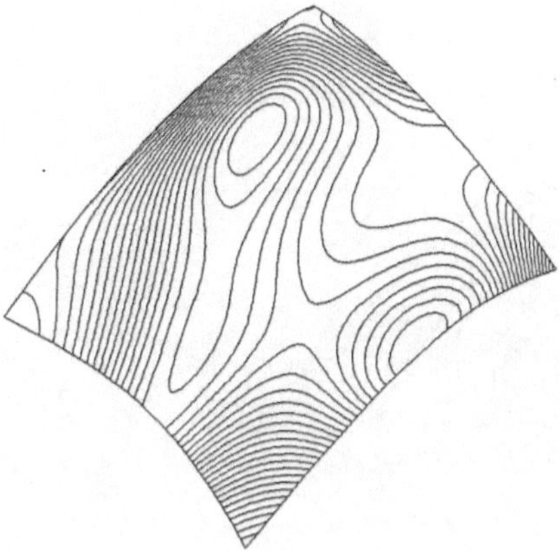

Figure 7.9 Equi-gradient curves of the patch shown in Fig.7.6

Figure 7.10 Curves orthogonal to the equi-gradient curves

shade boundary curves of the surface or represents its silhouette curves when the viewing point is at infinity. Differentiating eq. (7.16) and using Weingarten's equation, we obtain the differential equation of the curves:

$$(\alpha\phi_u + \beta\phi_v)du + (\alpha'\phi_u + \beta'\phi_v)dv = 0, \tag{7.17}$$

where $\phi = (\mathbf{n}_f \cdot \mathbf{s})$.

The equation of curves orthogonal to the equi-gradient curves is given by

$$\begin{aligned} \{(\alpha_u + \beta'\phi_v)E - (\alpha\phi_u + \beta\phi_v)F\}du+ & \\ \{(\alpha'\phi_u + \beta'\phi_v)F - (\alpha\phi_u + \beta\phi_v)G\}dv &= 0. \end{aligned} \tag{7.18}$$

As eq. (7.15) is deduced from eq. (7.13), so eq. (7.18) is obtained by ϕ_u and ϕ_v in eq. (7.15) being replaced with $(\alpha\phi_u + \beta\phi_v)$ and $(\alpha'\phi_u + \beta'\phi_v)$ of eq. (7.17). Accordingly, the inner product of tangent vectors derived from equations (7.17) and (7.18) is zero. Examples of the equi-gradient curves and their orthogonal curves are shown in Fig. 7.9 and Fig. 7.10.

An equi-gradient curve crosses an umbilic in a horizontal direction, because the shape of the surface becomes spherical at an umbilic. Accordingly, the equi-gradient curves take an extremum height at an umbilic.

The pattern of the equi-gradient curves can be used to evaluate surface smoothness including connecting conditions.

Figure 7.11(a) shows the equi-gradient curves of $C^{(1)}$ connected patches and (b) shows those of $C^{(2)}$ connected patches. In the former pattern, we can clearly recognize cusps at the patch boundaries.

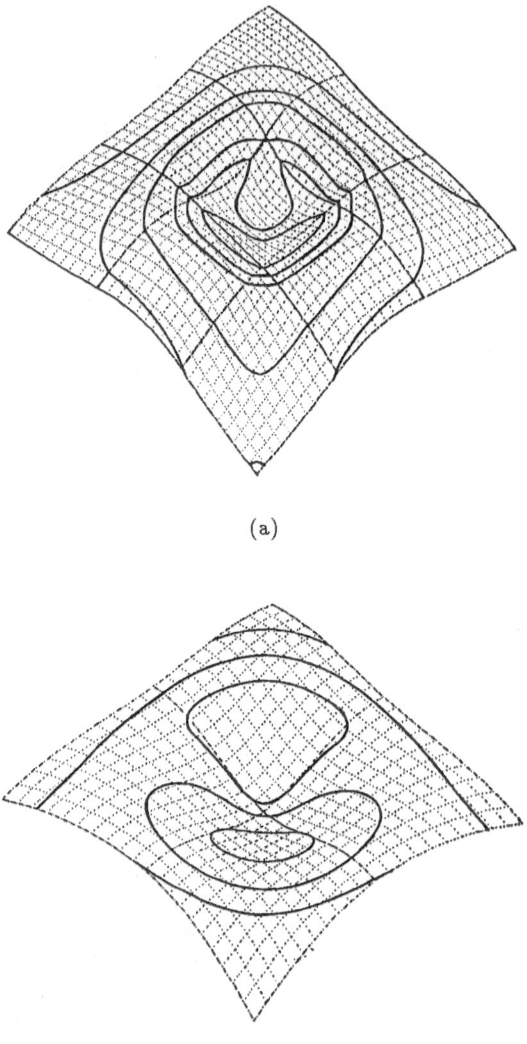

(a)

(b)

Figure 7.11 Equi-gradient curves for composite patches: (a) $C^{(1)}$ connection, (b) $C^{(2)}$ connection

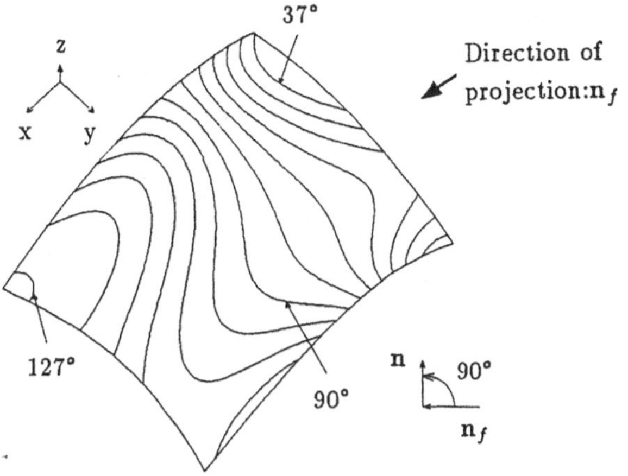

Figure 7.12 Example of a silhouette pattern

7.3.6 Silhouette Curve and Silhouette Pattern

In eq. (7.16) for $\sigma = 90$ degrees and $\mathbf{n}_f$ being the vector of the line of sight from infinity, eq. (7.17) gives silhouette curves. For a designed surface, the silhouette curve pattern, which is generated by repeated small angle rotations of $\mathbf{n}_f$, gives information on the smoothness of the surface and for the evaluation of its aesthetic quality. Figure 7.12 is an example of the silhouette pattern. In patches with $C^{(1)}$ connection, the silhouette curves show tangent discontinuity on the patch boundaries, though their contour curves are smooth. Figure 7.13(a) shows an example and Fig. 7.13(b) shows the contour curves of the same patches. The pattern of the lines of curvature of the same surface is shown in Fig. 7.5. Comparing these two curves the silhouette pattern seems more sensitive to discontinuity of the curvature on the patch boundaries than that of the lines of curvature.

When a viewing point is not at infinity, but at $\mathbf{p}_e$, the condition of the silhouette curve is given by

$$(\mathbf{p}_e - \mathbf{s}) \cdot \mathbf{n} = 0 \tag{7.19}$$

instead of eq. (7.16). Differentiating eq. (7.19), we have

$$(\mathbf{p}_e - \mathbf{s}) \cdot \mathbf{n}_u du + (\mathbf{p}_e - \mathbf{s}) \cdot \mathbf{n}_v dv = 0. \tag{7.20}$$

The solution of eq. (7.20) gives true silhouette curves. Figure 7.14(a) is a parallel projection of a patch from a direction other than the viewing direction to show the loci of the silhouette curves together with highlight equi-brightness curves described in the next section. Figure 7.14(b) is the perspective projection from the viewpoint $\mathbf{p}_e$. The silhouette curves bound the highlighted surface as well as the shaded part of the surface beside the visible boundary.

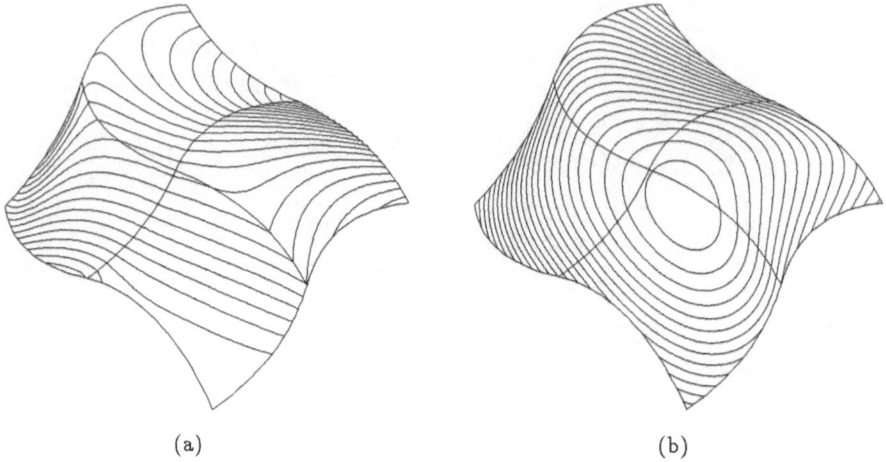

Figure 7.13 (a) Silhouette pattern and (b) contour curves, for $C^{(1)}$ connected patches

In most of the shaded pictures produced by computers, the quality of their silhouette curves is poor, for the computers do not draw the silhouette curves explicitly and the boundaries of the object are shown only in the scan lines.

7.3.7 Highlight Curves

The highlight equi-brightness curve is defined as the locus of points on a surface where the brightness of the surface to a human observer's eye is constant. The brightness is caused by uniform ambient light, diffusion of the incident light and also by rays specularly reflected to an observer's eye.

See Fig. 7.15. Let the angle between a normal vector of the surface and the incident ray be θ and the angle between the line of sight and the specularly reflected ray be ϕ, and $\mathbf{a}$, $\mathbf{a}'$ and $\mathbf{b}$ be the unit vectors of the incident ray, the reflected ray and the line of sight. There are relations:

$$-\mathbf{a}\cdot\mathbf{n} = \cos\theta, \quad \mathbf{a}'\cdot\mathbf{b} = \cos\phi = \mathit{const}. \tag{7.21}$$

We are to express $\mathbf{a}'$ by $\mathbf{a}$ and $\mathbf{n}$. The normal components of $\mathbf{a}$ and $\mathbf{a}'$ have the same magnitude and opposite directions: they are $(\mathbf{a}\cdot\mathbf{n})\mathbf{n}$ and $-(\mathbf{a}\cdot\mathbf{n})\mathbf{n}$. The tangential components of $\mathbf{a}$ and $\mathbf{a}'$ are the same: $\mathbf{a}-(\mathbf{a}\cdot\mathbf{n})\mathbf{n}$. So we have

$$\mathbf{a}' = \mathbf{a} - 2(\mathbf{a}\cdot\mathbf{n})\mathbf{n}.$$

One of the equations (7.21) becomes

$$-2(\mathbf{a}\cdot\mathbf{n})(\mathbf{b}\cdot\mathbf{n}) + (\mathbf{a}\cdot\mathbf{b}) = \cos\phi. \tag{7.22}$$

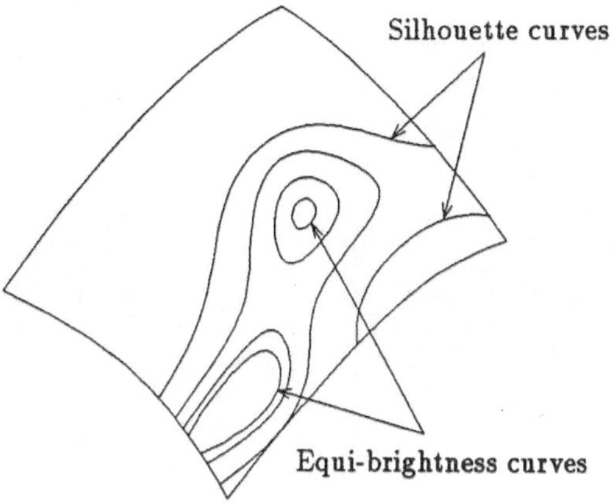

(a)

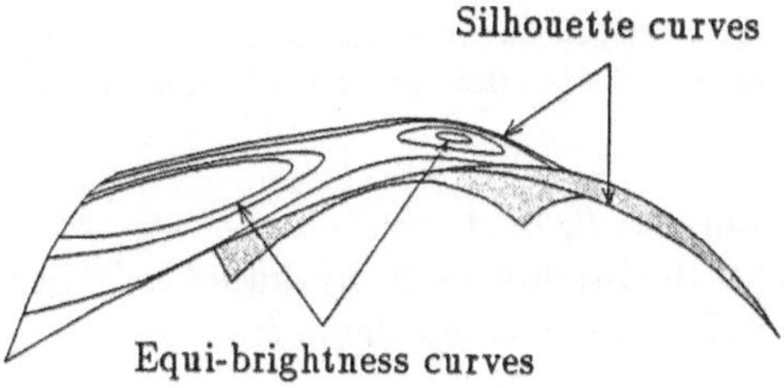

(b)

Figure 7.14 True silhouette curves and equi-highlight curves: (a) parallel projection to show curves clearly, (b) projection from the view point

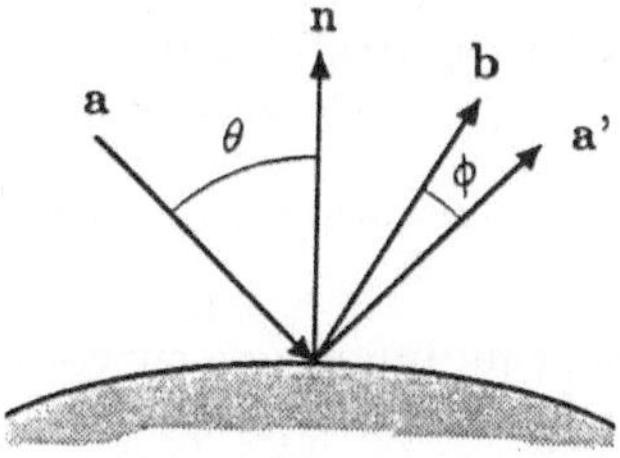

Figure 7.15 Model of specular reflection

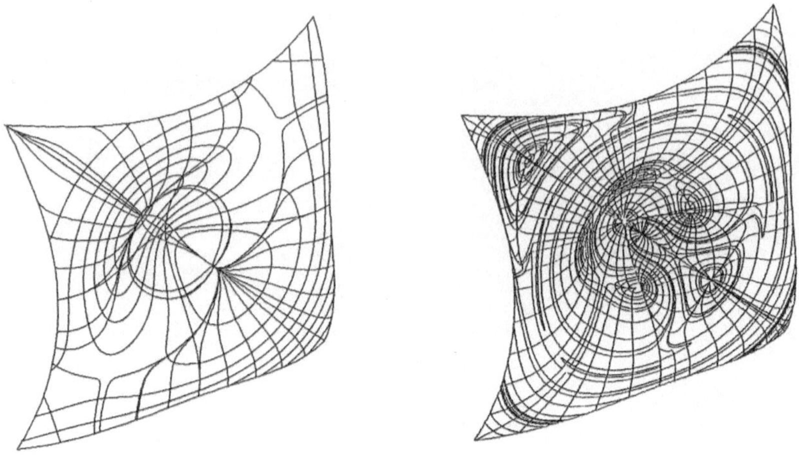

(a) (b)

Figure 7.16 Highlight equi-brightness curves and their orthogonal curves: (a) case of only diffusion reflection, (b) case of diffusion and specular reflection included

The perceived brightness B_p of the surface consists of the uniform brightness B_0 by the ambient light, the brightness B_d by diffuse reflection and the brightness B_s by specular reflection of the incident light I:

$$B_p = B_0 + I\{B_d \cos\theta + B_s(\theta)\cos^n \phi\}, \tag{7.23}$$

where B_d, $B_s(\theta)$ and n are factors and an exponent value dependent on the conditions and the material of the surface. If we assume they are constant, we can make a differential equation of the equi-brightness curves from eq. (7.23) using the relations (7.21) and (7.22),

$$-\mu d(\mathbf{a}\cdot\mathbf{n}) + n\nu d\{-2(\mathbf{a}\cdot\mathbf{n})(\mathbf{b}\cdot\mathbf{n}) + (\mathbf{a}\cdot\mathbf{b})\}^{n-1} = 0, \tag{7.24}$$

where μ and ν are relative constants to determine the ratio of the contributions of diffusion and specular reflections.

Figures 7.16(a) and (b) are examples of the equi-brightness curves together with their orthogonal curves for the same composite patch shown in Fig. 7.11(b). Figure 7.16(a) shows a case of only diffusion reflection, but Fig. 7.16(b) shows a case including both diffusion and specular reflection. Highlighted region appears clearly.

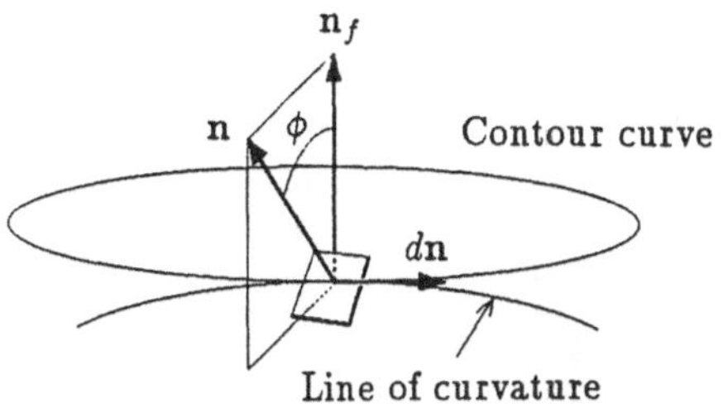

Figure 7.17 Vertical component of a unit normal and extremum slope

7.4 Characteristic Curves on a Surface 2

7.4.1 Gradient Extremum Curves or Ridge-Valley Curves

There are points of extremum height on an equi-gradient curve. The loci of these points on all equi-gradient curves are called here the gradient extremum curves, because at an intersecting point of this curve with a contour curve the downward gradient along the contour takes an extremum value. Their differential equation is obtained from eq. (7.17) and the condition of extremum height of the points, $d\mathbf{n}_f \cdot \mathbf{s}(u, v) = 0$, which becomes

$$\phi_u du + \phi_v dv = 0. \tag{7.25}$$

By eliminating du and dv from equations (7.17) and (7.25), and using equations (6.28) and (6.29), we obtain

$$(GM - FN)\phi_u^2 - (GL - EN)\phi_u\phi_v + (LF - EM)\phi_v^2 = 0. \tag{7.26}$$

The above equation can also be derived from the equation of the lines of curvature and the condition (7.25) of the extremum height of the points on it. And the condition (7.25) is the same as eq. (7.15).

It follows that at these extremum points the equi-gradient curve, the line of curvature and the contour curve have a common tangent. Accordingly, variation $d\mathbf{n}$ of the unit normal vector $\mathbf{n}$ along the contour curve becomes zero from Rodorigues' formula. So, its component in the $\mathbf{n}_f$ direction is zero at this point.

$$d(\mathbf{n}_f \cdot \mathbf{n}) = \mathbf{n}_f \cdot d\mathbf{n} = 0. \tag{7.27}$$

Since $(\mathbf{n}_f \cdot \mathbf{n})$ is $\cos\sigma$ (refer to Fig. 7.17) and this is a measure of the slope along the contour curve, at this point the slope takes an extremum value. This is another interpretation of the gradient extremum curve. An example of this curve is shown in Fig. 7.18 together with other characteristic curves.

The curvature k of a contour curve also takes an extremum value at this point because the tangent direction of the contour curve coincides with the principal

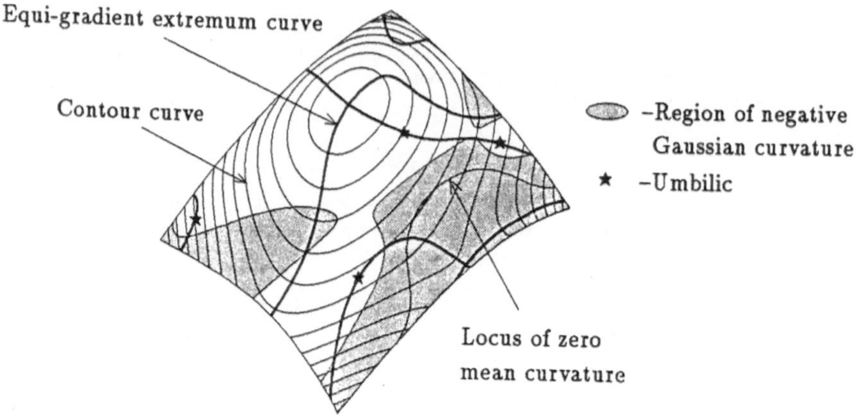

Figure 7.18 Equi-gradient extremum curves and other characteristic curves

direction of the surface. The shape of the surface around this point is represented by a quadric surface from eq. (6.42) given by

$$z = (1/2)(\kappa_1 x^2 + \kappa_2 y^2)$$

and the contour curve is a section curve given by a plane which includes the principal axis, for instance x axis, of this quadric surface. The section curve is symmetrical about the yz plane, that is, its curvature takes an extremum value.

The model surface in Fig. 7.18 is the same as that used in Figs. 7.5, 7.6, 7.9, and 7.13. The locus passes always through a point of maximum or minimum slope on each level, which coincides with a point of extremum curvature of a contour curve. This is an image of a ridge or a valley path in a topographical map. So we call the loci the ridge-valley curves. Their equation is given by eq. (7.26).

Since it is rather difficult to obtain all the curves from step-by-step solution of eq. (7.26), we convert it into the form of differential equations, which we solve with the help of a differential equation solver with adaptive stepsize control. Figure 7.19 is a more complicated case, which is the same surface as in Fig. 7.4. The curves correctly represent the ridge or valley paths of the terrain. The even number of curves start or end at points of the extremum heights, and the curves pass the umbilics which are the locations of the extremum heights on equi-gradient curves.

7.4.2 Loci of Zero Gaussian Curvature and Loci of Extremum Principal Curvatures

A locus of zero Gaussian curvature divides a surface into positive and negative Gaussian curvature regions. In Fig. 7.18, the regions of negative Gaussian

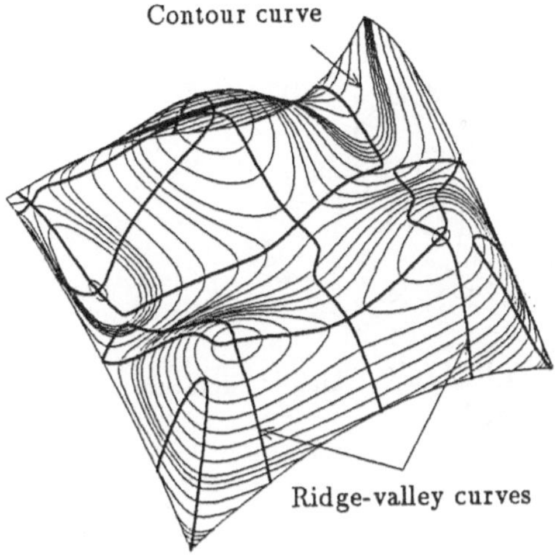

Figure 7.19 Ridge-valley curves and contour curves

curvature are shown in light grey. At any points in the negative Gaussian curvature regions there are two directions of zero curvature, which are given by $\theta = \pm \tan^{-1}\sqrt{(-\kappa_1/\kappa_2)}$. At a point where mean curvature is zero, the angle θ becomes 45 and 135 degrees. There is a net of curves whose tangent directions coincide with zero curvature directions. Currently, this net does not seem useful.

Along a line of curvature, the value of the principal curvature changes and there exists a point of extremum value of the curvature. The loci of these extremum curvature points for all lines of curvature can be considered. They are called here the principal curvature extremum curves. Their equations are obtained from the equation of the lines of curvature (7.8) and the curvature extremum condition $d\kappa = \kappa_u du + \kappa_v dv = 0$, that is,

$$(FL - EM)\kappa_v^2 - (GL - EN)\kappa_u\kappa_v + (GM - FN){\kappa_u}^2 = 0. \tag{7.28}$$

Since direct calculation or use of the differential equation form of eq. (7.28) is very complicated, we determine the extremum curvature points numerically while tracing each line of curvature, then we make their interpolation curves. An example is shown in Fig. 7.20.

7.5 Offset Surfaces

There is a set of points, each of which has a constant distance d from a corresponding point of a surface $\mathbf{s}(u, v)$ in the direction of its normal vector at that point. They constitutes a surface $\mathbf{s}^+(u, v)$. This surface is called the offset surface of the original one. A point of $\mathbf{s}^+(u, v)$ is located on the normal of the corresponding point of $\mathbf{s}(u, v)$, so we have

$$\mathbf{s}^+(u, v) = \mathbf{s}(u, v) + \mathbf{n}d. \tag{7.29}$$

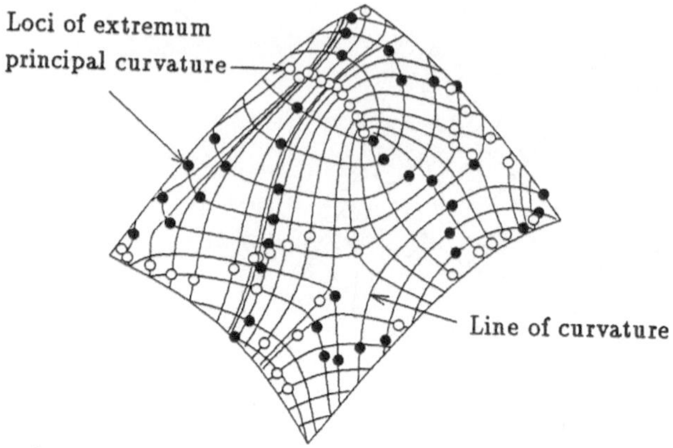

Figure 7.20 Loci of extremum principal curvatures

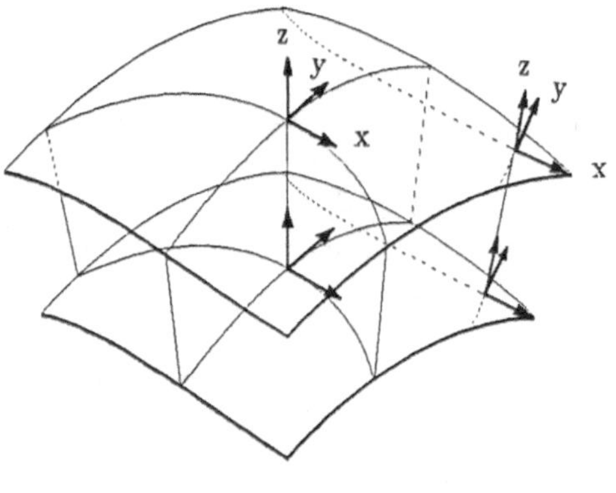

Figure 7.21 Normal vectors and lines of curvature between offset surfaces

The normal planes of both the original and the offset planes coincide because they include their common normals. Now let the parameters u and v be taken along the lines of curvature of $\mathbf{s}(u, v)$. See Fig. 7.21. Differentiating eq. (7.29) with respect to u and using Rodrigues' formula, we obtain

$$\mathbf{s}_u^+ = \mathbf{s}_u + \mathbf{n}_u d = \mathbf{s}_u - \kappa_1 \mathbf{s}_u d = (1 - \kappa_1 d)\mathbf{s}_u \tag{7.30}$$

and similarly,

$$\mathbf{s}_v^+ = \mathbf{s}_v + \mathbf{n}_v d = \mathbf{s}_v - \kappa_2 \mathbf{s}_v d = (1 - \kappa_2 d)\mathbf{s}_v \tag{7.31}$$

where κ_1 and κ_2 are the principal curvatures in the principal directions of $\mathbf{s}(u, v)$, which are the same as those of constant parameter curves in this case. Notice that sign of curvature of a surface obeys the definition given in eq. (5.6). For example, when the normal directs upwards and the normal section is convex upwards, the curvature is negative.

Any point on the lines of curvature of $\mathbf{s}(u,v)$ has a corresponding point on the offset surface $\mathbf{s}(u,v)^+$. The parameters u and v can be considered those of the offset surface $\mathbf{s}(u,v)^+$. Using equations (7.30) and (7.31), an inner product of its basic vectors is

$$\mathbf{s}_u^+ \cdot \mathbf{s}_v^+ = (1-\kappa_1 d)(1-\kappa_2 d)\mathbf{s}_u \cdot \mathbf{s}_v.$$

Since $\mathbf{s}_u \cdot \mathbf{s}_v = 0$, the above value is zero. It follows that the u, v parameter net of the offset surface is orthogonal and they constitute also the lines of curvature net. Since unit normal vectors $\mathbf{n}^+$ and $\mathbf{n}$ on the corresponding points are equal, their increments along a line of curvature are the same, that is, $d\mathbf{n}^+ = d\mathbf{n}$ and also the differentials of distances along the respective line of curvature have the relation $d\mathbf{s} = d\mathbf{s}^+/(1-\kappa_i d)$, which is derived from equations (7.30) and (7.31). Rodrigues' formula applied to $\mathbf{s}(u,v)$ is

$$d\mathbf{n} + \kappa_i d\mathbf{s} = 0.$$

Substituting $d\mathbf{n}$ and $d\mathbf{s}$ into the above equation, we obtain

$$d\mathbf{n}^+ + \frac{\kappa_i}{(1-\kappa_i d)} d\mathbf{s}^+ = 0. \tag{7.32}$$

This is Rodrigues' formula for $\mathbf{s}(u,v)^+$. Accordingly, the principal curvatures of $\mathbf{s}^+(u,v)$ are

$$\kappa_i^+ = \frac{\kappa_i}{(1-\kappa_i d)}, \quad for\ i = 1, 2. \tag{7.33}$$

So its Gaussian and mean curvatures are given by

$$\begin{aligned} K_g^+ &= \kappa_1^+ \kappa_2^+ = \frac{\kappa_1 \kappa_2}{(1-\kappa_1 d)(1-\kappa_2 d)} \\ &= \frac{K_g}{(1-2K_m d + K_g d^2)}, \end{aligned} \tag{7.34}$$

$$\begin{aligned} K_m^+ &= \frac{\kappa_1^+ + \kappa_2^+}{2} = \frac{1}{2}\{\frac{\kappa_1}{(1-\kappa_1 d)} + \frac{\kappa_2}{(1-\kappa_2 d)}\} \\ &= \frac{(K_m - K_g d)}{(1-2K_m d + K_g d^2)}. \end{aligned} \tag{7.35}$$

In the above description of the surface, the lines of curvature are taken as its parametric curves for easy understanding. Since values of the curvatures are independent of parameterization, so are the expressions of κ_1^+, κ_2^+, K_g^+ and K_m^+.

Next, in the usual parametric expression of a surface, we are to derive the fundamental magnitudes of the first and the second order of its offset surface. Since the basic vectors of the offset surface are given by

$$\begin{aligned} &\mathbf{s}_u^+ = \mathbf{s}_u + \mathbf{n}_u d, \quad \mathbf{s}_v^+ = \mathbf{s}_v + \mathbf{n}_v d \\ &\mathbf{s}_{uu}^+ = \mathbf{s}_{uu} + \mathbf{n}_{uu} d, \quad \mathbf{s}_{uv}^+ = \mathbf{s}_{uv} + \mathbf{n}_{uv} d, \quad \mathbf{s}_{vv}^+ = \mathbf{s}_{vv} + \mathbf{n}_{vv} d, \end{aligned}$$

we can derive their fundamental magnitudes from their definitions,

$$\left.\begin{array}{lcl} \mathbf{n}^+ &=& \mathbf{s}_u^+ \times \mathbf{s}_v^+ / |\mathbf{s}_u^+ \times \mathbf{s}_v^+| = \mathbf{n} \\ E^+ &=& E - 2Ld + \mathbf{n}_u^2 d^2 \\ F^+ &=& F - 2Md + (\mathbf{n}_u \cdot \mathbf{n}_v) d^2 \\ G^+ &=& G - 2Nd + \mathbf{n}_v^2 d^2 \\ H^+ &=& |\mathbf{s}_u^+ \times \mathbf{s}_v^+| = H(1 - 2K_m d + K_g d^2) \\ L^+ &=& L - \mathbf{n}_u^2 d \\ M^+ &=& M - (\mathbf{n}_u \cdot \mathbf{n}_v) d \\ N^+ &=& N - \mathbf{n}_v^2 d \end{array}\right\} . \tag{7.36}$$

For derivation of the above equations the following relations are used. By differentiating the relations $\mathbf{n} \cdot \mathbf{n} = 1$, $\mathbf{n} \cdot \mathbf{s}_u = 0$, we obtain

$$\begin{array}{lcl} \mathbf{n} \cdot \mathbf{n}_{uu} &=& -\mathbf{n}_u^2, \quad \mathbf{n} \cdot \mathbf{n}_{uv} = -\mathbf{n}_u \cdot \mathbf{n}_v, \quad \mathbf{n} \cdot \mathbf{n}_{vv} = -\mathbf{n}_v^2 \\ \mathbf{n}_u \cdot \mathbf{s}_u &=& -\mathbf{n} \cdot \mathbf{s}_{uu} = -L, \quad \mathbf{n}_v \cdot \mathbf{s}_v = -\mathbf{n} \cdot \mathbf{s}_{vv} = -N, \\ \mathbf{n}_v \cdot \mathbf{s}_u &=& \mathbf{n}_u \cdot \mathbf{s}_v = -\mathbf{n} \cdot \mathbf{s}_{uv} = -M \end{array}$$

and $\mathbf{n}_u^2$, $\mathbf{n}_v^2$ and $\mathbf{n}_u \cdot \mathbf{n}_v$ in eq. (7.36) are given by equations (6.31) and (6.32), which are:

$$\left.\begin{array}{rcl} \mathbf{n}_u^2 &=& (M^2E + L^2G - 2LMF)/H^2 = -K_gE + 2K_mL \\ \mathbf{n}_v^2 &=& (M^2G + N^2E - 2MNF)/H^2 = -K_gG + 2K_mN \\ \mathbf{n}_u \cdot \mathbf{n}_v &=& (LMG + MNE - M^2F - LNF)/H^2 = -K_gF + 2K_mM \end{array}\right\} . \tag{7.37}$$

Since we can determine the fundamental magnitudes of the offset surface, we can calculate the intersection curves and various characteristic curves as in the original surface.

So long as $\mathbf{s}^+(u,v)$ does not intersect with itself, a sphere of radius d, the center of which is on the offset surface, touches the original surface at one point, but if there are regions bounded by the self-intersection curve of the offset surface, the sphere whose center is in these regions not only touches the original surface, but also intersects with it at other places.

Since the fundamental magnitudes of an offset surface can be determined from the data of the original surface, the interference calculations for the offset surface are performed without having the explicit equation of the offset surface(refer to Sect. 16.5). Figure 7.22 shows a surface and its offset one. Problems of obtaining the intersection curves of offset surfaces are treated in Sect. 13.6 and Sect. 16.5.

Sets of the normal lines emanated from points on the net of the lines of curvature of a surface $\mathbf{s}(u,v)$ constitute the mutually orthogonal surfaces S_1 and S_2 in three-dimensional space. The surfaces $\mathbf{s}^+(u,v)$ and $\mathbf{s}(u,v)$ are orthogonal to them. Since the offset distance d can take any values, $\mathbf{s}(u,v)^+$ and $\mathbf{s}(u,v)$ can be considered to belong to a set of the surfaces S_0. Then the space can be divided by the three mutually orthogonal sets of the surfaces S_0, S_1 and S_2 .

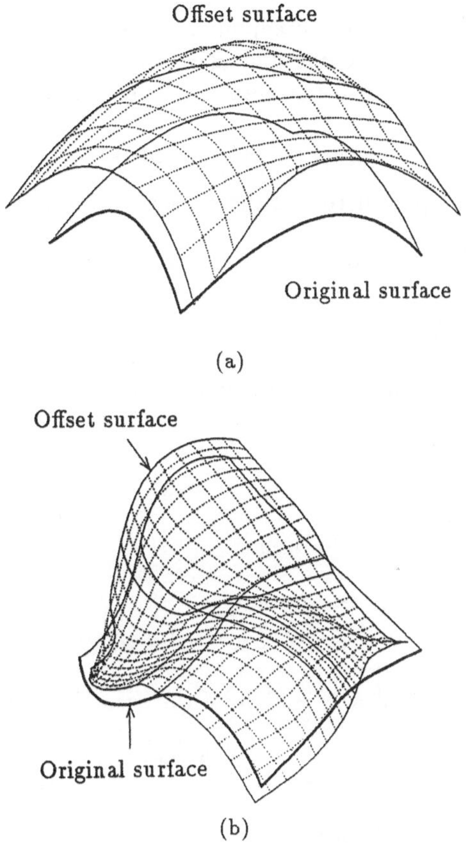

Figure 7.22 Original and offset surfaces: (a) 3 × 3 patch, (b) $C^{(1)}$ connected patches

Since the parametric expressions of surfaces do not describe explicitly the relation of their mutual locations, it is difficult to search for the global relations among them. The orthogonal set of surfaces explained above, which is generated from one of the parametric surfaces, serves to give the distance relation to the other surfaces. This can be used to obtain all the intersection curves between two parametric surfaces.

Exercise. Show that the expressions of the Gaussian and the mean curvatures of an offset surface derived from its fundamental magnitudes are equal to those given by equations (7.34) and (7.35).

7.6 Ruled Surfaces

A ruled surface is generated by a sweeping line which is guided by a space curve. Let $\mathbf{p}(u)$ be the given space curve and $\mathbf{a}(u)$ a unit vector of the sweeping line and let them be functions of a parameter u. An equation of the ruled surface $\mathbf{s}(u, v)$ is defined by

$$\mathbf{s}(u, v) = \mathbf{p}(u) + v\mathbf{a}(u), \tag{7.38}$$

where u and v are the parameters of the surface. From this we obtain

$$\begin{aligned} \mathbf{s}_u &= \mathbf{p}' + v\mathbf{a}', \quad \mathbf{s}_v = \mathbf{a}, \\ \mathbf{s}_{uu} &= \mathbf{p}'' + v\mathbf{a}'', \quad \mathbf{s}_{uv} = \mathbf{a}', \quad \mathbf{s}_{vv} = 0, \end{aligned}$$

where a prime mark indicates differentiation by u.

Its fundamental magnitudes of the first order are

$$\begin{aligned} E &= \mathbf{p}'^2 + 2v\mathbf{p}'\mathbf{a}' + v^2\mathbf{a}'^2, \\ F &= \mathbf{p}'\mathbf{a}, \quad G = \mathbf{a}^2. \end{aligned}$$

Its unit normal vector is by definition

$$\begin{aligned} \mathbf{n} &= (\mathbf{s}_u \times \mathbf{s}_v)/H \\ &= (\mathbf{p}' \times \mathbf{a} + v\mathbf{a}' \times \mathbf{a})/H. \end{aligned} \tag{7.39}$$

The fundamental magnitudes M and N of the second order of the surface are

$$M = \mathbf{n} \cdot \mathbf{s}_{uv} = \frac{[\mathbf{p}', \mathbf{a}, \mathbf{a}']}{H}, \quad N = 0. \tag{7.40}$$

Its Gaussian curvature is given by

$$K_g = -\frac{M^2}{H^2}, \tag{7.41}$$

which is negative or zero. If $\mathbf{a}$ is constant or parallel to $\mathbf{p}'$, which is the tangent direction of the curve, then M becomes zero, hence the Gaussian curvature is zero. The surface is developable.

A helical surface is developable, because it is produced by a sweeping tangent to a helical curve, but a hyperboloid of one sheet and a hyperbolic paraboloid are not developable, because their generators are not parallel to tangents of their guiding curves.

8 Curves Through Given Points, Interpolation and Extrapolation

8.1 Introduction

Methods of constructing a curve which passes through all the points of a given sequence are described. In CAD, input data for a curve are values of a point sequence supplied by automatic measuring instruments or values from digitizers used by a person or supplied from other processes. Since those data contain errors of various sorts, sometimes smoothing procedures such as methods of least squares or appropriate filtering procedures are required.

For storage and use of data by computers, input data finally have to be converted to mathematical expressions which preserve features and accuracy of the original data. There are several curve expressions, but except for parametric polynomial or rational expressions, they are not suitable for use in CAD for reasons of difficulty of application and non-favorable properties of the curves. Non-parametric expressions are not widely used, but since polynomial and rational interpolations and extrapolations are used in the latter sections for numerical methods of solving differential equations, these two representations with their calculation methods are included in this chapter [16].

In CAD, a curve segment is usually represented by a parametric polynomial expression for ease of manipulation. We introduce Hermite expressions for polynomials and then simulation methods of an elastic beam for smooth interpolation. The simulated method has to do with spline curves, but control point methods, which are a very useful technique to represent and control the shapes of curve segments, are mentioned. The detail is treated in Chaps. 9 – 12. Curve fitting by the rational expression of degree two or the conics in implicit forms is treated in Chap. 12.

8.2 Polynomial and Rational Interpolation and Extrapolation

8.2.1 Lagrange's Formula

Given $n + 1$ sets of coordinates values (x_i, y_i), where the values of x_i are in ascending or descending order, a polynomial of degree n,

$$p(x) = \sum_{j=0}^{n} a_j x^j, \tag{8.1}$$

which passes through points

$$y_i = p(x_i), \qquad i = 0 \ldots n\,, \tag{8.2}$$

is to be determined. The coefficients a_j of eq.(8.1) can be obtained by solving the simultaneous equations (8.2) in $n+1$ unknown a_j. But this method is not useful for applications, for it requires solving the simultaneous equations each time the data are changed, and the coefficients a_j do not give direct information on the shape of the curve.

Now, we introduce functions $g_j(x), \{j = 0 \ldots n\}$ of x, which have favorable properties for their determination. We express a curve by a linear combination of $g_i(x)$:

$$p(x) = \sum_{j=0}^{n} c_j g_j(x), \tag{8.3}$$

whose coefficients c_j are determined to satisfy the conditions (8.2). *Lagrange's interpolation formula* has the form of eq.(8.3), in which $c_j = y_j$ and

$$g_j(x) = \frac{(x-x_0)\cdots(x-x_{j-1})(x-x_{j+1})\cdots(x-x_n)}{(x_j-x_0)\cdots(x_j-x_{j-1})(x_j-x_{j+1})\cdots(x_j-x_n)} \tag{8.4}$$

Since the value of $g_j(x)$ becomes 1 at $x = x_j$ and is zero at the other x_i of the given points, changes of the values and the number of the given points do not require solving of the simultaneous equations. However, since the degree of the polynomial is proportional to the number of points to pass, it easily becomes a higher degree polynomial. Then an unwanted oscillation in the interpolated curve frequently appears.

Example. See Fig. 8.1. Values of $y = \sqrt{x}$ are given at $x = (1.0), (1/2), (1/4), (1/16), (1/16)^2, (1/16)^4$. Interpolation curves by the polynomials are shown in solid lines, whereas those in dotted lines are curves by the rational interpolations which are explained in the next section. Figure 8.1(a) shows a curve derived from the first four points, (b) a curve from the first five points, and (c) a curve from all six points. As the order of the curve increases, there appear increasing oscillations in the polynomial interpolations.

8.2.2 Numerical Methods of Interpolated Points

When additional input data (x_i, y_i) are introduced one by one to increase the accuracy of an interpolated or extrapolated value y at x, use of eq.(8.4) is not adequate for the calculation of the updated value, but the following method which utilizes the previous value of y is suitable.

Let $P_{i(i+1)}$ be the value at x of the polynomial of degree one, which passes through two points (x_i, y_i) and (x_{i+1}, y_{i+1}). Likewise let $P_{i(i+1)\ldots(i+n)}$ be the value at x of the polynomial of degree n which passes through $n+1$ points $(x_i, y_i), (x_{i+1}, y_{i+1}), \cdots, (x_{i+n}, y_{i+n})$. Analogously, since P_i is considered the value at x of the polynomial of degree zero, we define $P_i = y_i$.

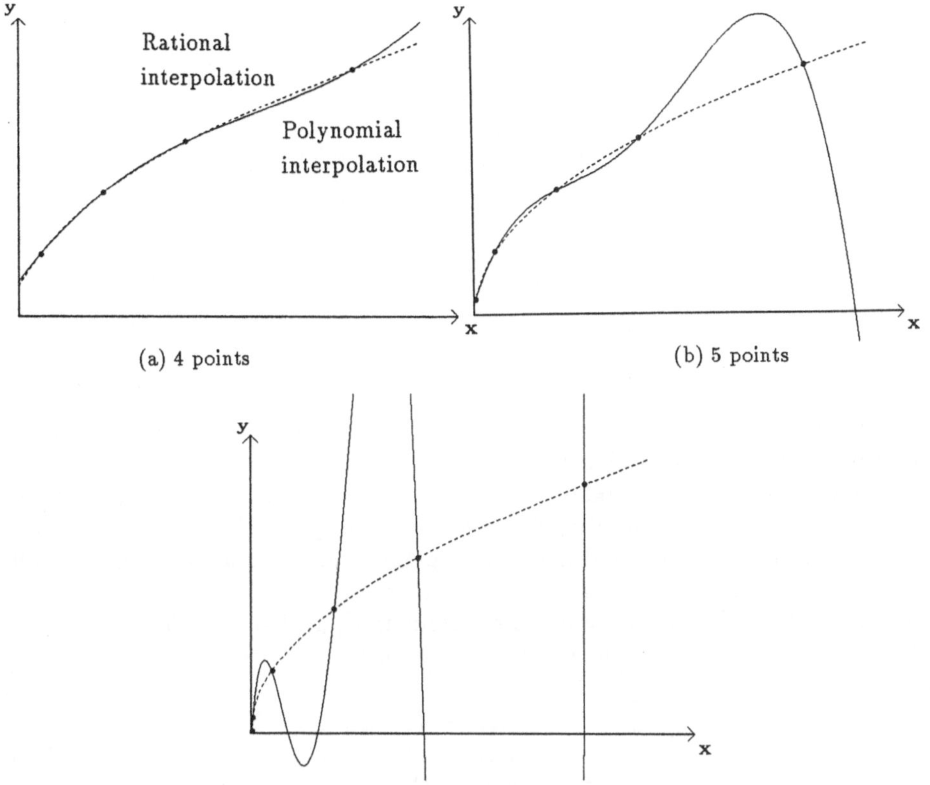

Figure 8.1 Polynomial and rational interpolation and extrapolation for $y = sqrt(x)$

A value of the polynomial of degree m at given x can be deduced from values of the polynomials of degree $m-1$ using the following recurrence formula:

$$P_{i(i+1)\ldots(i+m)} = \frac{(x - x_{i+m})P_{i\ldots(i+m-1)} + (x_i - x)P_{(i+1)\ldots(i+m)}}{(x_i - x_{i+m})} \quad . \tag{8.5}$$

It is easily proved from the definition of $P_{i\ldots(i+m)}$ that the right-hand side of eq.(8.5) takes values $y_i,\ y_{i+1},\cdots,y_{i+m},\ y_{i+m+1}$ at $x = x_i,\ x_{i+1},\cdots,\ x_{i+m+1}$ and is a polynomial of degree m. These values are produced systematically in the table below: for the case $i = 1$ and $m = 4$, From the first column of the above table the second column is produced using eq.(8.5). If the elements of the first column are only P_1 and P_2, the result value is P_{12}. If P_3 is added, then P_{23} is obtained and from P_{12} and P_{23} the final interpolated value P_{123} is determined. When this value is not so much different from P_{12} or P_{23}, we may not take in

Table 8.1 Value of each $P_{i(i+1)\cdots(i+m)}$

P_1				
	P_{12}			
P_2		P_{123}		
	P_{23}		P_{1234}	
P_3		P_{234}		P_{12345}
	P_{34}		P_{2345}	
P_4		P_{345}		
	P_{45}			
P_5				

new input data P_4 and P_5. But if we have to use all the data from P_1 to P_5 for interpolation or extrapolation, we have to calculate P_{12345}.

Our objective is to evaluate y at x from a given sequence of points (x_1, y_1), $\cdots$, (x_n, y_n). We assume that from y_1 , $\cdots$, y_{n-1}, a value $y = P_{1\cdots n-1}$ has already been determined and now a new point y_n at x_n is added. First we fill the entries of Table 8.1 to produce $P_{2..n}$, then we obtain $P_{12..n}$ from $P_{12..(n-1)}$ and $P_{2..n}$. To make the above process systematical, at first we define differences between $P_{i(i+1)...(i+m)}$ and the lower adjacent entries:

$$\left.\begin{aligned} C_{i,m} &= P_{i...(i+m)} - P_{i...(i+m-1)}, \\ D_{i,m} &= P_{i...(i+m)} - P_{(i+1)...(i+m)}, \\ C_{i,0} &= P_i = y_i, \quad D_{i,0} = P_i = y_i \end{aligned}\right\} \tag{8.6}$$

There are relations between these differences:

$$C_{i,m+1} = X_{i,m}(C_{i+1,m} - D_{i,m}), \tag{8.7}$$

$$D_{i,m+1} = (1 - X_{i,m})(C_{i+1,m} - D_{i,m}), \tag{8.8}$$

$$\text{where } X_{i,m} = \frac{x_i - x}{x_i - x_{i+m+1}}$$

We construct two triangular Tables 8.2 and 8.3, whose entries contain $C_{i,m}$ or $D_{i,m}$. The first index i indicates the i-th row of the difference tables and the second index m shows the m-th column of the tables. Entries in the tables are filled step by step as each new value $y_i = C_{i,0} = D_{i,0}$ is added in the both entries of the first columns $(i, 0)$. At each level m, the $C_{i,m}$ and $D_{i,m}$ are the correction terms for the interpolation value one order higher. The final value $P_{1,2} \ldots n$ is equal to the sum of any y_i plus $C_{i,m}$ or $D_{i,m}$ in the ascending order of m from $m = 2$ to $m = n$ according to eq.(8.6). Application of the tables appears in the Appendix of this book.

Table 8.2 Value of each $C_{i,j}$

$C_{1,0}$	$C_{1,1}$	$C_{1,2}$	$C_{1,3}$	$C_{1,4}$
$C_{2,0}$	$C_{2,1}$	$C_{2,2}$	$C_{2,3}$	
$C_{3,0}$	$C_{3,1}$	$C_{3,2}$		
$C_{4,0}$	$C_{4,1}$			
$C_{5,0}$				

Table 8.3 Value of each $D_{i,j}$

$D_{1,0}$	$D_{1,1}$	$D_{1,2}$	$D_{1,3}$	$D_{1,4}$
$D_{2,0}$	$D_{2,1}$	$D_{2,2}$	$D_{2,3}$	
$D_{3,0}$	$D_{3,1}$	$D_{3,2}$		
$D_{4,0}$	$D_{4,1}$			
$D_{5,0}$				

8.2.3 Rational Function Interpolation and Extrapolation

Approximations by polynomial functions are easily calculated and their derivatives and integrals are easily obtained, but they oscillate around the desired curves so that they frequently exceed given error bounds. Approximations by rational functions usually show smaller error bounds than by polynomials. In Fig. 8.2, polynomial and rational curves which pass through the same six points are shown. The rational curve seems good, but the polynomial one shows large oscillation. However, the rational interpolations do not always give good results;

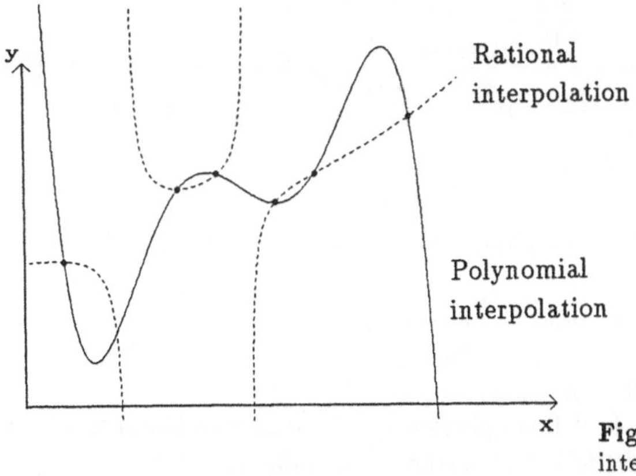

Figure 8.2 Polynomial and rational interpolation for six points

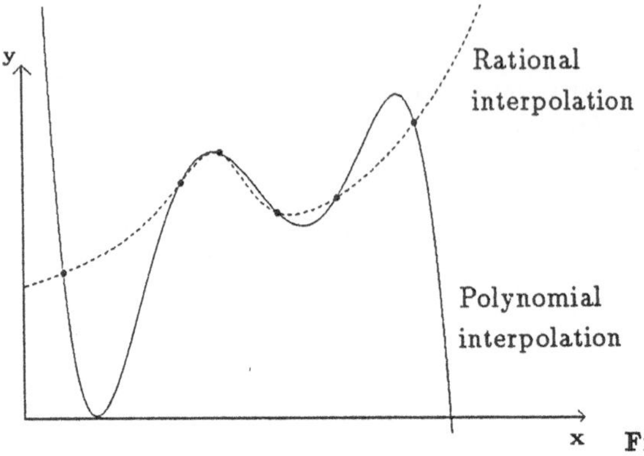

Figure 8.3 Polynomial and rational interpolation for six points — unfavorable case

for instance, Fig. 8.3 shows very bad interpolation curves for both polynomial and rational cases.

In numerical methods of solving ordinary differential equations, the Bulirsch-Stoer method [16][17] applies rational instead of polynomial extrapolation. It is an excellent method of solving ordinary differential equations and we use it in Chap. 16; its detail is explained in Appendix of this book. Since its procedure to obtain an interpolated or extrapolated value is similar to that of the polynomial procedure described above, we treat the case of a rational function here.

Let $R_{i(i+1)\ldots(i+n)}$ be a rational function which passes through $n+1$ points:

$$(x_i, y_i), (x_{i+1}, y_{i+1}), \cdots, (x_{i+n}, y_{i+n}).$$

A recurrence formula analogous to the polynomial case eq.(8.5) is

$$\begin{aligned} R_{i(i+1)\ldots(i+n)} &= R_{(i+1)\ldots(i+n)} + \\ & \frac{R_{(i+1)\ldots(i+n)} - R_{i\ldots(i+n-1)}}{\frac{x-x_i}{x-x_{i+n}}\{1 - \frac{R_{(i+1)\ldots(i+n)} - R_{i\ldots(i+n-1)}}{R_{(i+1)\ldots(i+n)} - R_{(i+1)\ldots(i+n-1)}}\} - 1}, \qquad (8.9) \\ R_i = y_i, \quad & R_{i(i-1)} \equiv 0. \end{aligned}$$

For $x = x_i$, the first term is $R_{(i+1)\ldots(i+n)}$ and the second term gives

$$-\{R_{(i+1)\ldots(i+n)} - R_{i\ldots(i+n-1)}\}.$$

The result is $R_{i\ldots(i+n-1)}$, which is y_i.

For $x = x_{i+1}, \ldots, x_{i+n}$, the first term gives $y_{i+1}, \ldots, y_{i+n}$ and the second term gives zero. Accordingly, eq.(8.9) is the recurrence formula we want.

For $n = 1$, we obtain a rational curve which passes two given points. From eq.(8.9) we have

$$R_{i(i+1)} = R_{i+1} + \frac{R_{i+1} - R_i}{\frac{x - x_i}{x - x_{i+n}}\{1 - \frac{R_{i+1} - R_i}{R_{i+1}}\} - 1}$$
$$= \frac{y_i y_{i+1}(x_{i+1} - x_i)}{\{(x - x_i)y_i - (x - x_{i+1})y_{i+1}\}}.$$

The denominator is a polynomial of degree one, whereas its numerator is constant, that is, a polynomial of degree zero. Since by definition $R_{i(i+1)}$ passes two points $(x_i y_i)$ and (x_{i+1}, y_{i+1}), we have the extrapolated value at $x = 0$:

$$R_{i(i+1)} = \frac{y_i y_{i+1}(x_{i+1} - x_i)}{x_{i+1} y_{i+1} - x_i y_i}. \tag{8.10}$$

For $n = 2$, a rational curve which passes given three points is obtained from eq.(8.9) by

$$R_{i(i+1)(i+2)} = R_{(i+1)(i+2)} + \frac{(x - x_{i+2})(R_{(i+1)(i+2)} - R_{i(i+1)})(R_{(i+1)(i+2)} - y_{i+1})}{(x_{i+2} - x)(R_{(i+1)(i+2)} - y_{i+1}) - (x - x_i)(R_{(i+1)(i+2)} - R_{i(i+1)})}.$$

We want to obtain $R_{12\ldots n}$ by constructing two triangular tables similar to Tables 8.2 and 8.3 of the polynomial case. For this purpose, we define the difference functions between the adjacent rational functions similar to eq.(8.6):

$$\begin{aligned} C_{i,m} &= R_{i\ldots(i+m)} - R_{i\ldots(i+m-1)}, \\ D_{i,m} &= R_{i\ldots(i+m)} - R_{(i+1)\ldots(i+m)}, \\ C_{i,0} &= R_i, \quad D_{i,0} = R_i. \end{aligned} \tag{8.11}$$

Instead of equations (8.7) and (8.8), we have the recurrence formulas for $C_{i,m}$ and $D_{i,m}$ for the rational case:

$$C_{i,m+1} = \frac{X_{i,m} D_{i,m}(C_{i+1,m} - D_{i,m})}{(X_{i,m} D_{i,m} - C_{i+1,m})}, \tag{8.12}$$

$$D_{i,m+1} = \frac{C_{i,m}(C_{i+1,m} - D_{i,m})}{(X_{i,m} D_{i,m} - C_{i+1,m})}, \tag{8.13}$$

$$\text{where} \quad X_{i,m} = \frac{(x - x_i)}{(x - x_{i+m+1})}.$$

Then we can construct a table similar to Tables 8.2 and 8.3 step by step each time a new pair (x_i, y_i) is added. If x is given, we can determine $y = R_{12\ldots n}$ by a similar method as for $y = P_{12\ldots n}$.

When x is outside of x_n, the corresponding value of y is an extrapolated value which is given by

$$R_{12\dots n} = R_n + D_{n-1,1} + D_{n-2,2} + \cdots + D_{1,n-1}. \tag{8.14}$$

Example. For the examples given in Fig. 8.1, we obtain extrapolated values of $y = \sqrt{x}$ at $x = 0$, when the correct y values are given at seven points of x_i:

$$x = 1.0,\ 1/2,\ 1/4,\ 1/16,\ (1/16)^2,\ (1/16)^4,\ (1/16)^8.$$

Relations between the extrapolated values at $x = 0$ and the number of the given points taken from the head of the above sequence are shown for the polynomial and the rational extrapolations.

number of points	4	5	6	7
polynomial	0.1417	0.04699	0.00366	0.000015
rational	0.1249	0.04096	0.00357	0.000015

(8.15)

Though the curves given by the polynomials oscillate extremely around the true curve as the order increases, the values of y given by the two extrapolations at $x = 0$ do not differ considerably for this sequence of values of x.

8.3 Polynomial Interpolation with Constraints of Derivatives

The Hermite interpolation method specifies points to pass as well as derivatives there. In the simplest case, a curve segment is to be determined by its two end points. A curve which passes through many points is composed from curve segments, each of which passes through the adjacent points.

Let $\mathbf{r}(t)$, $\{0 \le t \le 1\}$, be a curve segment with a parameter t between two points $\mathbf{r}(0) = \mathbf{r}_0$ and $\mathbf{r}(1) = \mathbf{r}_1$, where derivatives up to the m-th with respect to t are specified, then the degree of the curve is at least $2m+1$. We are to express $\mathbf{r}(t)$ in terms of these $2m+2$ derivatives with respect to the parameter at the end points, multiplied by suitable weighting functions. We use notations $\mathbf{r}_0^{(i)}$ and $\mathbf{r}_1^{(i)}$ for i-th derivatives of $\mathbf{r}(t)$ at $t = 0$ and $t = 1$ and the notations $g_i(t; 2m+1)$ and $h_i(t; 2m+1)$ for the weighting functions of $\mathbf{r}_0^{(i)}$ and $\mathbf{r}_1^{(i)}$.

The curve segments of degree 1, 3 and 5 are expressed by

$$m = 0 \ : \ \mathbf{r}(t) = \mathbf{r}_0^{(0)} g_0(t;1) + \mathbf{r}_1^{(0)} h_0(t;1), \tag{8.16}$$

$$m = 1 \ : \ \mathbf{r}(t) = \mathbf{r}_0^{(0)} g_0(t;3) + \mathbf{r}_1^{(0)} h_0(t;3) + \mathbf{r}_0^{(1)} g_1(t;3) + \mathbf{r}_1^{(1)} h_1(t;3), \tag{8.17}$$

$$\begin{aligned} m = 2 \ : \ \mathbf{r}(t) &= \mathbf{r}_0^{(0)} g_0(t;5) + \mathbf{r}_1^{(0)} h_0(t;5) \\ &+ \mathbf{r}_0^{(1)} g_1(t;5) + \mathbf{r}_1^{(1)} h_1(t;5) + \mathbf{r}_0^{(2)} g_2(t;5) + \mathbf{r}_1^{(2)} h_2(t;5), \end{aligned} \tag{8.18}$$

where the weighting functions $g_i(t; 2m+1)$ and $h_i(t; 2m+1)$ are given by

$$\begin{array}{lll} m=0 & : & g_0(t;1) = 1-t, \qquad h_0(t;1) = t, \\ m=1 & : & g_0(t;3) = (1-t)^3 + 3t(1-t)^2 = 1 - 3t^2 + 2t^3, \\ & & h_0(t;3) = t^3 + 3t^2(1-t) = 3t^2 - 2t^3, \\ & & g_1(t;3) = t(1-t)^2, \qquad h_1(t;3) = -t^2(1-t), \\ m=2 & : & g_0(t;5) = (1-t)^3(1+3t+6t^2), \quad h_0(t;5) = t^3(10-15t+6t^2), \\ & & g_1(t;5) = t(1-t)^3(1+3t), \qquad h_1(t;5) = -(1-t)t^3(4-3t), \\ & & g_2(t;5) = (1/2)t^2(1-t)^3, \qquad h_2(t;5) = (1/2)(1-t)^2t^3. \end{array} \tag{8.19}$$

The general form of r(t) of degree 2m+1 is [2]:

$$\mathbf{r}(t) = \sum_{j=0}^{m} \mathbf{r}_0^{(i)} g_i(t; 2m+1) + \mathbf{r}_1^{(i)} h_i(t; 2m+1), \tag{8.20}$$

where

$$g_i(t; 2m+1) = \frac{1}{i!} \sum_{j=0}^{m} C_j^{2m+1-i} \cdot t^{i+j}(1-t)^{2m+1-i-j}, \tag{8.21}$$

$$h_i(t; 2m+1) = \frac{1}{i!} \sum_{j=0}^{m} C_j^{2m+1-i} \cdot t^{2m+1-i-j}(1-t)^{i+j}, \tag{8.22}$$

and C_j^k represent the binomial coefficients.

These functions have the following relations,

$$g_0(t; 2m+1) + h_0(t; 2m+1) = (t+1-t)^{2m+1} = 1,$$
$$h_i(t; 2m+1) = (-1)^i g_i(1-t; 2m+1).$$

The shape of a curve segment is expressed by the weighted sum of the derivatives of $\mathbf{r}(t)$ from 0-th to m-th degree at both ends of the curve. Though values of the derivatives with respect to a parameter are difficult to give explicitly, the above expressions are used to connect with the adjacent curve segments.

8.4 Elastic Curves with Minimum Energy

The shape of an elastic wire, which is deformed by external constraints or forces, is determined to make its stored elastic energy minimum. So, its shape is considered natural and smooth. We want to use this elastic wire for interpolation of a given point sequence.

Let $n+1$ points $P_i(x_i, y_i)$ with $x_i < x_{i+1}$, $\{i = 0 \dots n\}$, be given and let $y_{i,i+1}$ be the shape of interpolated curve between x_i and x_{i+1}, then we are to determine all $y_{i,i+1}$ connected with continuous curvatures. Refer to Fig. 8.4. Using notations $d_{i,i+1} \equiv x_{i+1} - x_i$ and $t \equiv (x - x_i)/(d_{i,i+1})$, the slope in the i-th section is written as

$$\left(\frac{dy}{dx}\right)_i = \frac{1}{d_{i,i+1}} \left(\frac{dy_i}{dt}\right),$$

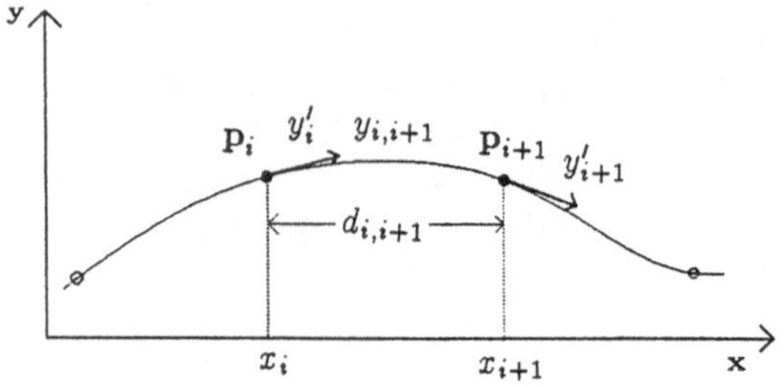

Figure 8.4 Elastic beam model for smoothing

Using the weighting functions given by eq.(8.19), we express the shape $y_{i,i+1}$ of each curve segment in the following form:

$$\begin{aligned} y_{i,i+1} &= y_i g_0(t) + y_{i+1} h_0(t) + y'_i d_{i,i+1} g_1(t) + y'_{i+1} d_{i,i+1} h_1(t), \\ &\quad \text{for } i = 0 \ldots n-1. \end{aligned} \tag{8.23}$$

There are $n+1$ unknown y'_i which are derivatives with respect to x at the junctions x_i. They are determined to minimize the stored elastic energy. This condition leads to the following equations (refer to Sect. 8.6):

$$\left. \begin{aligned} y''_{i-1,i}(x_i) &= y''_{i,i+1}(x_i), \qquad i = 1 \ldots n-1, \\ y''_{0,1}(x_0) &= y''_{n-1,n}(x_n) = 0, \end{aligned} \right\} \tag{8.24}$$

which stipulate that at each junction, the second derivatives of the curve segments are continuous and at the end points they are zero. Substituting eq.(8.23) into equations (8.24), we obtain the following three-term simultaneous equations:

$$\begin{aligned} & y'_{i-1}\left(\frac{1}{d_{i-1,i}}\right) + 2y'_i\left(\frac{1}{d_{i-1,i}} + \frac{1}{d_{i,i+1}}\right) + y'_{i+1}\left(\frac{1}{d_{i,i+1}}\right) \\ &= 3\frac{(y_i - y_{i-1})}{d^2_{i-1,i}} + 3\frac{(y_{i+1} - y_i)}{d^2_{i,i+1}}, \quad \text{for} \quad i = 1 \ldots n-1, \end{aligned} \tag{8.25}$$

and the conditions at both ends are

$$\left. \begin{aligned} 2y'_0 d_{0,1} + y'_1 d_{0,1} &= 3(y_1 - y_0), \\ y'_{n-1} d_{n-1,n} + 2y'_n d_{n-1,n} &= 3(y_n - y_{n-1}). \end{aligned} \right\} \tag{8.26}$$

Solving the equations, we determine values of y'_i and the shape of each segment from eq.(8.23).

If at both ends of the curve its slopes y'_0 and y'_n are fixed, the number of unknown y'_i is $n-1$, which is equal to that of equations (8.25). So, we can solve them and calculate the shape from eq.(8.23).

8.5 Interpolation by Parametric Curves

In the previous section, a curve is expressed in the form $y(x)$. Though this form is easy to treat, a limitation of the expression arises when y of the curve takes multivalues. Accordingly, for shape design, curve segments with parametric expressions are used. For a curve segment $\mathbf{r}_i(t)$, $\{0 \le t \le 1\}$, with its end points $\mathbf{p}_i$ and $\mathbf{p}_{i+1}$ and the tangent vectors $\dot{\mathbf{r}}_i(0)$ and $\dot{\mathbf{r}}_i(1)$ there, its expression is given from eq.(8.17) for $m = 1$ by

$$\mathbf{r}_i(t) = \mathbf{p}_i g_0(t) + \mathbf{p}_{i+1} h_0(t) + \dot{\mathbf{r}}_i(0) g_1(t) + \dot{\mathbf{r}}_i(1) h_1(t), \tag{8.27}$$

with the conditions

$$\dot{\mathbf{r}}_i(1) = k_{i+1}\dot{\mathbf{r}}_{i+1}(0), \quad i = 1 \ldots n-1, \tag{8.28}$$

which specify the continuity of the tangent at the junctions, where k_{i+1} is a constant which we take as the ratio of the chord lengths of two curves. Since in this section, the independent variable t of the curve expressions is normalized between 0 and 1, the ratios k_i have to be taken into consideration.

For each interval of the point sequence $\mathbf{p}_0, \mathbf{p}_1, \mathbf{p}_2, \cdots, \mathbf{p}_n$, a curve segment of the above form is assigned and connected to its adjacent curve segments with tangential continuity. To determine the value of $\dot{\mathbf{r}}_i(0)$ at the connecting points to make the whole shape as smooth as possible, pseudo-elastic energy stored in the curve is made a minimum. Let a chord vector of a segment be

$$\mathbf{a}_i = \mathbf{p}_{i+1} - \mathbf{p}_i, \tag{8.29}$$

and let its absolute value $|\mathbf{a}_i|$ be a_i and assume that shape of the curve does not deviate much from its chord line, then $ds \approx |\mathbf{a}_i|dt$ and the curvature is proportional to $\ddot{\mathbf{r}}_i/a_i^2$, and the stored pseudo-elastic energy is given by

$$E = \sum(1/a_i^3) \int_0^1 \ddot{\mathbf{r}}_i(t)^2 dt. \tag{8.30}$$

In the above equation $\ddot{\mathbf{r}}_i$ is the function of $\dot{\mathbf{r}}_i(0)$ and $\dot{\mathbf{r}}_{i+1}(1)$ by equations (8.27) and (8.28). From the minimum condition $\partial E/\partial \dot{\mathbf{r}}_i(0) = 0$, we obtain similar results as in the previous section; continuity of curvature at the intermediate connecting points of curve segments and zero curvatures at the end points,

$$\left.\begin{aligned} \ddot{\mathbf{r}}_{i-1}(1) &= k_i^2 \ddot{\mathbf{r}}_i(0), \quad i = 1 \ldots n-1, \\ \ddot{\mathbf{r}}_0(0) &= \ddot{\mathbf{r}}_{n-1}(1) = 0, \\ \text{where} \quad k_i &= a_{i-1}/a_i. \end{aligned}\right\} \tag{8.31}$$

From these equations, which correspond to equations (8.23) and (8.24), we get the simultaneous equations for the tangent vectors at the connecting points,

$$\dot{\mathbf{r}}_{i-1} + 2\dot{\mathbf{r}}_i(k_i + k_i^2) + \dot{\mathbf{r}}_{i+1} k_i^2 k_{i+1} = 3(\mathbf{p}_i - \mathbf{p}_{i-1}) + 3(\mathbf{p}_{i+1} - \mathbf{p}_i)k_i^2, \tag{8.32}$$

for $i = 1 \ldots n-1$,

and at the both ends

$$\left.\begin{aligned} 2\dot{\mathbf{r}}_0 + \dot{\mathbf{r}}_1 k_1 &= 3(\mathbf{p}_1 - \mathbf{p}_0), \\ \dot{\mathbf{r}}_{n-1} + 2\dot{\mathbf{r}}_n &= 3(\mathbf{p}_n - \mathbf{p}_{n-1}), \\ \text{where} \quad \dot{\mathbf{r}}_i = \dot{\mathbf{r}}_i(0), & \qquad \dot{\mathbf{r}}_n = \dot{\mathbf{r}}_{n-1}(1). \end{aligned}\right\} \tag{8.33}$$

When at the end points the directions of the tangents are specified instead of zero curvatures, the magnitudes of the tangents are unknown. So, $\partial E/\partial|\dot{\mathbf{r}}_0| = 0$ is needed instead of $\partial E/\partial \dot{\mathbf{r}}_0 = 0$, that is,

$$2\dot{\mathbf{r}}_0^2 + \dot{\mathbf{r}}_0 \cdot \dot{\mathbf{r}}_1 k_1 = -3\dot{\mathbf{r}}_0 \cdot (\mathbf{p}_1 - \mathbf{p}_0). \tag{8.34}$$

Similarly for the other end,

$$\dot{\mathbf{r}}_n \cdot \dot{\mathbf{r}}_{n-1} + 2\dot{\mathbf{r}}_n^2 = 3\dot{\mathbf{r}}_n \cdot (\mathbf{p}_n - \mathbf{p}_{n-1}). \tag{8.35}$$

These boundary conditions are used instead of eq.(8.33). Since we can use the Bézier expression (refer to Chap. 9) in place of eq.(8.23), a curve segment $\mathbf{r}_i(t)$ of degree three is represented by its control polygon with three edges $\mathbf{e}_{3i}$, $\mathbf{e}_{3i+1}$ and $\mathbf{e}_{3i+2}$ (in Chap. 9, a symbol $\mathbf{a}$ is used instead of the symbol $\mathbf{e}$ for an edge of a control polygon). The tangent vectors at the ends of the segment are given by its edges vectors:

$$\dot{\mathbf{r}}_i(0) = 3\mathbf{e}_{3i}, \quad \dot{\mathbf{r}}_i(1) = 3\mathbf{e}_{3i+2} = k_{i+1}\dot{\mathbf{r}}_{i+1}(0) = 3k_{i+1}\mathbf{e}_{3i+3}.$$

It follows that equations (8.32) and (8.33) are converted to the relations among the polygon edges as below,

$$\left.\begin{aligned} &\mathbf{e}_{3i-3} + 2\mathbf{e}_{3i}(k_i + k_i^2) + \mathbf{e}_{3i+3}k_i^2 k_{i+1} = (\mathbf{p}_i - \mathbf{p}_{i-1}) + (\mathbf{p}_{i+1} - \mathbf{p}_i)k_i^2, \\ &\quad i = 1 \ldots n-1, \\ &2\mathbf{e}_0 + \mathbf{e}_3 k_1 = (\mathbf{p}_1 - \mathbf{p}_0), \qquad \mathbf{e}_{3n-3} + 2\mathbf{e}_{3n-1} = (\mathbf{p}_n - \mathbf{p}_{n-1}). \end{aligned}\right\} \tag{8.36}$$

Equations (8.32) and (8.33) or (8.34) and (8.35) are three-term simultaneous equations whose solution is easily obtained. When a point sequence is given, the chord vectors of adjacent point pairs and the ratios of their absolute values $k_i = a_{i-1}/a_i$ are all the data for the coefficients of the equations for determining $\dot{\mathbf{r}}_i$'s. Each curve segment is calculated from eq.(8.27). Their connection continuity is $C^{(2)}$.

When the distances between adjacent points of the sequence are almost the same, then constants k_i may well be taken equal to 1, but when their variation is great, k_i should be taken as defined. In Fig. 8.5 two examples are shown with the same given points: (a) constants k_i are taken as ratios of the chord lengths, (b) all k_i are equal to 1. The resulting shapes are quite different.

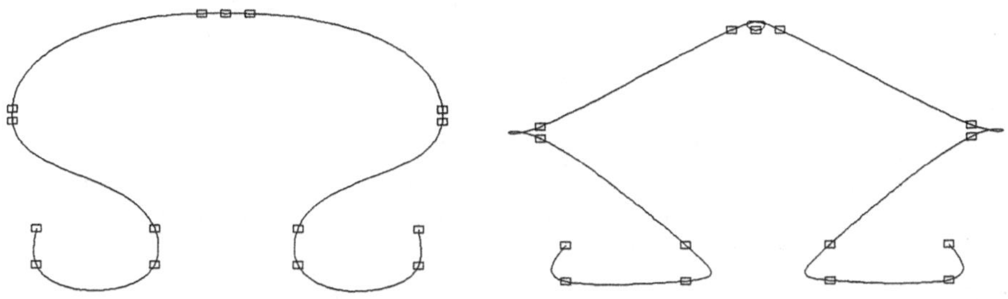

(a) k_i = ratio of adjacent chord lengths (b) $k_i = 1$

Figure 8.5 Curve shapes and scale ratios: (a) k_i = ratio of adjacent chord lengths, (b) $k_i = 1$

8.6 Appendix. Derivation of Equations by Elastic Beam Analogy

Equations (8.24) and (8.25) are derived in the following way. For deflection of a wire or slender beam, a bending moment applied to its axis contributes most to the stored elastic energy, and any other elastic energy such as that caused by uniform compression need not be considered for the present purpose. The stored elastic energy of a wire by bending is expressed by$(1/2) \int G\kappa^2 ds$, where G is the rigidity of bending and κ is the curvature of the wire axis, and the integral is taken along its axis.

Minimizing this elastic energy under given constraints is difficult, because of the nonlinearity of the curvature expression and the dependence of the bending moment on the deformed shape of the wire. This is the kind of problem of Elastica [19], and even if it could be solved any way it may not be practical.

Since its energy is used for a shape criterion only, its expression need not be very accurate so long as its characteristics are preserved. Accordingly, instead of a flexible wire, an elastic beam of small deflection is assumed. In this case, its curvature is approximated by d^2y/dx^2 with the initial beam axis in the x direction and its deflection in the y direction. This means the reaction forces from the supports are directed vertically, so the bending moments at a section of the beam between the supports is a linear function of the distance from a support point. Since the bending moment divided by the bending rigidity of the beam is equal to the curvature at that point, its deflection y is expressed by a polynomial of third degree in x, because of the double integration of the linear function of x representing the bending moment of a section of the beam.

This polynomial is expressed by the vertical positions (y_i, y_{i+1}) and the slopes y'_i and y'_{i+1} at the end points x_i and x_{i+1}, multiplied by their corresponding weighting function g_0, h_0, g_1 and h_1 given in Sect. 8.4. So we have eq.(8.23). We construct a continuous elastic beam which consists of n curve segments connected at $n-1$ supports. The total elastic energy stored in the deformed beams is

$$E = (1/2) \sum G \int {y''_{i,i+1}}^2 dx,$$

where $y''_{i,i+1}$ is the second derivative of the deflection between the i-th and $(i+1)$-th supports, which represents the curvature there. The total energy is a function of the slopes y'_i at the supports. Differentiating it with respect to y'_i and setting the result to zero, we have

$$\begin{aligned} \frac{\partial E}{\partial y'_i} &= \sum G \int y''_{i,i+1} \frac{\partial y''_{i,i+1}}{\partial y'_i} dx \\ &= 0, \qquad i = 0 \ldots n. \end{aligned}$$

By evaluating the above equations, we obtain equations (8.25) and (8.26).

9 Bézier Curves and Control Points

9.1 Introduction

There are various expressions for describing a shape, but the number of practical ones is limited. A polynomial expression in parametric form with control points is most useful in most applications, for it is not only mathematically simple, but also it has favorable properties for engineering design; its expression does not depend on coordinate systems, geometrical properties are easily grasped and manipulated by control points, unwanted waviness usually does not appear in construction and modification of shapes, and long curves and large areas can be designed by connecting short or small segments.

Accordingly, we treat single segments of Bézier curves and surfaces which have the above stated characteristics. Their expressions are a little different from those in other textbooks, but because of our adoption of shifting operators for manipulation of control points, our expressions for Bézier curves are simpler and easier to process [5][8]. Then in the next chapter, B spline curves and surfaces which have similar favorable properties in design are introduced as the special connection of segments of Bézier curves and surfaces. We think this approach is the easiest for design engineers to understand and use. The extension of the approach to special applications is explained in later chapters.

9.2 Curve Segment and Its Control Points

We explain the relation between the shape of a curve segment and its control points together with their weighting functions. Let a curve segment be denoted by $\mathbf{r}(t)$ whose components are $x(t), y(t), z(t)$, where the parameter t takes values $0 \leq t \leq 1$. A curve segment is expressed by the summation of its control points $\mathbf{p}_0$, $\mathbf{p}_1$, $\mathbf{p}_2$, $\mathbf{p}_3 \cdots \mathbf{p}_n$ multiplied by the corresponding weighting functions $f_0(t)$, $f_1(t)$, $f_2(t), \cdots, f_n(t)$:

$$\mathbf{r}(t) = \mathbf{p}_0 f_0(t) + \mathbf{p}_1 f_1(t) + \mathbf{p}_2 f_2(t) + \cdots + \mathbf{p}_n f_n(t). \tag{9.1}$$

A polyline made of the control points $\mathbf{p}_0, \mathbf{p}_1, \mathbf{p}_2, \mathbf{p}_3 \cdots, \mathbf{p}_n$ is called a *control polygon*. For the expression (9.1) to be useful, it needs to have the following properties:

- The expression is independent of coordinate systems used.
- The shape of the control polygon approximates to that of $\mathbf{r}(t)$.
- The geometric characteristics of the curve are obtained from the control polygon only.

To meet these requirements, the following relation must hold for the weighting functions independently of t:

$$f_0(t) + f_1(t) + f_2(t) + \cdots + f_n(t) = 1. \tag{9.2}$$

When all the control points converge to a same point $\mathbf{p}$, the expression (9.1) becomes

$$\mathbf{r}(t) = \mathbf{p}\{f_0(t) + f_1(t) + f_2(t) + \cdots + f_n(t)\}. \tag{9.3}$$

Since both sides of the above expression must be independent of t and equal to $\mathbf{p}$, eq. (9.2) is derived. This is also the condition for independence of coordinate transformation. A transformation applied to the left-hand side of eq. (9.1) must be the same as the curve with the transformed control points and the same weighting functions. Since a transformation consists of a rotation and a translation, the form of eq. (9.1) is invariant for a rotational transformation and together with the condition eq. (9.2) it is invariant for translation of the coordinates.

Properties of the weighting functions $f_i(t)$ are described next. When all the control points except an arbitrary $\mathbf{p}_j$ converge to one point $\mathbf{p}$, the expression (9.1) becomes

$$\mathbf{r}(t) = \mathbf{p}\{f_0(t)+f_1(t)+f_2(t)+\cdots+f_n(t)\}+(\mathbf{p}_j-\mathbf{p})f_j(t) = \mathbf{p}+(\mathbf{p}_j-\mathbf{p})f_j(t), \tag{9.4}$$

because of eq. (9.2). If $\mathbf{r}(t)$ is to be within the line segment joining $\mathbf{p}$ and $\mathbf{p}_j$, a weighting function $f_j(t)$ must have a value between 0 and 1: hence for all j

$$0 \leq f_j(t) \leq 1. \tag{9.5}$$

The equality holds only for $t = 0$ or $t = 1$, where only one weighting function is 1 and the others must be all zero. Without these conditions, the curve is not within the line segment $(\mathbf{p}, \mathbf{p}_j)$. And all the weighting functions must have a single peak in the range, otherwise the relation between the control points and the curve shape becomes complicated and the shape cannot be inferred from the control polygon. Then the roles of the control points cannot be satisfied.

With these properties of $f_j(t)$, the curve $\mathbf{r}(t)$ exists within the convex hull made by all the control points $(\mathbf{p}_0\ \mathbf{p}_1\ \mathbf{p}_2\ \mathbf{p}_3\ \cdots \mathbf{p}_n)$. See Fig. 9.1. If we take the x-axis to coincide with one of the edges of the convex hull, the signs of y values of all the control points are the same or zero, consequently the value of the curve generated by the control points multiplied by their weighting functions has the same sign. This means that the curve is on one side of the edge of the convex hull. For all edges a similar property holds, hence the curve cannot go out of the convex hull. It follows that if the control polygon is planar, the curve is planar, and if the edges of the control polygon are collinear, so is the curve.

9.3 Bézier Curve and Its Operator Form

We explain a curve segment with its control points expressed in a compact operator form and the merits of using the operator. To indicate explicitly the degree

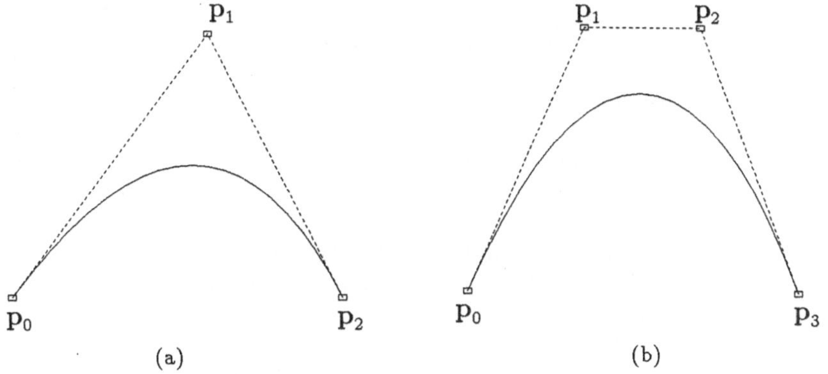

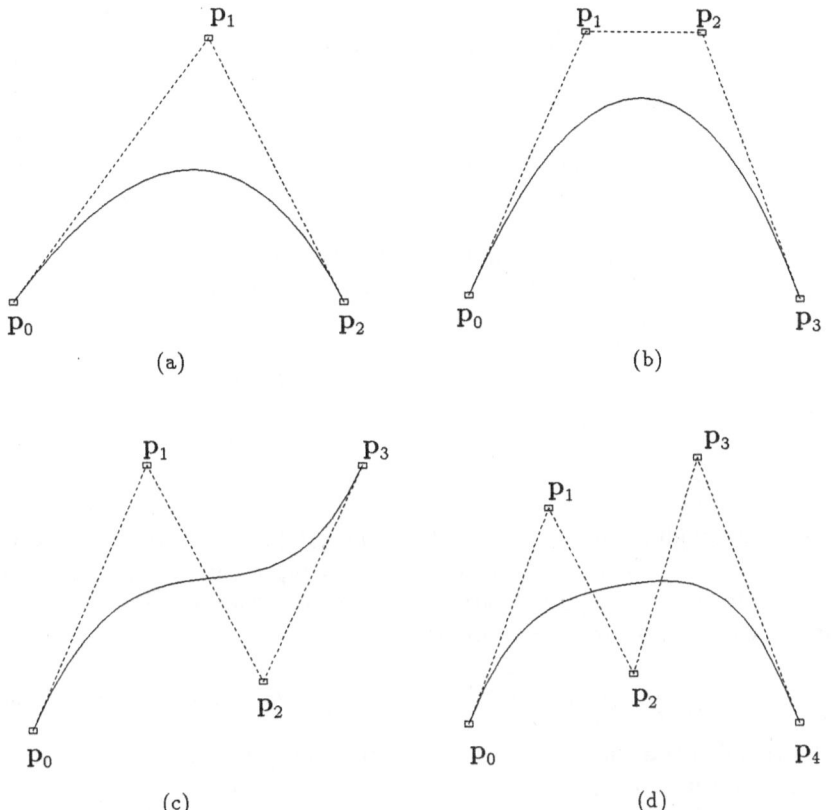

Figure 9.1 Control polygons and their generating curves

n of the polynomial of t and the first control point $\mathbf{p}_i$ of $\mathbf{r}(t)$, we use a notation

$$\mathbf{r}_i(t;n) \tag{9.6}$$

for representing the curve segment. The control points of this curve begin from $\mathbf{p}_i$ and end at $\mathbf{p}_{i+n}$.

A curve for $n = 0$ is considered a single point, so the following expressions can be used:

$$\mathbf{r}_0(t;0) = \mathbf{p}_0 \ , \qquad \mathbf{r}_i(t;0) = \mathbf{p}_i \ . \tag{9.7}$$

For $n = 1$, since the number of control points of a curve segment is two, a curve $\mathbf{r}_0(t;1)$ is considered a line segment connecting them, and the degree of its two weighting functions is 1. Owing to the conditions imposed on them, their simplest forms are:

$$f_0(t) = (1-t), \qquad f_1(t) = t. \tag{9.8}$$

Then, we have the expression

$$\mathbf{r}_0(t;1) = \mathbf{p}_0(1-t) + \mathbf{p}_1 t. \tag{9.9}$$

This is rewritten using eq. (9.7) as

$$\mathbf{r}_0(t;1) = \mathbf{r}_0(t;0)(1-t) + \mathbf{r}_1(t;0)t. \tag{9.10}$$

A point given by the left-hand side of the above expression is the point dividing the line segment joining $\mathbf{r}_0(t;0)$ and $\mathbf{r}_1(t;0)$ in the ratio $t : 1-t$.

For $n = 2$, increasing all the degree parameters of the curve expressions in eq. (9.10) by 1, we get

$$\mathbf{r}_0(t;2) = \mathbf{r}_0(t;1)(1-t) + \mathbf{r}_1(t;1)t. \tag{9.11}$$

The expression $\mathbf{r}_1(t;1)$ is the line segment connecting $\mathbf{p}_1$ and $\mathbf{p}_2$, which is formally derived from $\mathbf{r}_0(t;1)$ by increasing its subscript by one or from $\mathbf{r}_1(t;0)$ by increasing its degree parameter by one. For fixed t, eq. (9.11) is considered to give a point dividing the line segment joining $\mathbf{r}_0(t;1)$ and $\mathbf{r}_1(t;1)$ in the ratio $t : 1-t$. When t changes, the locus of $\mathbf{r}_0(t;2)$ represents a curve segment of degree two, whose control points are $\mathbf{p}_0$, $\mathbf{p}_1$ and $\mathbf{p}_2$. Generalizing eq. (9.11) for a curve segment of degree n, we obtain

$$\mathbf{r}_0(t;n) = \mathbf{r}_0(t;n-1)(1-t) + \mathbf{r}_1(t;n-1)t. \tag{9.12}$$

The left-hand side of the above equation is the locus of a point dividing the line segment connecting $\mathbf{r}_0(t;n-1)$ and $\mathbf{r}_1(t;n-1)$ in the ratio $t : 1-t$. The curves $\mathbf{r}_0(t;n-1)$ and $\mathbf{r}_1(t;n-1)$ are determined respectively from the control points $(\mathbf{p}_0, \mathbf{p}_1, \cdots \mathbf{p}_{n-1})$ and $(\mathbf{p}_1, \mathbf{p}_2, \cdots \mathbf{p}_n)$.

Now, to simplify the manipulation of mathematical expressions of curve segments generated from the control points, we introduce the shift operator E which operates on identifiers of the sequentially ordered objects of a group such as the control points of a curve. Its basic property is to increase or decrease the number which is used as a subscript of an object in the group in the following way:

$$\mathsf{E}\mathbf{p}_i = \mathbf{p}_{i+1}, \qquad \mathsf{E}^{-1}\mathbf{p}_{i+1} = \mathbf{p}_i. \tag{9.13}$$

From these, we obtain the rules of E for integers m and n,

$$\mathsf{E}^m\mathsf{E}^n = \mathsf{E}^{m+n}, \qquad \mathsf{E}\mathsf{E}^{-1} = \mathsf{E}^0 = \mathsf{I} = 1, \tag{9.14}$$

where I is the identity operator which does not change its operand, just like 1 in arithmetic multiplication. The rules (9.14) indicate that the shift operator E obeys the law of multiplication in algebra, hence we can treat E as an algebraic constant.

Since $\mathbf{r}_i(t;n)$ is expressed by summation of terms each of which consists of one of the consecutive control points $\mathbf{p}_i, \mathbf{p}_{i+1}, \cdots, \mathbf{p}_{i+n}$ multiplied by its corresponding weighting function, the shift operator E can operate on $\mathbf{r}_i(t;n)$ as well. So we can express eq. (9.12) as

$$\begin{aligned} \mathbf{r}_0(t;n) &= \mathbf{r}_0(t;n-1)(1-t) + \mathsf{E}\mathbf{r}_0(t;n-1)t \\ &= (1-t+\mathsf{E}t)\mathbf{r}_0(t;n-1). \end{aligned} \tag{9.15}$$

By applying the above equation to its right-hand side factor $\mathbf{r}_0(t;n-1)$ repeatedly, we finally obtain the following form for a curve segment of degree n:

$$\mathbf{r}_0(t;n) = (1-t+\mathsf{E}t)^n\mathbf{r}_0(t;0) \tag{9.16}$$

$$\text{or} \quad \mathbf{r}_0(t;n) = (1-t+\mathsf{E}t)^n\mathbf{p}_0\,. \tag{9.17}$$

To express the above equation by the last control point $\mathbf{p}_n$, the right-hand side of eq. (9.17) is multiplied by $\mathsf{E}^{-n}\mathsf{E}^n = \mathsf{I}$ and then it is converted to the desired form.

$$\begin{aligned} \mathbf{r}_0(t;n) &= (1-t+\mathsf{E}t)^n\mathsf{E}^{-n}\mathsf{E}^n\mathbf{p}_0 \\ &= \{(1-t)\mathsf{E}^{-1}+t\}^n\mathbf{p}_n. \end{aligned} \tag{9.18}$$

In the right-hand side, $\mathbf{p}_n$ appears explicitly instead of $\mathbf{p}_0$, so we denote this expression as $\grave{\mathbf{r}}_n(t;n)$ instead of $\mathbf{r}_0(t;n)$,

$$\grave{\mathbf{r}}_n(t;n) = \{(1-t)\mathsf{E}^{-1}+t\}^n\mathbf{p}_n. \tag{9.19}$$

Equations (9.17) and (9.19) are the operator expressions of a curve segment of degree n with $n+1$ control points. This curve segment is the same as what is called a Bézier curve, whose form is given by eq. (9.20) in the next section. This operator form of a Bézier curve is convenient not only for its conciseness, but also for its mathematical manipulation. We call the control points $\mathbf{p}_i$ Bézier points or B points for brevity. Instead of B point sequence, we use the term B polygon with the same meaning, but when its edges are emphasized, the term B polygon is more appropriate than B point sequence.

9.4 Different Expressions of B Curve

Expanding eq. (9.17) and noticing that $\mathsf{E}^i\mathbf{p}_0 = \mathbf{p}_i$, we get

$$\mathbf{r}_0(t;n) = \sum_{i=0}^{n} C_i^n(1-t)^{n-i}t^i\mathbf{p}_i \tag{9.20}$$

where C_i^n are the binomial coefficients, which are equal to $n!/\{i!(n-i)!\}$. The weighting function of eq. (9.1) in the B curve is

$$f_i(t) = C_i^n(1-t)^{n-i}t^i. \tag{9.21}$$

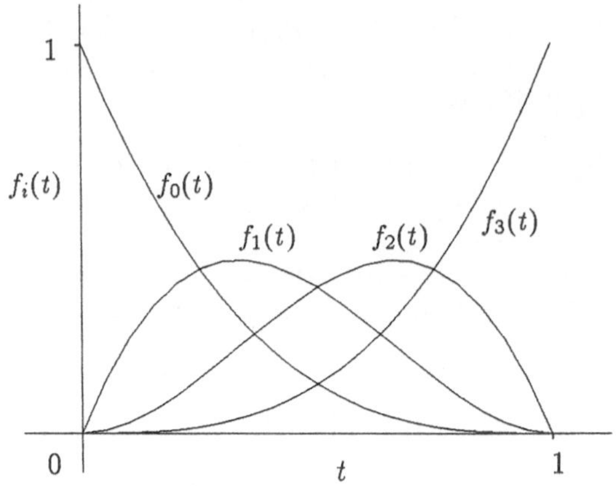

Figure 9.2 Shapes of weighting functions $f_i(t)$ for $n = 3$

The shapes of functions $f_i(t)$ are shown in Fig. 9.2.

The expression (9.20) is a standard form of a Bézier curve or a form using Bernstein polynomials [3]. The original expression by Bézier was a rather complicated one; see Sect. 9.13. In this book we use eq. (9.17) for the Bézier curve, because it is easy for algebraic manipulation. For numerical calculation, the polynomial form of t is sometimes convenient, which is obtained by expanding eq. (9.17) in t,

$$\begin{aligned} \mathbf{r}_0(t;n) = (1-t+\mathsf{E}t)^n \mathbf{p}_0 &= \{1+(\mathsf{E}-1)t\}^n \mathbf{p}_0 = \sum_{i=0}^{n} C_i^n \mathbf{q}_i t^i, \qquad (9.22) \\ \text{where} \quad \mathbf{q}_i &\equiv (\mathsf{E}-1)^i \mathbf{p}_0 = \sum_{j=0}^{i} (-1)^i C_j^i \mathbf{p}_j. \end{aligned}$$

9.5 Derivatives at Ends of a Segment and Hodographs

The tangent direction of a B curve is obtained by differentiation of eq. (9.17) with respect to t :

$$\begin{aligned} \frac{d\mathbf{r}_0(t;n)}{dt} &= \dot{\mathbf{r}}_0(t;n) = n(1-t+\mathsf{E}t)^{n-1}(\mathsf{E}-1)\mathbf{p}_0 \\ &= n(1-t+\mathsf{E}t)^{n-1}(\mathbf{p}_1-\mathbf{p}_0) \\ &= n\{\mathbf{r}_1(t;n-1)-\mathbf{r}_0(t;n-1)\}. \qquad (9.23) \end{aligned}$$

This shows that the tangent direction at a point on $\mathbf{r}_0(t;n)$ is that of a vector from $\mathbf{r}_0(t;n-1)$ to $\mathbf{r}_1(t;n-1)$, while according to eq. (9.12) the point $\mathbf{r}_0(t;n)$ divides this vector in the ratio $(t : 1-t)$.

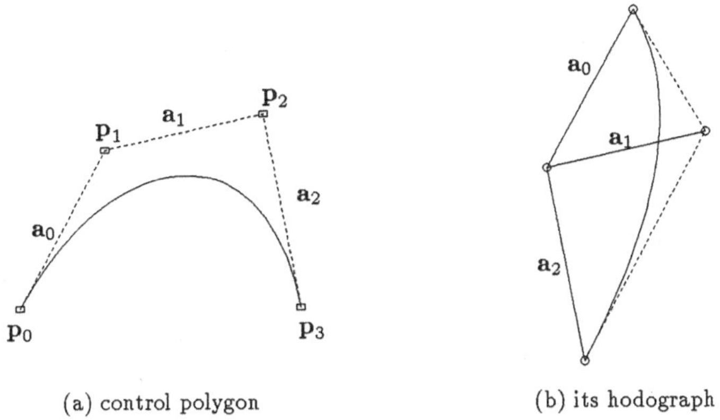

Figure 9.3 Construction of a hodograph

Next we explain the *hodograph* of a curve, which provides detailed information on its derivative vectors. We use notations $\mathbf{a}_0, \mathbf{a}_1, \mathbf{a}_2, \cdots \mathbf{a}_{n-1}$ for edges of a control polygon of degree n. Since $(\mathsf{E}-1)\mathbf{p}_0$ is equal to $\mathbf{a}_0$, we can rewrite eq. (9.23) with edges $\mathbf{a}_i$ of the polygon:

$$\dot{\mathbf{r}}_0(t;n) = n(1-t+\mathsf{E}t)^{n-1}\mathbf{a}_0 \ . \tag{9.24}$$

If we consider the edge vectors $\mathbf{a}_i$ as position vectors, a B polygon can be constructed from $\mathbf{a}_i$, such as shown in Fig. 9.3. This B polygon produces a curve proportional to $\dot{\mathbf{r}}_0(t;n)$ which is called a *hodograph* of the original curve $\mathbf{r}_0(t;n)$.

The hodograph is useful for examining the behavior of its original curve. For example in Fig. 9.4, whether each of three curves (a), (b) and (c) of degree three generated from almost similar B polygons has a loop or not can be determined by using their hodographs (d), (e) and (f). The hodograph (d) of the curve (a) passes to the right-hand side of the origin, that of (b) passes to the left-hand side, whereas that of (c) passes through the origin. The direction of the tangent vector of the curve (a) changes by far less than 180 degrees clockwise, that of (b) by far more than 180 degrees counter-clockwise, and the curve (c) has a singular point where its tangent vector vanishes.

The second derivative of $\mathbf{r}_0(t;n)$ is given by

$$\frac{1}{n(n-1)}\ddot{\mathbf{r}}_0 = (1-t+\mathsf{E}t)^{n-2}(\mathsf{E}-1)^2\mathbf{p}_0 = (1-t+\mathsf{E}t)^{n-2}(\mathsf{E}-1)\mathbf{a}_0.$$

Just as the hodograph of $\mathbf{r}_0(t;n)$ is constructed from edges of the B polygon of $\mathbf{r}_0(t;n)$, the hodograph of $\dot{\mathbf{r}}_0(t;n)$ is made from edges of the B polygon of $\dot{\mathbf{r}}_0(t;n)$. It shows graphically the behavior of $\ddot{\mathbf{r}}_0(t;n)$. At inflection points of

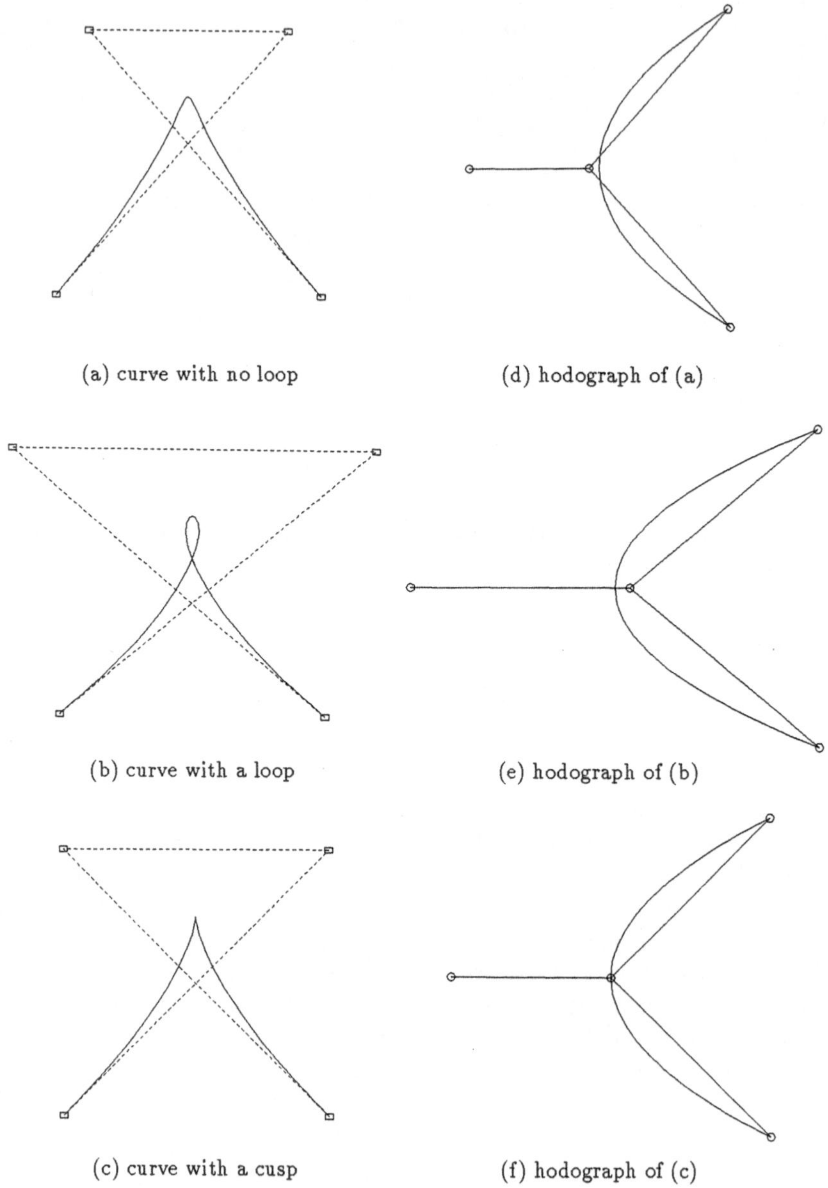

Figure 9.4 Examples of three hodographs

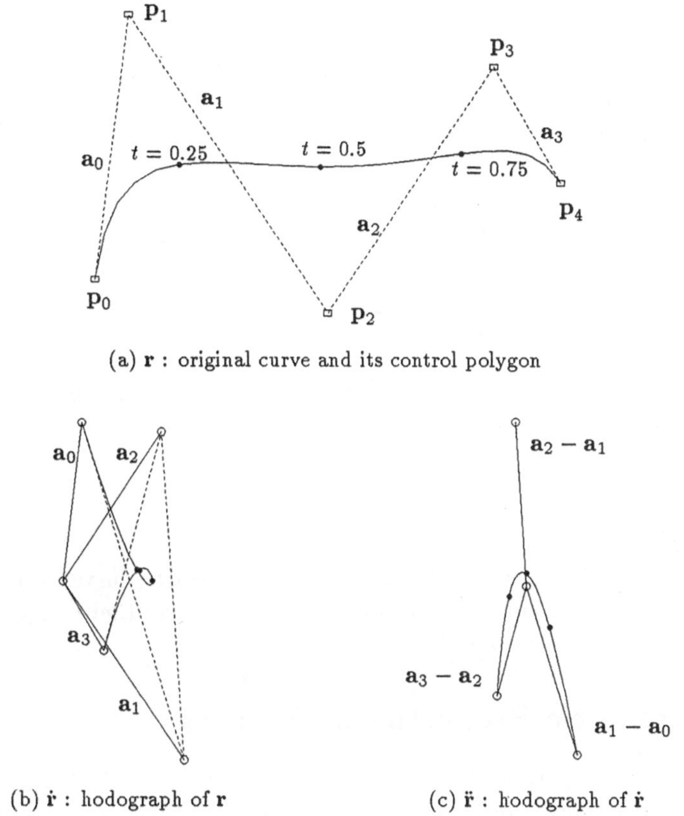

Figure 9.5 Behavior of $\dot{\mathbf{r}}$ and $\ddot{\mathbf{r}}$

$\mathbf{r}_0(t;n)$ two vectors $\dot{\mathbf{r}}_0(t;n)$ and $\ddot{\mathbf{r}}_0(t;n)$ become parallel, because its curvature vector is proportional to the normal component of $\ddot{\mathbf{r}}_0(t;n)$ (refer to eq. (5.24)) and it becomes zero. Figure 9.5 shows B polygons and vectors of $\mathbf{r}_0(t;4)$, $\dot{\mathbf{r}}_0(t;4)$ and $\ddot{\mathbf{r}}_0(t;4)$ from t=0 to t=1. There are two values of t where $\dot{\mathbf{r}}_0(t;4)$ and $\ddot{\mathbf{r}}_0(t;4)$ become parallel. These values correspond to inflection points of $\mathbf{r}_0(t;4)$.

The higher-order derivatives are obtained similarly:

$$\frac{(n-j)!}{n!}\frac{d^j\mathbf{r}_0}{dt^j} = \mathbf{r}_0^{(j)}(t;n) = (1-t+\mathsf{E}t)^{n-j}(\mathsf{E}-1)^j\mathbf{p}_0 \,. \tag{9.25}$$

They also have the form of the B curve expression.

The derivatives at the end points of the curve are given by setting $t=0$ or $t=1$ in the above equation:

$$\mathbf{r}_0^{(j)}(0;n) \;=\; \frac{n!}{(n-j)!}(\mathsf{E}-1)^j\mathbf{p}_0, \tag{9.26}$$

$$\mathbf{r}_0^{(j)}(1;n) = \frac{n!}{(n-j)!}(1-\mathsf{E}^{-1})^j \mathbf{p}_n. \tag{9.27}$$

The j-th derivative at an end point is determined by the $j+1$ consecutive control points from the end point. Conversely, the j-th control points $\mathbf{p}_j$ and $\mathbf{p}_{n-j}$ from the end points $\mathbf{p}_0$ and $\mathbf{p}_n$ are determined by the $j+1$ derivatives from zero-th to j-th at the respective ends:

$$\begin{aligned}\mathbf{p}_j &= (1+\mathsf{E}-1)^j \mathbf{p}_0 = \sum_{k=0}^{j}(C_k^j)(\mathsf{E}-1)^k \mathbf{p}_0 \\ &= \frac{1}{n!}\sum_{k=0}^{j}(C_k^j)(n-k)!\mathbf{r}_0^{(k)}(0;n),\end{aligned} \tag{9.28}$$

$$\begin{aligned}\mathbf{p}_{n-j} &= \{1-(1-\mathsf{E}^{-1})\}^j \mathbf{p}_n = \sum_{k=0}^{j}(C_k^j)(-1)^k(1-\mathsf{E}^{-1})^k \mathbf{p}_n \\ &= \frac{1}{n!}\sum_{k=0}^{j}(C_k^j)(-1)^k(n-k)!\mathbf{r}_0^{(k)}(1;n),\end{aligned} \tag{9.29}$$

for $j = 0, \cdots n.$

The relation between the control points of connecting curves at their junction is determined by the continuity conditions of derivatives there. Equations (9.26) and (9.27) are also used for this purpose.

9.6 Geometric Properties of B Curve

The geometric properties of a B curve can be expressed by the edges of its control polygon. At an end point of a curve segment, we can express important values of geometric features of the curve in terms of the first three edges of the polygon. From eq. (5.22) in Chap. 5, its unit tangent vector $\mathbf{t}$ is expressed by

$$\mathbf{t} = \frac{\dot{\mathbf{r}}}{|\dot{\mathbf{r}}|} = \frac{\mathbf{a}_0}{|\mathbf{a}_0|}. \tag{9.30}$$

Since the osculating plane is made by $\mathbf{t}$ and $\mathbf{t} + d\mathbf{t}$, it coincides with the plane made by the first and the second edges of the polygon. The unit normal vector $\mathbf{n}$ is in the osculating plane and 90 degrees anti-clockwise from the first edge vector. This is derived from eq. (5.26) and the definition of the unit normal. From equation (5.24), the curvature at the end point is given by

$$\kappa = \frac{n-1}{n}\frac{(\mathbf{a}_1 \cdot \mathbf{n})}{\mathbf{a}_0^2}. \tag{9.31}$$

The sign of κ is determined by the sign of $(\mathbf{a}_1 \cdot \mathbf{n})$. The unit binormal vector $\mathbf{b}$ is determined by $\mathbf{t} \times \mathbf{n}$ and the torsion at the end point is given from eq. (5.27) by

$$\tau = \frac{n-2}{n}\frac{(\mathbf{a}_2 \cdot \mathbf{b})}{(\mathbf{a}_1 \cdot \mathbf{n}) \cdot |\mathbf{a}_0|}. \tag{9.32}$$

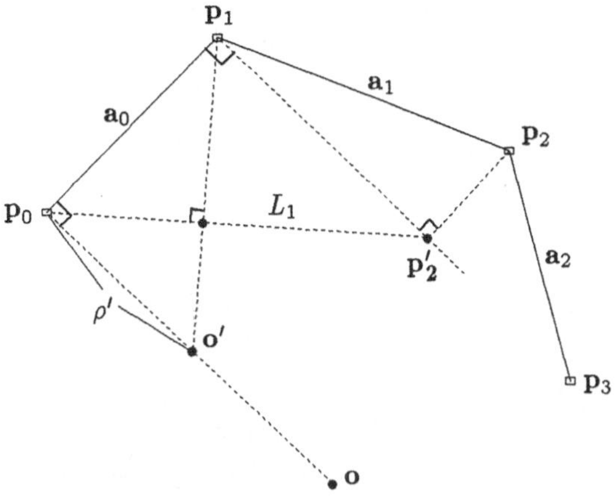

Figure 9.6 Locating the center of curvature

At the starting point of a curve segment, its tangent direction, curvature and torsion are represented by the first three edges of its Bézier polygon with forms very similar to their original definitions: $\mathbf{t}$ has the direction of $\mathbf{a}_0$, the change of the tangent direction $d\theta$ of the curve is considered to be proportional to the normal component of $\mathbf{a}_1$ divided by the length of $\mathbf{a}_0$, and the arc length ds is represented by the length of $\mathbf{a}_0$, that is, $d\theta/ds$ which defines the curvature is given by eq. (9.31). The change of torsion angle $d\mathbf{b}$ is represented in the B polygon as the binormal component of $\mathbf{a}_2$ divided by the normal component of $\mathbf{a}_1$, accordingly, $d\mathbf{b}/ds$ which defines the torsion is given by eq. (9.32).

When the B polygon of a curve is given, its tangent, curvature and torsion at its end points are obtained graphically. For instance, a procedure for determining the center of curvature is as follows. Letting ρ be the radius of curvature and ρ' be equal to $(n-1)\rho/n$, we get from eq. (9.31) the following relation:

$$\rho' : |\mathbf{a}_0| = |\mathbf{a}_0| : (\mathbf{a}_1 \cdot \mathbf{n}),$$

from which ρ' is determined graphically. Referring to Fig. 9.6,

1. Draw a perpendicular at $\mathbf{p}_1$ to the edge $\mathbf{a}_0$, to which draw a perpendicular from $\mathbf{p}_2$ to $\mathbf{p}_2'$. Draw a line L_1 joining its foot $\mathbf{p}_2'$ and $\mathbf{p}_0$.
2. Draw a perpendicular from $\mathbf{p}_1$ to the line L_1 and extend it to intersect with the normal line from $\mathbf{p}_0$. Let the intersection point be $\mathbf{o}'$.
3. Extend the vector $\mathbf{p}_0\mathbf{o}'$ by $n/(n-1)$ times from $\mathbf{p}_0$ to determine $\mathbf{o}$. Then $\mathbf{o}$ is the center of the curvature at the point $\mathbf{p}_0$ and the length of $\mathbf{p}_0\mathbf{o}$ is the radius of curvature.

A similar procedure can be applied to locate the center of torsion graphically by using the following relation:

$$(n-2)\rho_\tau/n : |\mathbf{a}_0| = (\mathbf{a}_1 \cdot \mathbf{n}) : (\mathbf{a}_2 \cdot \mathbf{b}),$$

where ρ_τ is the radius of torsion. The geometric properties stated above are those at the end points of a curve segment. Those at the other point on the curve can be expressed in the same form, because any point on the curve can be converted to an end point of a curve segment made by division of the original segment, and the B polygon of the divided segment can be obtained from that of the original B polygon.

9.7 Division of a Curve Segment and Its B Polygon

Equation (9.17) is the expression for the B curve segment for $t = 0$ to 1. Dividing it into two segments, we can give them similar expressions, that is, we can construct two B polygons for the divided segments.

Suppose we divide the curve $\mathbf{r}_0(t;n)$ at $t = t^*$ into two divided segments $\mathbf{r}_0^a(t;n)$ with a B polygon $B^a = \{\mathbf{p}_0^a, \cdots, \mathbf{p}_n^a\}$ and $\mathbf{r}_0^b(t;n)$ with a B polygon $B^b = \{\mathbf{p}_0^b, \cdots, \mathbf{p}_n^b\}$. To deduce $\mathbf{r}_0^a(t;n)$ in the standard operator form from eq. (9.17), we change the parameter t to τ (this is not the torsion of the curve) so that

$$\tau = t/t^*, \tag{9.33}$$

and we get an intermediate expression:

$$\mathbf{r}_0(t;n) = (1 - \tau t^* + \mathsf{E}\tau t^*)^n \mathbf{p}_0 = \{1 - \tau + (1 - t^* + \mathsf{E}t^*)\tau\}^n \mathbf{p}_0. \tag{9.34}$$

Here, we introduce a shift operator E_a for the control points $\mathbf{p}^a{}_i$ of the divided first segment. The relation between E and E_a is defined by

$$(1 - t^* + \mathsf{E}t^*)^i \mathbf{p}_0 \equiv \mathsf{E}_a^i \mathbf{p}_0^a = \mathbf{p}_i^a. \tag{9.35}$$

Using this and denoting the left-hand side of eq. (9.34) by $\mathbf{r}_0^a(\tau;n)$, we have an expression

$$\mathbf{r}_0^a(\tau;n) = (1 - \tau + \mathsf{E}_a\tau)^n \mathbf{p}_0^a. \tag{9.36}$$

Since τ takes a value between 0 and 1, a parameter t is used in place of τ and the above eq. (9.36) becomes

$$\mathbf{r}_0^a(t;n) = (1 - t + \mathsf{E}_a t)^n \mathbf{p}_0^a. \tag{9.37}$$

This is the equation of the first part of the divided curves.

Next, we treat the second part. We use eq. (9.19) in which the last control point appears.

$$\mathbf{r}_n(t;n) = \{(1 - t)\mathsf{E}^{-1} + t\}^n \mathbf{p}_n. \tag{9.38}$$

Changing the parameter t to τ, which is given by

$$\tau = (t - t^*)/(1 - t^*), \tag{9.39}$$

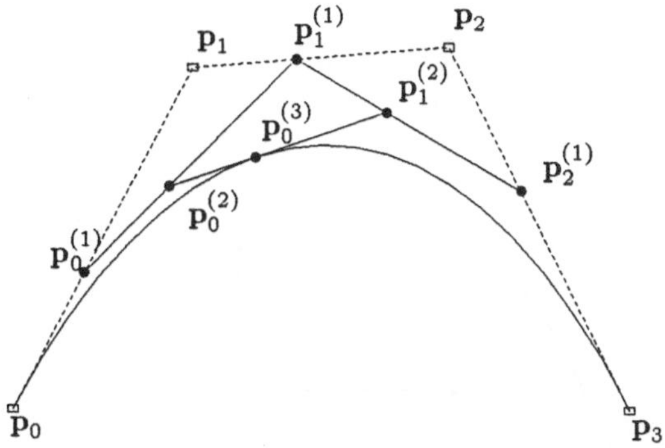

Figure 9.7 Division of a B polygon

we obtain from eq. (9.38) an intermediate expression:

$$\begin{aligned} \grave{\mathbf{r}}_n(\tau;n) &= [\{1-(1-t^*)\tau - t^*\}\mathsf{E}^{-1} + (1-t^*)\tau + t^*]^n \mathbf{p}_n \\ &= [\{(1-t^*)\mathsf{E}^{-1} + t^*\}(1-\tau) + \tau]^n \mathbf{p}_n. \end{aligned}$$

Here, we define a new shift operator E_b for the control points $\mathbf{p}_i^b$. The relation between $\mathbf{p}_{n-i}^b$ and $\mathbf{p}_{n-i}$ is determined by

$$\{(1-t^*)\mathsf{E}^{-1} + t^*\}^i \mathbf{p}_n \equiv \mathsf{E}_b^{-i}\mathbf{p}_n^b = \mathbf{p}_{n-i}^b. \tag{9.40}$$

Substituting this in the right-hand side of $\grave{\mathbf{r}}_n(\tau;n)$ and changing τ to t , we obtain

$$\grave{\mathbf{r}}_n^b(t;n) = \{(1-t)(\mathsf{E}_b)^{-1} + t\}^n \mathbf{p}_n^b. \tag{9.41}$$

This is the equation of the second part of the divided curves.

A Bézier curve $\mathbf{r}_0(t;n)$ is divided at $t = t^*$ into two B curves $\mathbf{r}_0^a(t;n)$ and $\mathbf{r}_0^b(t;n) = \grave{\mathbf{r}}_n^b(t;n)$, the locations of whose B points are given respectively by

$$\mathbf{p}_i^a = (1-t^* + \mathsf{E}t^*)^i \mathbf{p}_0, \tag{9.42}$$

$$\mathbf{p}_{n-i}^b = \{(1-t^*)\mathsf{E}^{-1} + t^*\}^i \mathbf{p}_n, \quad \text{for } i = 0, \cdots n. \tag{9.43}$$

From eq. (9.42) we can see that the location of i-th B point $\mathbf{p}_i^a$ of B^a is at a point $\mathbf{r}_0(t^*;i)$, whose B polygon vertices are $(\mathbf{p}_0, \mathbf{p}_1, \cdots, \mathbf{p}_i)$. In the same way, the location of $(n-i)$-th B point $\mathbf{p}^b{}_{n-i}$ of B^b is a point $\grave{\mathbf{r}}_n(t^*;i)$, whose B polygon vertices are $(\mathbf{p}_{n-i}, \mathbf{p}_{n-i+1}, \cdots, \mathbf{p}_n)$.

We call this process of producing B polygons B^a and B^b from the polygon B_0 the *division of a B polygon* in the dividing ratio $\{t^* : (1-t^*)\}$.

We explain a graphical method of determining the location of the new B points $\mathbf{p}^a{}_i$ and $\mathbf{p}^b{}_{n-i}$. Refer to Fig. 9.7.

1. Divide each edge $\mathbf{a}_i$ of B_0 at $\mathbf{p}_i^{(1)}$ in the ratio $(t^* : 1-t^*)$, which we call the dividing point of level one. A number in the superscript indicates the level. The original vertices are denoted by $\mathbf{p}_i^{(0)}$. A case for $n = 3$ is shown in parentheses as $(\mathbf{p}_0^{(1)}, \mathbf{p}_1^{(1)}, \mathbf{p}_2^{(1)})$.

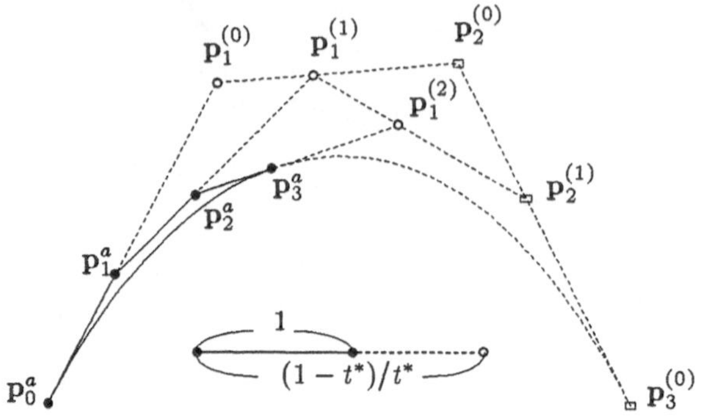

Figure 9.8 Reconstruction of an original B polygon from its divided one

2. Connect $p_i^{(1)}$ and $p_{i+1}^{(1)}$ by a line segment, which we call the dividing line of level one, and divide it at $p_i^{(2)}$ in the ratio $(t^* : 1-t^*)$. Then $p_i^{(2)}$ is called a dividing point of level two. (For $n = 3$, $p_0^{(2)}$ is determined on the dividing line $p_0^{(1)}p_1^{(1)}$, and $p_1^{(2)}$ is on $p_1^{(1)}p_2^{(1)}$.)
3. Connect $p_i^{(2)}$ and $p_{i+1}^{(2)}$ by the dividing line of level two and divide it at $p_i^{(3)}$ in the ratio$(t^* : 1-t^*)$. (For $n = 3$, $p_0^{(3)}$ is the only dividing point of level three, and this is final.)
4. Repeat this procedure until division of the dividing line of level $(n-1)$ at the dividing point $p_0^{(n)}$ of level n occurs. Then we determine locations of the control points of $B^a = (p_0^a, \cdots, p_n^a)$ and $B^b = (p_0^b, \cdots p_n^b)$ as follows.

$$\begin{array}{lllllll} B^a & : & p_0^{(0)} & p_0^{(1)} & p_0^{(2)} & \cdots & p_0^{(n-1)} & p_0^{(n)}, \\ B^b & : & p_0^{(n)} & p_1^{(n-1)} & p_2^{(n-2)} & \cdots & p_{n-1}^{(1)} & p_n^{(0)}. \end{array} \tag{9.44}$$

The location of the dividing point of level n is a point on the original curve:

$$p_n^a = p_0^b = p_0^{(n)} = r_0(t^*; n) = (1 - t^* + \mathrm{E}t^*)^n p_0.$$

When one of the divided B polygons B^a or B^b and a dividing ratio $(t^* : 1-t^*)$ are given, we can construct the other B polygon B^b or B^a and also the original polygon B_0 uniquely. The graphical procedure is just the inverse of the division procedure of the polygon B_0. For instance, referring to Fig. 9.8,

1. By extending each edge of B^a by $(1-t^*)/t^*$times, we determine the locations $p_1^{(0)}, p_1^{(1)}$ and $p_1^{(2)}$.
2. By extending edges $p_1^{(0)}p_1^{(1)}$ and $p_1^{(1)}p_1^{(2)}$ by $(1-t^*)/t^*$ times, we determine the locations $p_2^{(0)}$ and $p_2^{(1)}$,
3. Finally, $(1-t^*)/t^*$ times extension of a vector $p_2^{(0)}p_2^{(1)}$ fixes the location of $p_3^{(0)}$.

Thus, all the locations of the polygons B^b and B_0 are determined from B^a.

The dividing lines and points of a B polygon are closely connected to those of a B spline polygon, which is explained in the next chapter.

9.8 Continuity Conditions of Connection of B Polygons

At a dividing point $\mathbf{r}_0(t^*; n)$ of a Bézier curve segment of degree n the condition of continuity is naturally $C^{(n)}$; continuous up to n-th derivatives, for $i = 0, \cdots n$,

$$(d/dt)^i \mathbf{r}_0(t^* - 0; n) = (d/dt)^i \mathbf{r}_0(t^* + 0; n). \tag{9.45}$$

If these conditions are to be expressed by the divided segments $\mathbf{r}_0^a(1; n)$ and $\mathbf{r}_0^b(0; n)$, the differentiation with respect to t has to be converted to that by their own normalized parameters τ which are defined by equations (9.33) and (9.39). Accordingly, when comparing their differentiated values at their junction we have to consider the scale ratio of the parameters, which is $t^* : (1-t^*)$. Then eq. (9.45) becomes

$$\begin{aligned} (\frac{d}{dt})^i \mathbf{r}_0^a \mid_{t=1} &= k^i (\frac{d}{dt})^i \mathbf{r}_0^b \mid_{t=0}, \ i = 0, \cdots n, \\ \text{where} \quad k &= \frac{t^*}{1-t^*}. \end{aligned} \tag{9.46}$$

We call k the scale ratio of the adjacent parameters.

If eq. (9.46) holds for $i = 0$ to $i = j$, the $j+1$ control points from $\mathbf{p}_n^a$ to $\mathbf{p}_{n-j}^a$ of $\mathbf{r}_0^a(t; n)$ and the $j+1$ control points from $\mathbf{p}_0^b(= \mathbf{p}_n^a)$ to $\mathbf{p}_j^b$ of $\mathbf{r}_0^b(t; n)$ are constrained. The other control points of $\mathbf{r}_0^a(t; n)$ and $\mathbf{r}_0^b(t; n)$ are free. These constrained control points can be derived from the division of a B polygon of degree j in the ratio $t^* : (1 - t^*)$. For instance, in Fig. 9.7, the control points $(\mathbf{p}_1^a, \mathbf{p}_2^a, \mathbf{p}_3^a = \mathbf{p}_0^b, \mathbf{p}_1^b, \mathbf{p}_2^b)$ are determined by the division of a B polygon $(\mathbf{p}_0^{(1)} \mathbf{p}_1^{(1)} \mathbf{p}_2^{(1)})$ in the ratio $t^* : (1 - t^*)$. This assures the connection of $\mathbf{r}_0^a(1; n)$ and $\mathbf{r}_0^b(1; n)$ to be $C^{(2)}$.

Though a curve segment $\mathbf{r}_0(t; n)$ is divided into many segments, each of which has $n + 1$ control points, they are equivalent to the original curve segment so long as the continuity constraints at their connecting points are not reduced. If they are reduced by one degree at each junction, then the amount of freedom increases, that is, control points which can be moved freely are generated. This kind of problem is treated in Chap. 10.

9.9 Elevation of Degree of a Curve Segment

Increasing the number of control points of a curve segment without changing its shape is attained by dividing the curve segment as well as by increasing degree of the curve. If one wants to get more freedom for control of a shape consisting

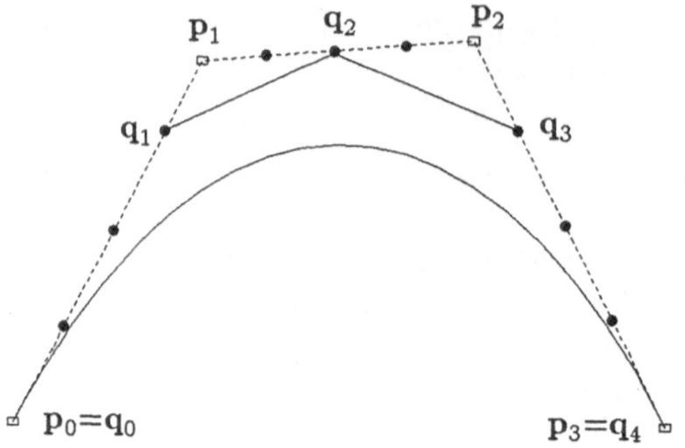

Figure 9.9 Degree elevation of B polygon

of connecting segments, the method by division is accompanied by decrease of degree of the continuity condition, but the method of degree elevation increases the degree of the segment, that is, increases the number of control points.

Referring to Fig. 9.9, let $(\mathbf{p}_0\ \mathbf{p}_1\ \mathbf{p}_2\ \mathbf{p}_3)$ be a B polygon of degree three, divide each of its edges into four equal sub-edges and let the third dividing point of the first edge be $\mathbf{q}_1$, the second dividing point of the second edge be $\mathbf{q}_2$ and the first dividing point of the third edge be $\mathbf{q}_3$, then the polygon $(\mathbf{p}_0\ \mathbf{q}_1\ \mathbf{q}_2\ \mathbf{q}_3\ \mathbf{p}_3)$ is of degree four, but generates a curve of the same shape as that derived from the original B polygon.

We prove the above rule in general form. Multiplying the right-hand side of eq. (9.17) by $\{(1-t)+t\}$, we make its degree increase apparently by one and denote the left-hand side by $\tilde{\mathbf{r}}_0(t; n+1)$, which has new control points $(\mathbf{q}_0, \mathbf{q}_1, \mathbf{q}_2, \cdots \mathbf{q}_n, \mathbf{q}_{n+1})$ with their new shift operator $\tilde{\mathsf{E}}$. Then this new expression and its expansion are given by

$$\tilde{\mathbf{r}}_0(t; n+1) = (1 - t + \tilde{\mathsf{E}}t)^{n+1}\mathbf{q}_0 = \sum_{i=0}^{n+1}(C_i^{n+1})(1-t)^{n+1-i}t^i\mathbf{q}_i. \tag{9.47}$$

To obtain the relation between points $\mathbf{p}_i$ and $\mathbf{q}_i$, we expand $\{(1-t)+t\} \times \mathbf{r}_0(t;n)$ in terms of $\{(1-t)^{n+1-i}t^i\}$ for $i = 0 \cdots (n+1)$. Then it becomes

$$\begin{aligned}
&\{(1-t)+t\} \times \mathbf{r}_0(t;n) \\
&= \sum_{i=0}^{n}(C_i^n)(1-t)^{n-i+1}t^i\mathbf{p}_i + \sum_{i=0}^{n}(C_i^n)(1-t)^{n-i}t^{i+1}\mathbf{p}_i \\
&= \sum_{i=0}^{n+1}(C_i^{n+1})(1-t)^{n+1-i}t^i\{(C_i^n)/(C_i^{n+1})\mathbf{p}_i\delta'_{n+1,i} \\
&\quad +(C_{i-1}^n)/(C_i^{n+1})\mathbf{p}_{i-1}\delta'_{0,i}\},
\end{aligned} \tag{9.48}$$

where

$$\delta'_{i,j} = \begin{cases} 0 & \text{for } i = j \\ 1 & \text{for } i \neq j \end{cases}. \tag{9.49}$$

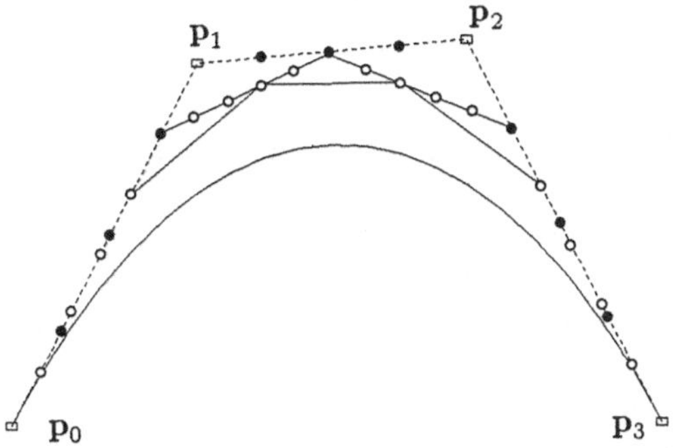

Figure 9.10 B polygon by repeated degree elevations and a curve segment

Comparing equations (9.47) and (9.48), we get the locations of the new control points in terms of the old control points:

$$\mathbf{q}_i = \{\frac{n+1-i}{n+1}\}\mathbf{p}_i\delta'_{n+1,i} + \{\frac{i}{n+1}\}\mathbf{p}_{i-1}\delta'_{0,i}. \tag{9.50}$$

The relation is that both of the first control points $\mathbf{p}_0$ and $\mathbf{q}_0$ are the same point, $\mathbf{q}_1$ is located at the dividing point of the first edge $\mathbf{a}_0$ of the original polygon in the ratio $n : 1$, $\mathbf{q}_i$ is located at the dividing point of the i-th edge $\mathbf{a}_i$ in the ratio $(n+1-i) : i$ and the last point $\mathbf{q}_{n+1}$ coincides with the last point $\mathbf{p}_n$ of the original polygon.

By repeating this degree increasing process indefinitely, the corresponding B polygon gradually approaches the shape of the curve segment it defines. See Fig. 9.10.

9.10 Expression for a Surface Patch

We introduce an expression for a Bézier surface patch in an operator form by extending that of a B curve. Instead of a B polygon, a control point net is used to define a four-sided surface called a Bézier patch, B patch for brevity. A B patch, whose parameters are u and v and degrees are m and n, is denoted by $\mathbf{s}_{00}(u, v; m, n)$ and its control point net is given by an $m \times n$ matrix $\{\mathbf{p}_{ij}\}$:

$$\begin{bmatrix} \mathbf{p}_{00} & \mathbf{p}_{01} & \cdots & \mathbf{p}_{0,n-1} & \mathbf{p}_{0,n} \\ \mathbf{p}_{10} & \mathbf{p}_{11} & \cdots & \mathbf{p}_{1,n-1} & \mathbf{p}_{1,n} \\ \vdots & \vdots & \ddots & \vdots & \vdots \\ \mathbf{p}_{m-1,0} & \mathbf{p}_{m-1,1} & \cdots & \mathbf{p}_{m-1,n-1} & \mathbf{p}_{m-1,n} \\ \mathbf{p}_{m,0} & \mathbf{p}_{m,1} & \cdots & \mathbf{p}_{m,n-1} & \mathbf{p}_{m,n} \end{bmatrix}. \tag{9.51}$$

To deduce an operator expression for $\mathbf{s}_{00}(u, v; m, n)$, we use a tensor product method. Considering each column of (9.51) as an independent B polygon of degree m, we have $(n+1)$ B curves

$$\mathbf{r}_j(u:m) = (1-u+\mathsf{E}u)^m \mathbf{p}_{0j}, \quad j = 0, \cdots n. \tag{9.52}$$

For fixed u, we make a row of $(n+1)$ consecutive points from eq. (9.52):

$$\left\{ \mathbf{r}_0(u;m) \quad \mathbf{r}_1(u;m) \quad \cdots \quad \mathbf{r}_n(u;m) \right\}. \tag{9.53}$$

Considering this $1 \times (n+1)$ array as a B polygon of degree n, we construct a B curve of degree n with a parameter v,

$$(1-v+\mathsf{F}v)^n \mathbf{r}_0(u;m), \tag{9.54}$$

where F is the shift operator used for the elements of the row (9.53), which are expressed by eq. (9.52), and operates on the second subscript of the control points $\mathbf{p}_{ij}$. The role of F is just the same as that of E for the first subscript. Since eq. (9.54) is a function of parameters u and v, whose degrees are m and n, and represents a patch, we denote it by $\mathbf{s}_{00}(u,v;m,n)$. Substituting eq. (9.52) into eq. (9.54) we obtain

$$\begin{aligned} \mathbf{s}_{00}(u,v;m,n) &= (1-v+\mathsf{F}v)^n(1-u+\mathsf{E}u)^m \mathbf{p}_{00} \\ &= (1-u+\mathsf{E}u)^m(1-v+\mathsf{F}v)^n \mathbf{p}_{00}, \quad 0 \le u, v \le 1. \end{aligned} \tag{9.55}$$

This is an operator expression for a Bézier surface patch constructed from the $(m+1) \times (n+1)$ control points net $\{\mathbf{p}_{ij}\}$, for $i = 0 \cdots n, \ j = 0 \cdots m$. An example of $\mathbf{s}_{00}(u,v;3,3)$ is shown in Fig. 9.11.

Since the equation of a B patch has a similar structure to that of a B curve, it has similar characteristics:

$$\begin{aligned} \mathbf{s}_{00}(u,v;0,0) &= \mathbf{p}_{00}, & (9.56)\\ \mathbf{s}_{00}(u,v;1,1) &= (1-u+\mathsf{E}u)(1-v+\mathsf{F}v)\mathbf{p}_{00} & (9.57)\\ &= (1-u)(1-v)\mathbf{p}_{00} + (1-u)v\mathbf{p}_{01} + u(1-v)\mathbf{p}_{10} + uv\mathbf{p}_{11}, & \\ \mathbf{s}_{00}(u,v;2,2) &= (1-u+\mathsf{E}u)^2(1-v+\mathsf{F}v)^2\mathbf{p}_{00} & (9.58)\\ &= (1-u+\mathsf{E}u)(1-v+\mathsf{F}v)\mathbf{s}_{00}(u,v;1,1) & \\ &= (1-u)(1-v)\mathbf{s}_{00}(u,v;1,1) + (1-u)v\mathbf{s}_{01}(u,v;1,1) & \\ &\quad + u(1-v)\mathbf{s}_{10}(u,v;1,1) + uv\mathbf{s}_{11}(u,v;1,1). & \end{aligned}$$

Equation (9.57) is a surface patch of degree 1×1, whose control points are at the four corners of a general twisted quadrilateral. A surface patch of degree 2×2 is equivalent to a special surface patch of degree 1×1, whose four control points are on the four patches $\mathbf{s}_{00}, \mathbf{s}_{01}, \mathbf{s}_{10}$ and $\mathbf{s}_{11}$ of degree 1×1 respectively. Similarly a Bézier surface patch of degree $m \times n$ is made from a patch of degree 1×1, whose four control points are replaced by expressions for four patches $\mathbf{s}_{00}, \mathbf{s}_{01}, \mathbf{s}_{10}$

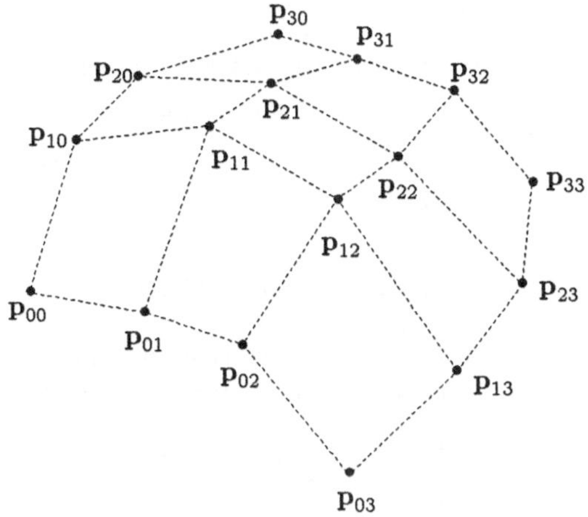

(a) control net

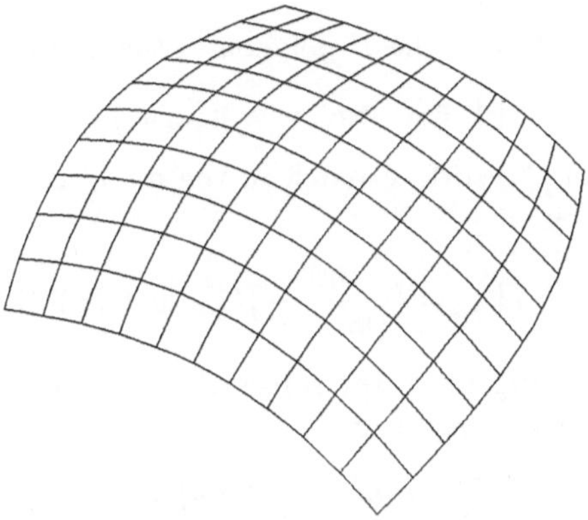

(b) B patch

Figure 9.11 B patch and its control net

and $\mathbf{s}_{11}$ of degree $(m-1) \times (n-1)$. The general form of eq. (9.57) is

$$\begin{aligned} \mathbf{s}_{00}(u,v;m,n) = (1-u)(1-v)&\mathbf{s}_{00}(u,v;m-1,n-1) \\ +(1-u)v&\mathbf{s}_{01}(u,v;m-1,n-1) \\ +u(1-v)&\mathbf{s}_{10}(u,v;m-1,n-1) \\ +uv&\mathbf{s}_{11}(u,v;m-1,n-1). \end{aligned} \quad (9.59)$$

The expression for a B patch in operator form is concise and useful for theoretical manipulation, but if we want to write all the control points explicitly, for instance in numerical calculation, then the following matrix form is used (refer to Sect. 1.8, *Example 2*):

$$\begin{aligned} s_{00}(u,v;m,n) &= F(u;m)\{p_{ij}\}F(v;n)^T, && (9.60)\\ \text{where } F(u;m) &= \{f_0(u;m)\ f_1(u;m)\cdots f_m(u;m)\}\\ F(v;n) &= \{f_0(v;n)\ f_1(v;n)\cdots f_n(v;n)\}\\ f_i(t;m) &= C_i^m(1-t)^{m-i}t^i. && (9.61)\end{aligned}$$

9.11 Geometric Properties of a Patch

One of the four boundaries of $s_{00}(u,v;m,n)$ is obtained by setting one of the parameter values $u=0$ or 1, or $v=0$ or 1, for example,

$$s_{00}(0,v;m,n) = (1-v+\mathsf{F}v)^n p_{00} \tag{9.62}$$

This is the boundary curve of $u=0$. Let the edges of the B polygon of this curve be

$$\begin{pmatrix} b_{00} & b_{01} & \cdots & b_{0,n-1} \end{pmatrix},$$

and similarly, let the edges of the B polygon of the boundary curve $v=0$ be

$$\begin{pmatrix} a_{00} & a_{10} & \cdots & a_{n-1,0} \end{pmatrix}.$$

The i-th derivative of the surface patch in the direction u along the boundary curve $u=0$ is given by

$$(\frac{\partial}{\partial u})^i s_{00}(u,v;m,n)\mid_{u=0} = \{\frac{m!}{(m-i)!}\}(1-v+\mathsf{F}v)^n(\mathsf{E}-1)^i p_{00}. \tag{9.63}$$

For $i=1$, this is the tangent vector across the boundary $u=0$, which we call the *cross-boundary tangent* vector $\mathbf{T}_u$.

$$\mathbf{T}_u = m(1-v+\mathsf{F}v)^n a_{00}\ . \tag{9.64}$$

Similarly, we get the cross-boundary tangent vector at the boundary $v=0$,

$$\mathbf{T}_v = n(1-u+\mathsf{E}u)^m b_{00}\ . \tag{9.65}$$

The normal of the surface patch at the corner point p_{00} is

$$\mathbf{n} = \frac{a_{00}\times b_{00}}{|a_{00}\times b_{00}|}. \tag{9.66}$$

The cross-boundary tangent vectors $\mathbf{T}_u$ and $\mathbf{T}_v$ are not constant along the boundaries and their rates of variation along the boundaries at the corner point p_{00} are given by

$$\begin{aligned} \frac{\partial \mathbf{T}_u}{\partial v}\mid_{v=0} &= mn(a_{01}-a_{00}) = mn\{(p_{11}-p_{01})-(p_{10}-p_{00})\}, && (9.67)\\ \frac{\partial \mathbf{T}_v}{\partial u}\mid_{u=0} &= mn(b_{10}-b_{00}) = mn\{(p_{11}-p_{10})-(p_{01}-p_{00})\}. && (9.68)\end{aligned}$$

They are equal to $\partial^2 \mathbf{s}_{00}(0,0;m,n)/\partial u \partial v$ and $\partial^2 \mathbf{s}_{00}(0,0;m,n)/\partial v \partial u$ and are called *twist vectors* at the corner, and take the same value when the corner point is not singular.

As stated in the surface theory, the fundamental magnitudes of the first and the second order at $\mathbf{p}_{00}$ are given by

$$\begin{aligned} E &= \mathbf{T}_u^2 = m^2 \mathbf{a}_{00}^2, \ F = \mathbf{T}_u \cdot \mathbf{T}_v = mn\mathbf{a}_{00}\mathbf{b}_{00}, \ G = \mathbf{T}_v^2 = n^2 \mathbf{b}_{00}^2, && (9.69) \\ L &= \mathbf{n} \cdot \frac{\partial^2 \mathbf{s}_{00}(0,0;m,n)}{\partial u^2} = m(m-1)d_{02}, && (9.70) \\ M &= \mathbf{n} \cdot \frac{\partial^2 \mathbf{s}_{00}(0,0;m,n)}{\partial u \partial v} = mnd_{11}, && (9.71) \\ N &= \mathbf{n} \cdot \frac{\partial^2 \mathbf{s}_{00}(0,0;m,n)}{\partial v^2} = n(n-1)d_{20}, && (9.72) \end{aligned}$$

where d_{02}, d_{11} and d_{20} are distances from $\mathbf{p}_{02}, \mathbf{p}_{11}$ and $\mathbf{p}_{20}$ to the tangential plane at $\mathbf{p}_{00}$. The Gaussian curvature, the mean curvature and the principal directions can be calculated from the locations of the control points. As in the case of B curves, any point on a B surface patch can be made a corner point of a divided patch; the control points around it are determined and the fundamental magnitudes at that point can be obtained accordingly.

As in the case of B curve, either of derivatives at the boundary or the control points of the B patch can be expressed by the others. For instance, $\mathbf{p}_{11}$ is expressed by the derivatives at the corner,

$$\begin{aligned} \mathbf{p}_{11} &= \mathsf{EF}\mathbf{p}_{00} = \{((\mathsf{E}-1)+1)((\mathsf{F}-1)+1)\}\mathbf{p}_{00} \\ &= \{(1/mn)\mathbf{s}_{uv} + (1/m)\mathbf{s}_u + (1/n)\mathbf{s}_v + \mathbf{s}\}\,|_{u=0,v=0}\,. && (9.73) \end{aligned}$$

9.12 Division and Degree Elevation of a Patch

Since each control polygon corresponding to a column of the control point matrix $\{\mathbf{p}_{ij}\}$ can be divided at $u = u^*$ and becomes two control polygons of the same degree connected in $C^{(m)}$ without changing the shape of the curves generated, the matrix $\{\mathbf{p}_{ij}\}$ is changed into two matrices $\{\mathbf{p}_{ij}^a\}$ and $\{\mathbf{p}_{ij}^b\}$ of the same magnitude, which generate two patches $\mathbf{s}_{00}^a$ and $\mathbf{s}_{00}^b$ according to the division of $\mathbf{s}_{00}(u,v;m,n)$ at $u = u^*$. Similarly two matrices $\{\mathbf{p}_{ij}^a\}$ and $\{\mathbf{p}_{ij}^b\}$ can be divided at $v = v^*$ into four matrices $\{\mathbf{p}_{ij}^{ac}\}, \{\mathbf{p}_{ij}^{ad}\}, \{\mathbf{p}_{ij}^{bc}\}$ and $\{\mathbf{p}_{ij}^{bd}\}$. Their corresponding patches divide $\mathbf{s}_{00}(u,v;m,n)$ at $u = u^*$ and $v = v^*$.

Each control point in the above matrices is derived from the original control points, such as

$$\begin{aligned} \mathbf{p}_{ij}^{ac} &= (1-u^*+\mathsf{E}u^*)^i(1-v^*+\mathsf{F}v^*)^j\mathbf{p}_{00}, && (9.74) \\ \mathbf{p}_{(m-i)j}^{bc} &= \{(1-u^*)\mathsf{E}^{-1}+u^*)\}^i(1-v^*+\mathsf{F}v^*)^j\mathbf{p}_{m0}, && (9.75) \\ \mathbf{p}_{i(n-j)}^{ad} &= (1-u^*+\mathsf{E}u^*)^i\{(1-v^*)\mathsf{F}^{-1}+v^*\}^j\mathbf{p}_{0n}, && (9.76) \\ \mathbf{p}_{(m-i)(n-j)}^{bd} &= \{(1-u^*)\mathsf{E}^{-1}+u^*\}^i\{(1-v^*)\mathsf{F}^{-1}+v^*\}^j\mathbf{p}_{mn}\,. && (9.77) \end{aligned}$$

They have also tensor product forms.

A surface patch of $(m+1)\times(n+1)$ degree, which is the same shape as the patch of degree $m \times n$, is obtained by the same degree increasing process as that adopted in a curve. Its new control points $\mathbf{q}_{ij}$ are given by the tensor product form of eq. (9.50),

$$\mathbf{q}_{ij} = \{1/(m+1)\}\{1/(n+1)\}\{(m+1-i)\delta'_{m+1,i+1} + i\delta'_{0,i}\mathsf{E}^{-1}\} \times\{(n+1-j)\delta'_{n+1,j+1} + j\delta'_{0,j}\mathsf{F}^{-1}\}\mathbf{p}_{ij}, \tag{9.78}$$

where $\mathbf{p}_{ij}$ is the control point of the original patch, and $\delta'_{i,j} = 0$ for $i = j$, otherwise $\delta'_{i,j} = 1$, which is the same as eq. (9.49). The control points at patch corners before and after the degree elevation are the same, but their indices are different:

$$\mathbf{q}_{00} = \mathbf{p}_{00}, \quad \mathbf{q}_{m+1,0} = \mathbf{p}_{m,0}, \quad \mathbf{q}_{0,n+1} = \mathbf{p}_{0,n}, \quad \mathbf{q}_{m+1,n+1} = \mathbf{p}_{m,n}. \tag{9.79}$$

9.13 Appendix. The Original Form of the Bézier Curve

The original form of the expression for the Bézier curve [3], which Bézier gave using polygon edges $(\mathbf{a}_1, \mathbf{a}_2, \cdots \mathbf{a}_n)$, was

$$\sum_{i=1}^{n} \frac{(-t)^i}{(i-1)!}\left(\frac{d}{dt}\right)^i\left\{\frac{(1-t)^n-1}{t}\mathbf{a}_i\right\}. \tag{9.80}$$

This is a very difficult expression, from which R. Forrester deduced the Bernstein polynomial form (9.20) in 1970 [3]. The form (9.17) was introduced by M. Hosaka in 1976 [5].

Equation (9.80) can be obtained by conversion from eq. (9.17) as follows: by putting

$$\mathbf{a}_1 = (\mathsf{E}-1)\mathbf{p}_0, \quad (1-\mathsf{E})t \equiv x, \tag{9.81}$$

we get from eq. (9.17)

$$\mathbf{r}_0(t;n) - \mathbf{p}_0 = \{(1-t+\mathsf{E}t)^n - 1\}\mathbf{p}_0 = -t[\{(1-x)^n - 1\}/x]\mathbf{a}_1.$$

Inside [ ] in the above equation is a polynomial of degree $(n-1)$ in x, hence from eq. (9.81) it is a polynomial of degree $(n-1)$ in E. So its Taylor expansion in E gives

$$[\cdots] = \sum_{i=0}^{n-1}\left[\left(\frac{d}{dx}\right)^i\left\{\frac{(1-x)^n-1}{x}\right\}\right]_{\mathsf{E}=0}\left(\frac{dx}{d\mathsf{E}}\right)^i\frac{\mathsf{E}^i}{i!}$$

Since from eq. (9.81) x becomes equal to t for $\mathsf{E}=0$, we have

$$\left[\left(\frac{d}{dx}\right)^i\left\{\frac{(1-x)^n-1}{x}\right\}\right]_{\mathsf{E}=0} = \left(\frac{d}{dt}\right)^i\left\{\frac{(1-t)^n-1}{t}\right\}. \tag{9.82}$$

And since $(dx/d\mathsf{E})^i = (-t)^i$, eq. (9.82) becomes

$$\mathbf{r}_0(t;n) - \mathbf{p}_0 = -t\sum_{i=0}^{n-1}\left(\frac{d}{dt}\right)^i\left\{\frac{(1-t)^n - 1}{t}\right\}(-t)^i\frac{\mathsf{E}^i\mathbf{a}_1}{i!}. \tag{9.83}$$

Expressing $\mathsf{E}^i\mathbf{a}_1$ as $\mathbf{a}_{i+1}$ and changing the summation range from (0 to $n-1$) to (1 to n), we obtain the equation (9.80) which Bézier gave without explanation.

10 Connection of Bézier Curves and Relation to Spline Polygons

10.1 Introduction

In design of free-form shapes, use of low-degree curves is desirable rather than adopting higher-degree ones or other complex expressions, so long as the objective of the design is attained by connecting low-degree curves. Curves of low degree are easy in analysis, synthesis and calculation of their geometry and also in detailed control of designing shapes, though there are some problems with their connections.

Bézier curve segments have favorable features for CAD application as we have explained in the previous chapter. But when a curve of complex shape is required to be represented by a single Bézier curve segment, it becomes necessarily of high degree and moreover local shape control by its polygon is difficult, because of the global nature of the influence of each of its control points. So, if we can connect B polygons with appropriate constraints which meet a slightly lower continuity condition at each junction than that expected from a single segment, we will be able to limit the influence of one segment over others. And if we can find a geometric relation between these constraints, we will be able to control local shape without violating the continuity conditions and can synthesize a favorable shape.

The vector representation of a basis spline curve, which is usually called B spline (refer to Sect. 11.7), has the characteristics stated above. In ordinary textbooks, the basis spline functions are introduced mathematically at first [22][33]. Hence, some amount of mathematical preparation and background are needed. Moreover numerical calculation of the basis spline curves and surfaces, due to their recurrence formula, is far more complicated when calculation of interference or evaluation of shapes is required, compared to that of Bézier curves and surfaces which have simple analytical expressions as well as a simple graphical interpretation.

Since a basis spline curve consists of piecewise polynomial curves, each curve segment can be expressed as a Bézier curve which is defined by a control polygon. So in this book, we take a different approach for synthesizing a curve or a surface, in which they are composed of Bézier curves or surface patches of degree n connected in $C^{(n-1)}$ [12]. For automatic adjustment of control points of adjacent Bézier curves to satisfy the connecting condition, we introduce constraints by a polygon with a scale ratio list, which determines connections and scale ratios of tangent vectors at junctions of connecting Bézier curves. We call this polygon an S polygon (S stands for spline), from which B polygons (B stands for Bézier) connected in $C^{(n-1)}$ are deduced.

We give formulas for obtaining B points from the S polygon and also the method of inserting vertices of the S polygon without changing the shape of its generating curve. The reason for adopting B polygons is that they are geometrically clear and simple and their expressions are easy to manipulate compared with the recursive formulas used in B splines. Especially in advanced design activities, calculation problems such as interference and quality evaluation of surfaces require analytical expressions of shapes, so simple closed forms of Bézier type are suitable.

The description of spline curves and surfaces is divided into Chaps. 10 and 11. In this chapter, we first treat a constraint on the connection of two Bézier curves and then introduce the S polygon from which a row of B polygons connected in $C^{(n-1)}$ can be generated. Omitting the theoretical aspects of the S polygon, which are treated in the next chapter, we explain practical methods of locating B polygons from a given S polygon and of inserting new vertices into it.

Then we loosen the connecting condition, from $C^{(2)}$, which requires continuity up to the second derivative vectors at the junction, to $G^{(2)}$ which satisfies continuity only of curvature vectors there. This allows a little freedom to move locations of the B points without changing the S polygon of degree three. The spline curve satisfying this condition is usually called the Beta spline, but the method of deriving it adopted in this chapter is far simpler and the result is easy to use.

In the last part of this chapter, we treat a problem of effects of scale ratios on generated shapes and that of control of curvature distribution along a curve.

The reader with good knowledge of Bézier curves and surfaces, who just wants to make shapes by connecting them, need not know the theoretical aspects treated in the next chapter, but may refer to Sects. 10.3 and 10.4 for curve synthesis and Sect. 11.1 for surface synthesis.

10.2 Connection of B Curve Segments

10.2.1 Scale Ratios

For various curve segments of parametric expressions, there may be great differences in increments of position vectors for the same incremental value of the respective parameters, because usually the parameters are normalized between 0 and 1 irrespective of the real magnitude of the objects they represent. This does not cause a problem so long as curve segments are independent. But when connecting them, we need to take account of these differences of scale for each segment, because their derivatives with respect to the parameter cannot be compared with each other without correction for their scale differences. Accordingly, we introduce a list of ratios of the scale factors for connecting segments. For convenience, we consider that each constituent segment has its own scale factor λ.

Let there be a row of Bézier curve segments of degree n, each connected in $C^{(i)}$.

$$\cdots, \mathbf{r}_{-1}(t;n), \mathbf{r}_0(t;n), \mathbf{r}_1(t;n), \cdots.$$

Instead of a curve segment $\mathbf{r}_j(t;n)$, we treat its B polygon $B_j^{(n)}$:

$$(\mathbf{p}_{jn}, \mathbf{p}_{jn+1}, \cdots, \mathbf{p}_{jn+n}), \tag{10.1}$$

$$\text{or} \quad (\mathbf{a}_{jn}, \mathbf{a}_{jn+1}, \cdots, \mathbf{a}_{jn+n-1}) \tag{10.2}$$

where $\mathbf{p}_{jn}$ and $\mathbf{a}_{jn}$ indicate the first control point and the first control edge of the control polygon $B_j^{(n)}$. Its superscript indicates the degree of the curve, and if this is clear it may be omitted. Two curve segments $\mathbf{r}_{j-1}(t;n)$ and $\mathbf{r}_j(t;n)$ are connected continuously up to the i-th derivative at their ends.

Though we use the same symbol t for the parameter of each curve segment, they are substantially independent each other. However, when two curves are connected smoothly, it is natural to consider that the element ds of arc length of both segments take the same value at the junction. Therefore, we postulate (refer to Sect. 5.2) a condition, setting $j = 0$ for simplicity:

$$ds = |\dot{\mathbf{r}}_{-1}(1;n)|(dt)_{-1} = |\dot{\mathbf{r}}_0(0;n)|(dt)_0, \tag{10.3}$$

where $(dt)_{-1}$ and $(dt)_0$ indicate that they belong to $\mathbf{r}_{-1}(t;n)$ and $\mathbf{r}_0(t;n)$ at the junction respectively.

Introducing constants λ_{-1} and λ_0 which are proportional to the magnitudes of tangent vectors $\dot{\mathbf{r}}_{-1}(1;n)$ and $\dot{\mathbf{r}}_0(0;n)$ at the junction, we define the ratio $k_0 = \lambda_{-1}/\lambda_0$ and obtain the following relation from eq. (10.3):

$$k_0 = \frac{\lambda_{-1}}{\lambda_0} = \frac{|\dot{\mathbf{r}}_{-1}(1;n)|}{|\dot{\mathbf{r}}_0(0;n)|} = \frac{(dt)_0}{(dt)_{-1}}. \tag{10.4}$$

This ratio is considered the scale factor ratio of the parameters at the junction. For connecting curve segments $(\cdots, \mathbf{r}_{-1}, \mathbf{r}_0, \mathbf{r}_1, \mathbf{r}_2, \cdots)$, a list of scale factors $(\cdots, \lambda_{-1}, \lambda_0, \lambda_1, \lambda_2, \cdots)$ must be provided to ensure a natural shape. When curve segments are of almost equal magnitude, all factors λ_i may well take the value 1.

10.2.2 Conditions of $C^{(i)}$ Connection

If we interpret the connection of continuity $C^{(i)}$ geometrically, the following conditions must hold at the junction of two curve segments $\mathbf{r}_{-1}(t;n)$ and $\mathbf{r}_0(t;n)$:

$$\left(\frac{d}{ds}\right)^j \mathbf{r}_{-1}(t;n)\,|_{t=1} = \left(\frac{d}{ds}\right)^j \mathbf{r}_0(t;n)\,|_{t=0}, \quad \text{for } j = 0, \cdots i. \tag{10.5}$$

Substituting eq. (10.4) into eq. (10.5), we rewrite eq. (10.5) as

$$(\frac{d}{dt})^j \mathbf{r}_{-1}(t;n)\mid_{t=1} = k_0^j (\frac{d}{dt})^j \mathbf{r}_0(t;n)\mid_{t=0} . \tag{10.6}$$

Let the two connected curve segments $\mathbf{r}_{-1}(t;n)$ and $\mathbf{r}_0(t;n)$ be represented by Bézier curves in the operator forms(refer to equations (9.20) and (9.17)):

$$\mathbf{r}_{-1}(t;n) = \{(1-t)\mathsf{E}^{-1} + t\}^n \mathbf{p}_0, \tag{10.7}$$

$$\mathbf{r}_0(t;n) = (1-t+\mathsf{E}t)^n \mathbf{p}_0. \tag{10.8}$$

Substituting these into eq. (10.6), we have, for $j = 0, \cdots i$,

$$(1-E^{-1})^j \mathbf{p}_0 = k_0^j (E-1)^j \mathbf{p}_0. \tag{10.9}$$

This is the condition of $C^{(i)}$.

For example,

$$j = 0 \ : \ \mathbf{p}_0 = \mathbf{p}_0,$$
$$j = 1 \ : \ \mathbf{p}_0 - \mathbf{p}_{-1} = k_0(\mathbf{p}_1 - \mathbf{p}_0) \quad \text{or} \quad \mathbf{a}_{-1} = k_0\mathbf{a}_0,$$
$$j = 2 \ : \ (\mathbf{p}_0 - \mathbf{p}_{-1}) - (\mathbf{p}_{-1} - \mathbf{p}_{-2}) = k_0^2\{\mathbf{p}_2 - \mathbf{p}_1) - (\mathbf{p}_1 - \mathbf{p}_0\},$$
$$\text{or } \mathbf{a}_{-1} - \mathbf{a}_{-2} = k_0^2(\mathbf{a}_1 - \mathbf{a}_0) \tag{10.10}$$

and so on.

These relations are the geometric constraints, as shown in Fig. 10.1 for $C^{(1)}$, $C^{(2)}$, $C^{(3)}$ and $C^{(4)}$. The important fact is that in the $C^{(i)}$ connection, on both sides of the junction the i consecutive control points of one segment are completely determined by the i consecutive control points of the other segment. This sequence of $2i+1$ control points defines a B polygon of degree i. Application of this fact is explained in the following subsection.

10.2.3 $C^{(n-1)}$ Connection and Control Points

Let there be $C^{(n-1)}$ connected curve segments of degree n. We examine the influence of displacement of one control point of one of its constituent curve segments on the shape of the whole curve.

For $C^{(i)}$ connection, there are $i+1$ equations (10.9) for the $i+1$ consecutive control points of both sides of one junction. Accordingly, for $C^{(n)}$ connection of these segments, all the $n+1$ control points of one segment are completely determined by the $n+1$ control points of the adjacent segment. Consequently, displacement of one control point of a segment influences all the control points of the other segments connected one after another to keep the $C^{(n)}$ condition at all the junctions. So, there is no freedom of local deformation of shape of the curve.

But for $C^{(n-1)}$ connection of n+1 consecutive segments, the number of equations (10.9) at all the n junctions is equal to n^2 while the number of control points

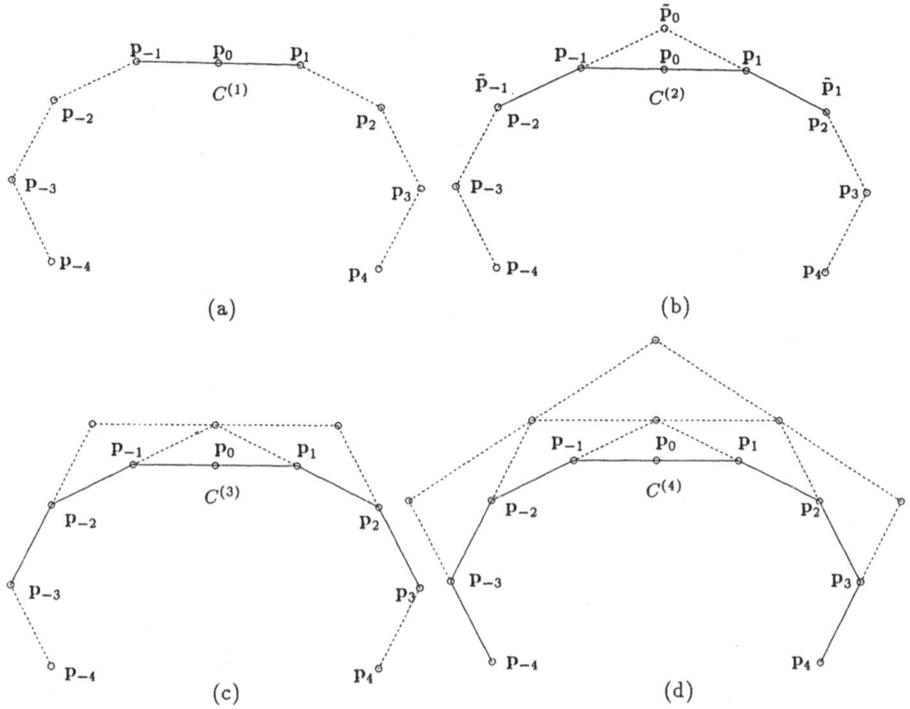

Figure 10.1 Constraints of control points on both sides of a junction of two B polygons

of all the $n+1$ segments is equal to $(n+1)^2$, so their difference of $2n+1$ control points are free. Therefore, the $n+1$ control points of the first segment and the n last control points of the $(n+1)$-th segment can be considered independent.

Accordingly, displacement of the control points of the first segment does not influence the $(n+2)$-th segment or further ones. For instance, let $B_0\ B_1\ B_2\ B_3, \cdots$ be Bézier control polygons of curve segments of degree three and let them connect in $C^{(2)}$. See Fig. 10.2. Then any displacement of control points of B_0 does not influence the location of the last three control points of B_2, while those of B_0 and B_1 have to be adjusted to keep the $C^{(2)}$ condition at the junctions, and B_3 does not change at all. We are to find a method of automatic adjustment of the control points of B_0, B_1 and B_2 to keep them in $C^{(2)}$ connection. Thus the $C^{(n-1)}$ connection enables local control of shape of connected curve segments of degree n.

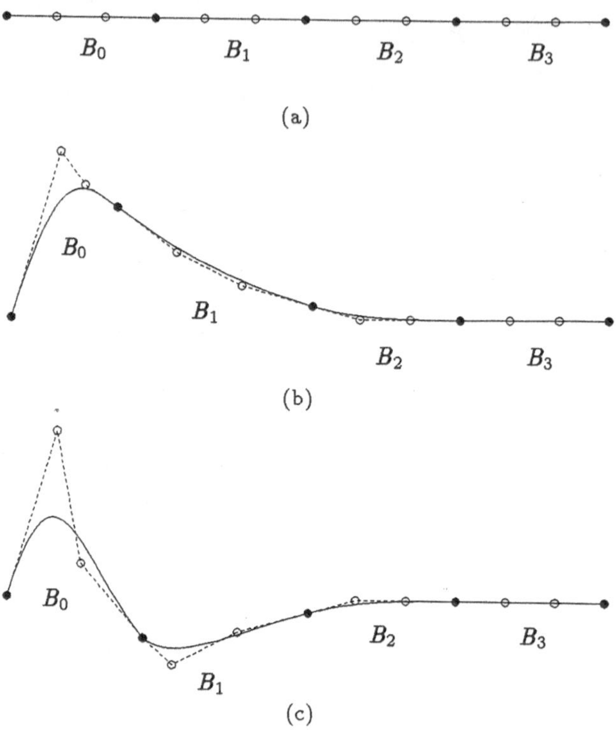

Figure 10.2 Propagation of displacement over other segments connected in $C^{(2)}$

10.2.4 Connection Defining Polygon

We introduce a special B polygon which stipulates the condition of connection of two B polygons. We call it the *Connection Defining Polygon* (*CDP*), whose properties are explained in this section.

Referring to Fig. 10.1(b), for connection of $B_{-1}^{(3)} = (\mathbf{p}_{-3}, \mathbf{p}_{-2}, \mathbf{p}_{-1}, \mathbf{p}_0)$ and $B_0^{(3)} = (\mathbf{p}_0, \mathbf{p}_1, \mathbf{p}_2, \mathbf{p}_3)$ in $C^{(2)}$, their five control points from $\mathbf{p}_{-2}$ to $\mathbf{p}_2$ have the two relations of eq. (10.9) for $j = 1$ and $j = 2$, so two of the five control points are determined from the other three. Or there exists a B polygon $\tilde{B}_{-1,0}^{(2)}$ which has three control points :

$$(\tilde{\mathbf{p}}_{-1}, \tilde{\mathbf{p}}_0, \tilde{\mathbf{p}}_1)^{(2)}, \tag{10.11}$$

division of which in the ratio $(\lambda_{-1} : \lambda_0)$ produces two B polygons

$$\acute{B}_{-1}^{(2)} = (\mathbf{p}_{-2}, \mathbf{p}_{-1}, \mathbf{p}_0), \qquad \acute{B}_0^{(2)} = (\mathbf{p}_0, \mathbf{p}_1, \mathbf{p}_2). \tag{10.12}$$

Their control points are from $\mathbf{p}_{-2}$ to $\mathbf{p}_2$.

We call this B polygon $\tilde{B}^{(2)}_{-1,0}$ the connection defining polygon or the *CDP* of $B^{(3)}_{-1}$and $B^{(3)}_{0}$, because the polygon $\tilde{B}^{(2)}_{-1,0}$ determines the connection of $\acute{B}^{(2)}_{-1}$ and $\acute{B}^{(2)}_{0}$ in $C^{(2)}$.

The control points of $\acute{B}^{(2)}_{-1}$ are the same as those of the control points of $B^{(3)}_{-1}$ except for the leftmost $\mathbf{p}_{-3}$, and those of $\acute{B}^{(2)}_{0}$ are the same as those of $B^{(3)}_{0}$ except for the rightmost $\mathbf{p}_3$. So we use notations ${}^*B^{(3)}_{-1}$ and $B^{*(3)}_{0}$ in place of $\acute{B}^{(2)}_{-1}$ and $\acute{B}^{(2)}_{0}$ when emphasizing their missing leftmost or rightmost control point. We call these B polygons the left-end or right-end *incomplete* B polygons.

Refer to Sects. 9.7 and 9.8. Since these control points are determined by the division of $\tilde{B}^{(2)}_{-1,0}$ in the ratio $(\lambda_{-1} : \lambda_0)$, their locations are given, for $0 \le j \le 2$, by

$$\mathbf{p}^{(2)}_{-2+j} = (1 - t + \mathrm{E}t)^j \tilde{\mathbf{p}}^{(2)}_{-1}\ , \tag{10.13}$$

$$\mathbf{P}^{(2)}_{2-j} = \{(1 - t)\mathrm{E}^{-1} + t\}^j \tilde{\mathbf{p}}^{(2)}_{1}, \tag{10.14}$$

$$\text{where} \quad t = \frac{\lambda_{-1}}{\lambda_{-1} + \lambda_0} = \frac{k_0}{1 + k_0}. \tag{10.15}$$

According to the dividing procedure for a B polygon explained in Sect. 9.7, it is understood that each edge of the *CDP* is considered to correspond to $(\lambda_{-1} + \lambda_0)$, each edge of its divided polygon ${}^*B^{(3)}_{-1}$ to λ_{-1} and that of $B^{*(3)}_{0}$ to λ_0.

The above stated relations in $C^{(2)}$ connection can be extended to those in $C^{(n-1)}$ connection. For the connection of two B polygons $B^{(n)}_{-1}$ and $B^{(n)}_{0}$ in $C^{(n-1)}$ with the scale ratio $(\lambda_{-1} : \lambda_0)$, there is a *CDP* $\tilde{B}^{(n-1)}_{-1,0}$, each edge of which corresponds to $(\lambda_{-1} + \lambda_0)$. By division of the *CDP* in the ratio $(\lambda_{-1} : \lambda_0)$, the incomplete B polygons ${}^*B^{(n)}_{-1}$ and $B^{*(n)}_{0}$ are obtained, where ${}^*B^{(n)}_{-1}$ indicates the control points of $B^{(n)}_{-1}$ with its first one missing and $B^{*(n)}_{0}$ the control points of $B^{(n)}_{0}$ with its last one missing.

If there exists a series of *CDP*s of degree $(n - 1)$:

$$\cdots \tilde{B}_{-1,0}\ \tilde{B}_{0,1}\ \tilde{B}_{1,2}\ \tilde{B}_{2,3} \cdots\ , \tag{10.16}$$

by dividing each *CDP* of (10.16) in the ratio

$$\cdots(\lambda_{-1} : \lambda_0), (\lambda_0 : \lambda_1), (\lambda_1 : \lambda_2)\cdots,$$

we obtain a series of incomplete B polygons:

$$(\cdots {}^*B_{-1}B^*_0, {}^*B_0B^*_1, {}^*B_1B^*_2 \cdots).$$

By removing duplicate B points from the above series, we obtain a row of B polygons of degree n connected in $C^{(n-1)}$:

$$(\cdots B_{-1}\ B_0\ B_1\ B_2 \cdots) \tag{10.17}$$

whose scale factors are

$$(\cdots \lambda_{-1}, \lambda_0, \lambda_1, \lambda_2 \cdots). \tag{10.18}$$

If we can make a polygon which controls the disposition of a series of *CDPs* and directly gives a series of B polygons (10.17), we are freed from problems of individually connecting B polygons in $C^{(n-1)}$, yet may utilize favorable characteristics of Bézier expressions. We introduce such a polygon in the next section, and call it the S polygon for brevity.

10.3 Introduction of S Polygon

10.3.1 Locating B Points from an S Polygon

Since an S polygon stipulates the disposition of *CDPs*, we can provide methods of directly determining its constituent B polygons without considering each *CDP* at a junction. We explain such practical methods in this section.

Since two adjacent *CDPs* $\tilde{B}_{-1,0}$ and $\tilde{B}_{0,1}$ are resolved into $({}^*B_{-1}B_0^*)$ and $({}^*B_0B_1^*)$, from their common part we can extract a (both-ends) *truncated* B polygon ${}^*B_0^*$. Similarly we have adjacent truncated B polygons ${}^*B_{-1}^*$ and ${}^*B_1^*$. From the condition (10.10) for $j = 1$, the location of their common control point $\mathbf{p}_0$ is the dividing point between $\mathbf{p}_{-1}$ and $\mathbf{p}_0$ in the ratio $(\lambda_{-1} : \lambda_0)$ and that of $\mathbf{p}_n$ is the dividing point between $\mathbf{p}_{n-1}$ and $\mathbf{p}_{n+1}$ in the ratio $(\lambda_0 : \lambda_1)$. Accordingly, if we can determine all the both-end-truncated B polygons, their missing common B points can be supplied easily.

Now we introduce a polygon S which constrains the row of *CDPs* given by eq.(10.16) with the list of scale factors. Let the vertices of an S polygon be denoted by

$$(\cdots \mathbf{q}_{-1}\ \mathbf{q}_0\ \mathbf{q}_1\ \mathbf{q}_2 \cdots) \tag{10.19}$$

and its edges by

$$(\cdots \mathbf{b}_{-1}\ \mathbf{b}_0\ \mathbf{b}_1\ \mathbf{b}_2 \cdots). \tag{10.20}$$

Each of its edges $\mathbf{b}_j$ is divided into n sub-edges whose dividing ratio is equal to the ratio of scale factors of n consecutive B polygons (10.17). The *dividing ratio list* for an edge $\mathbf{b}_j$ is given by a list of n elements λ denoted by

$$L_j^{(n)} = (\lambda_{j-n+1}, \cdots, \lambda_j). \tag{10.21}$$

Accordingly, each sub-edge has its own scale factor. The value j of the index of its last element of $L_j^{(n)}$ is the same as that of the edge $\mathbf{b}_j$ to be divided. This rule is determined logically; the reason is explained in Sect. 11.3. When terms a 'scale list' or a 'λ' list are used, they simply refer to a list such as eq.(10.18). A set of consecutive n elements of a scale list belonging to an S polygon of degree n becomes a dividing ratio list of an edge of the polygon.

Next, a *unit* S polygon of degree n, which is denoted by $S_i^{(n)}$, is defined as the n consecutive edges starting at the vertex $\mathbf{q}_i$ of an S polygon. Since all the dividing ratio lists of $S_i^{(n)}$ are $(L_i^{(n)}, \cdots, L_{i+n-1}^{(n)})$, the scale factor λ_i is the only common element in them. Accordingly, there is only one sub-edge, which has the same common scale factor λ_i, in each edge of a unit S polygon $S_i^{(n)}$. We call the common scale factor the *main scale factor* of the unit S polygon.

Leaving the theoretical aspect of the S polygon to Sects. 11.3–11.5, we give its features useful for derivation of B polygons.

1. A unit S polygon $S_i^{(n)}$ generates three successive B polygons denoted by $({}^*B_{i-1}\ B_i\ B_{i+1}^*)^{(n)}$: the central one, whose scale is λ_i, is complete and both side ones are incomplete B polygons of degree n with scale factors λ_{i-1} and λ_{i+1} respectively.
2. An *incomplete* S polygon $S_i^{*(n)}$ which misses the right-end edge of $S_i^{(n)}$ generates a *CDP* $\tilde{B}_{i-1,i}$, and its adjacent incomplete S polygon ${}^*S_i^{(n)}$ which misses the left-end edge of $S_i^{(n)}$ generates $\tilde{B}_{i,i+1}$.
3. From a both-end-truncated S polygon ${}^*S_i^{*(n)}$, which misses both end edges of $S_i^{(n)}$, an S polygon of degree $(n-2)$ with its main scale λ_i which is the same as that of $S_i^{(n)}$, can be deduced. The method is explained in Chap. 11. It is called the *reduced-truncated* S polygon of $S_i^{(n)}$, for which we use a symbol $\breve{S}_i^{(n-2)}$. The dividing ratio list of its first edge is denoted by
$$\breve{L}_i^{(n-2)} = (\lambda_{(i-n+3)}, \cdots, \lambda_i), \tag{10.22}$$
instead of $L_i^{(n-2)}$ for clarity. The reduced-truncated S polygon finally produces a truncated B polygon ${}^*B_i^{*(n)}$.
4. Location of a common control point $\mathbf{p}_{n(i+1)}$ of two adjacent B polygons $B_i^{(n)}$ and $B_{(i+1)}^{(n)}$ whose scales are λ_i and λ_{i+1} is determined at the $(\lambda_i : \lambda_{i+1})$ dividing point of the distance between the end control points $\mathbf{p}_{\{n(i+1)-1\}}$ and $\mathbf{p}_{\{n(i+1)+1\}}$ of the adjacent truncated B polygons ${}^*B_i^{*(n)}$ and ${}^*B_{(i+1)}^{*}{}^{(n)}$.

Applying the above properties, especially (3), we can determine locations of B polygons connected in $C^{(n-1)}$ from a given S polygon. We give examples. See Fig. 10.3.

$n = 1$: By the definition the following relations are determined.

$$\begin{aligned} S_0^{(1)} &= (\mathbf{q}_0\mathbf{q}_1) = \mathbf{b}_0, & L_0^{(1)} &= (\lambda_0) \\ S_0^{*(1)} &= \mathbf{q}_0, \quad {}^*S_0{}^{(1)} = \mathbf{q}_1, & {}^*S^*{}_0^{(1)} &= null. \end{aligned}$$

Each *CDP* coincides with each vertex of the S polygon, because the incomplete S polygons coincide with the vertices. Hence, each edge of S is a B polygon of degree one, which connects with adjacent B polygons at the vertices.

$$B_0^{(1)} = (\mathbf{p}_0\mathbf{p}_1) = \mathbf{b}_0 = (\mathbf{q}_0\mathbf{q}_1), \quad \breve{S}_0^{(-1)} = null, \quad {}^*B_0^{*(1)} = null.$$

Since the dividing ratio list is $L_0^{(1)} = (\lambda_0)$ and the number of its element is one, there are no dividing points on the edges.

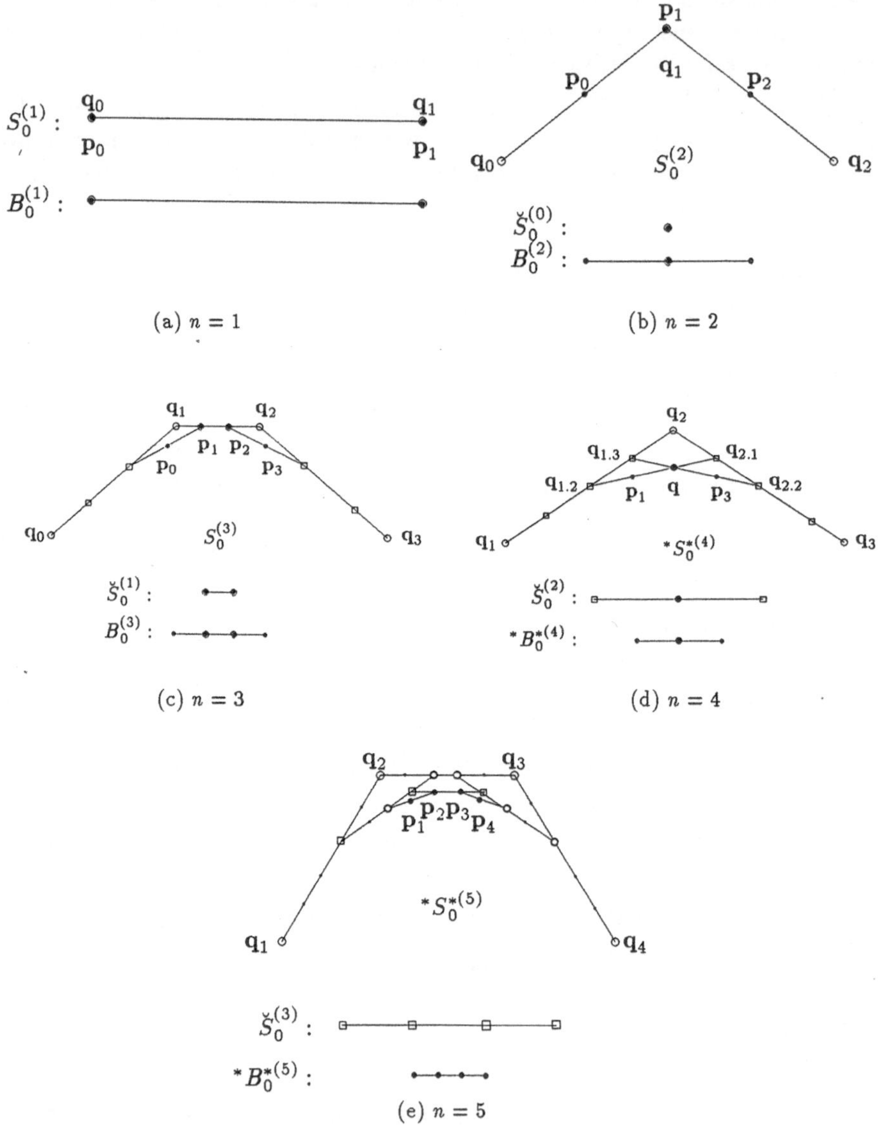

Figure 10.3 Locations of B points for $n = 1, \ldots 5$

$n = 2$: The following relations are easily obtained.

$$\begin{aligned} S_0^{(2)} &= (\mathbf{q}_0\ \mathbf{q}_1\ \mathbf{q}_2) = (\mathbf{b}_0\ \mathbf{b}_1), & L_2^{(2)} &= (\lambda_1\ \lambda_0), \\ S_0^{*(2)} &= (\mathbf{q}_0\ \mathbf{q}_1), \quad {}^*S_0^{(2)} = (\mathbf{q}_1\ \mathbf{q}_2), & {}^*S_0^{*(2)} &= \mathbf{q}_1, \\ \check{S}_0^{(0)} &= \mathbf{q}_1, & \check{L}_0^{(0)} &= null, \\ {}^*B_0^{*(2)} &= \mathbf{p}_1 = \mathbf{q}_1, & {}^*B_{-1}^{*\ (2)} &= \mathbf{p}_{-1} = \mathbf{q}_0, \\ {}^*B_1^{*(2)} &= \mathbf{p}_3 = \mathbf{q}_2, \end{aligned}$$

Since the reduced-truncated S polygon is a vertex of an S polygon, the truncated B polygon becomes the same point. The common B control point $\mathbf{p}_0$ between B_1 and B_0 is located at the dividing point between $\mathbf{q}_0$ and $\mathbf{q}_1$ in the ratio $(\lambda_{-1} : \lambda_0)$. Similarly the common B control point $\mathbf{p}_2$ between B_0 and B_1 is located at the dividing point between $\mathbf{q}_1$ and $\mathbf{q}_2$ in the ratio $(\lambda_0 : \lambda_1)$. We denote their locations symbolically as $\mathbf{q}_{0.1}$ and $\mathbf{q}_{1.1}$.

$n = 3$: By the definition we have

$$\begin{aligned} S_0^{(3)} &= (\mathbf{q}_0\ \mathbf{q}_1\ \mathbf{q}_2\ \mathbf{q}_3) = (\mathbf{b}_0\ \mathbf{b}_1\ \mathbf{b}_2), & L_0^{(3)} &= (\lambda_2\ \lambda_1\ \lambda_0), \\ {}^*S_0^{*(3)} &= (\mathbf{q}_1\ \mathbf{q}_2) = \mathbf{b}_1, & \check{L}_0^{(1)} &= \lambda_0. \end{aligned}$$

Its reduced-truncated S polygon $\check{S}_0^{(1)}$ is necessarily the λ_0 sub-edge of $\mathbf{b}_1$ which is ${}^*B_0^{*(3)} = (\mathbf{p}_1\ \mathbf{p}_2)$. Accordingly, locations of $\mathbf{p}_1$ and $\mathbf{p}_2$ are the dividing points of the edge $\mathbf{b}_1$ with the ratio $L_1^{(3)} = (\lambda_{-1}\ \lambda_0\ \lambda_1)$. Similarly, locations of $\mathbf{p}_{-2}$ and $\mathbf{p}_{-1}$ and those of $\mathbf{p}_4$ and $\mathbf{p}_5$ are determined at the dividing points of $\mathbf{b}_0$ and $\mathbf{b}_2$ by the ratios $L_0^{(3)}$ and $L_2^{(3)}$ respectively. The location of $\mathbf{p}_0$ is the $(\lambda_{-1} : \lambda_0)$ dividing point between locations $\mathbf{p}_{-1}$ and $\mathbf{p}_1$, and the location of $\mathbf{p}_3$ is the $(\lambda_0 : \lambda_1)$ dividing point of $\mathbf{p}_2$ and $\mathbf{p}_4$. We denote the locations of the control points $(\mathbf{p}_0\ \mathbf{p}_1\ \mathbf{p}_2\ \mathbf{p}_3)$ of $B_0^{(3)}$ as

$$\mathbf{q}_{0.2.1}\ \mathbf{q}_{1.1}\ \mathbf{q}_{1.2}\ \mathbf{q}_{1.2.1}. \tag{10.23}$$

These symbolic notation for locations of the dividing points are explained in Sect. 11.4, but they are not needed in this chapter.

$n = 4$:

$$\begin{aligned} S_0^{(4)} &= (\mathbf{q}_0\ \mathbf{q}_1\ \mathbf{q}_2\ \mathbf{q}_3\ \mathbf{q}_4) = (\mathbf{b}_0\ \mathbf{b}_1\ \mathbf{b}_2\ \mathbf{b}_3), & L_0^{(4)} &= (\lambda_0\ \lambda_1\ \lambda_2\ \lambda_3) \\ {}^*S_0^{*(4)} &= (\mathbf{q}_1\ \mathbf{q}_2\ \mathbf{q}_3) = (\mathbf{b}_1\ \mathbf{b}_2)\ , & \check{L}_0^{(2)} &= (\lambda_1\ \lambda_0). \end{aligned}$$

Its reduced-truncated S polygon $\check{S}_0^{(2)}$ is constructed as follows. Refer to Fig. 10.3(d). Make two new edges: one by connecting the second dividing point $\mathbf{q}_{1.2}$ of $\mathbf{b}_1$ to the first dividing point $\mathbf{q}_{2.1}$ of $\mathbf{b}_2$, and the other from the third dividing point $\mathbf{q}_{1.3}$ of $\mathbf{b}_1$ to the second dividing point $\mathbf{q}_{2.2}$ of $\mathbf{b}_2$. Let their intersection be $\mathbf{q}$. Then the polygon $(\mathbf{q}_{1.2}\ \mathbf{q}\ \mathbf{q}_{2.2})$ is the reduced-truncated S polygon $\check{S}_0^{(2)}$, from which the truncated B polygon ${}^*B_0^{*(4)} = (\mathbf{p}_1\ \mathbf{p}_2\ \mathbf{p}_3)$ is determined by the procedure for $n = 2$. In the lower part of the figure, relation between $\check{S}_0^{(2)}$ and ${}^*B_0^{*(4)}$ is shown schematically.

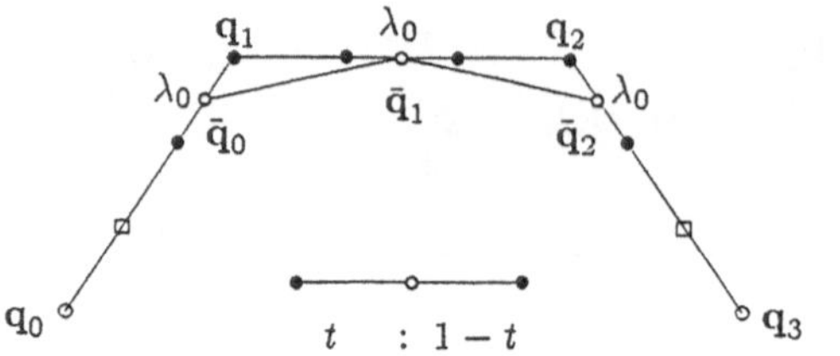

Figure 10.4 Relation between locations of old vertices ($\mathbf{q}_1$, $\mathbf{q}_2$) and those of new vertices ($\bar{\mathbf{q}}_0$, $\bar{\mathbf{q}}_1$ $\bar{\mathbf{q}}_2$)

$n = 5$:

$$^*S_0^{*(5)} = (\mathbf{q}_1\ \mathbf{q}_2\ \mathbf{q}_3\mathbf{q}_4) = (\mathbf{b}_1\ \mathbf{b}_2\mathbf{b}_3), \qquad \check{L}_0^{(3)} = (\lambda_2\ \lambda_1\ \lambda_0) \qquad (10.24)$$

Make a reduced-truncated S polygon $\check{S}$ of degree three, whose vertices are produced from the dividing points (indicated by small blank squares in Fig. 10.3(e)) of the edges of the truncated S polygon of degree five. From $\check{S}_0^{(3)}$ with $\check{L}_0^{(3)}$ and using the procedure for $n = 3$, we determine four B points indicated by small black circles, which are the locations of $^*B_0^{*(5)} = (\mathbf{p}_1\ \mathbf{p}_2\ \mathbf{p}_3\ \mathbf{p}_4)$. In the lower part of the figure, relation between $\check{S}_0^{(3)}$ and $^*B_0^{*(5)}$ is shown.

The above stated methods can be extended to degree n: from an S polygon and its edge dividing ratio lists we can obtain its reduced-truncated polygon $\check{S}$ of degree $(n-2)$. Repeating the degree reduction process, we can determine locations of the B polygons of degree n from the lower-degree S polygons. B polygons thus obtained from an S polygon of degree n are connected in $C^{(n-1)}$.

10.3.2 Increase of Vertices of an S Polygon

From an $S_0^{(n)}$ polygon with a scale list $(\cdots, \lambda_{-1}, \lambda_0\ \lambda_1, \cdots)$ we obtain a series of B polygons connected in $C^{(n-1)}$:

$$(\cdots, B_{-1}^{(n)}, B_0^{(n)}, B_1^{(n)}, \cdots),$$

whose scale ratios are $(\cdots : \lambda_{-1} : \lambda_0 : \lambda_1 : \cdots)$. If we divide $B_0^{(n)}$ in the ratio $(t : 1-t)$ into two B polygons $B^a{}_0^{(n)}$ and $B^b{}_0^{(n)}$, we can consider that their scales are $\lambda_0 t$ and $\lambda_0(1-t)$. They connect with their adjacent B polygons $B_{-1}^{(n)}$ and $B_1^{(n)}$ in $C^{(n-1)}$, because the original $B_0^{(n)}$ connects with the same adjacent B polygons in $C^{(n-1)}$.

Now, from the original unit S polygon $S_0^{(n)}$ we produce two new unit S polygons $S^a{}_0^{(n)}$ and $S^b{}_0^{(n)}$ by the following procedure: see Fig. 10.4, which is a case of $n = 3$.

In each edge of $S_0^{(n)}$ there is a sub-edges of the dividing ratio λ_0 which is the main scale of the polygon. We divided each sub-edge into two parts in the ratio

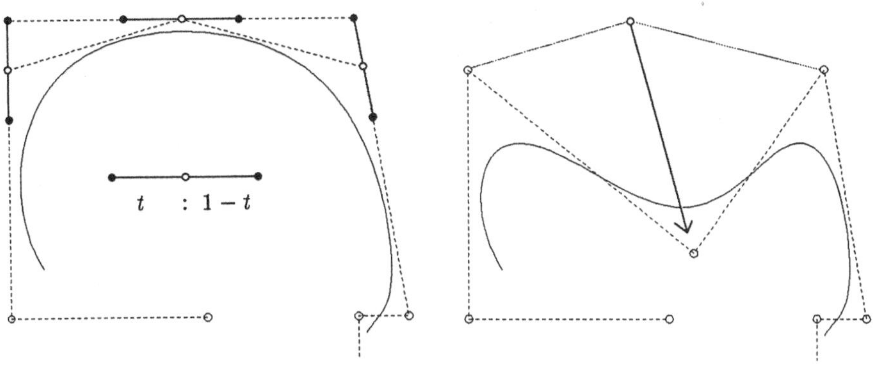

(a) increase of vertex of S polygon (b) displacement of a vertex and deformation of curve shape

Figure 10.5 Insertion of a vertex

$(t : 1-t)$, then we have n dividing points $(\bar{\mathbf{q}}_0 \cdots \bar{\mathbf{q}}_{n-1})$. We replace the original $n-1$ vertices $(\mathbf{q}_1 \cdots \mathbf{q}_{n-1})$ by these new n vertices and consider there generate two unit S polygons:

$$S^{a}{}_{0}^{(n)} = (\mathbf{q}_0\, \bar{\mathbf{q}}_0, \cdots, \bar{\mathbf{q}}_{n-1}),$$
$$S^{b}{}_{0}^{(n)} = (\bar{\mathbf{q}}_0, \cdots, \bar{\mathbf{q}}_{n-1}, \mathbf{q}_n).$$

We can prove these two unit S polygons have the main scales $\lambda_0 t$ and $\lambda_0(1-t)$ respectively and generate the B polygons $B^{a}{}_{0}^{(n)}$ and $B^{b}{}_{0}^{(n)}$, which are the same B polygons generated from the division of $B_0^{(n)}$ in the ratio $(t : 1-t)$. (This is treated in Sec. 11.5.4).

These B polygons $B^{a}{}_{0}^{(n)}$ and $B^{b}{}_{0}^{(n)}$ connect in $C^{(n)}$ if no displacement of the vertices occurs. Even when vertices of the new S polygon are moved, the condition of $C^{(n-1)}$ connection of the generated curve segments is not violated. An example is shown in Fig. 10.5. We call this process the division of an S polygon or the insertion of a vertex.

10.4 B points under Geometric Connecting Condition $G^{(2)}$

So far we have treated $C^{(n-1)}$ connection of B polygons of degree n, considering scale ratios of the parameters at their junctions. Now we are to loosen the condition of $C^{(2)}$. Instead of continuity of the second derivative vectors at the junctions, we use the continuity condition of curvature there, which is called the geometric continuity condition $G^{(2)}$. We restrict the degree of curve segments to three. This makes the constraints stipulated by the Connection Defining Polygon

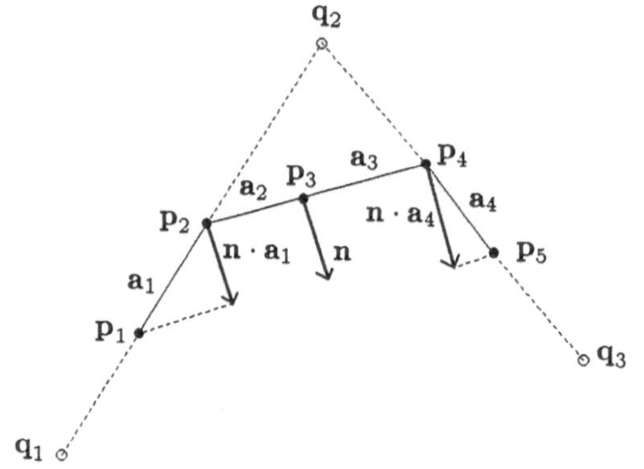

Figure 10.6 Control vectors at a junction in $G^{(2)}$ connection

of degree two hold only in the normal direction at the junction. It follows that the dividing ratio list of each edge of an S polygon can be adjusted to modify the shape of connected curve segments without breaking the condition of $G^{(2)}$.

Refer to Fig. 10.6. Let the connecting B polygons of degree three be expressed by their edges

$$B_0 = (\mathbf{a}_0\ \mathbf{a}_1\ \mathbf{a}_2), \qquad B_1 = (\mathbf{a}_3\ \mathbf{a}_4\ \mathbf{a}_5),$$

and their CDP be

$$\tilde{B}^{(2)}_{0,1} = (\mathbf{p}_1\ \mathbf{q}_2\ \mathbf{p}_5).$$

In $C^{(2)}$ connection, division of the CDP in the ratio $(\lambda_0 : \lambda_1)$ produces two incomplete B polygons ${}^*B_0$ and B_1^* and the following relations hold:

$$|\mathbf{a}_1| : |\overline{\mathbf{q}_2\mathbf{p}_2}| = |\mathbf{a}_2| : |\mathbf{a}_3| = |\overline{\mathbf{q}_2\mathbf{p}_4}| : |\mathbf{a}_4| = \lambda_0 : \lambda_1 = k_1 : 1. \tag{10.25}$$

In $G^{(2)}$ connection, instead of the continuity condition of the second derivatives at the junction, which is given by

$$\mathbf{a}_2 - \mathbf{a}_1 = k_1^2(\mathbf{a}_4 - \mathbf{a}_3), \tag{10.26}$$

only the continuity of the curvatures there is allowed. Referring to the expression (9.31) for curvature at the left end of B_0, which is proportional to $(\mathbf{n} \cdot \mathbf{a}_1)/\mathbf{a}_0^2$, we have the equation of equality of the curvatures at the junction B_0 and B_1:

$$-\frac{(\mathbf{n} \cdot \mathbf{a}_1)}{\mathbf{a}_2^2} = \frac{(\mathbf{n} \cdot \mathbf{a}_4)}{\mathbf{a}_3^2}. \tag{10.27}$$

This is converted to

$$-(\mathbf{n} \cdot \mathbf{a}_1) = k_1^2(\mathbf{n} \cdot \mathbf{a}_4) \tag{10.28}$$

by the condition of continuity of the tangent vectors $\mathbf{a}_2 = k_1\mathbf{a}_3$. Equation (10.28) is the normal component of eq. (10.26). Using eq. (10.25), from eq. (10.28), we have

$$-\mathbf{n} \cdot (\mathbf{a}_1\lambda_1/\lambda_0) = \mathbf{n} \cdot (\mathbf{a}_4\lambda_0/\lambda_1). \tag{10.29}$$

Accordingly, in $G^{(2)}$ connection, the tangential component of eq. (10.26) need not be satisfied, but only its normal component eq. (10.29) is the condition. Here we introduce an arbitrary positive constant σ, which we call the stretch factor, in the above equation thus:

$$-\mathbf{n}\cdot(\mathbf{a}_1\sigma\lambda_1/\lambda_0) = \mathbf{n}\cdot(\mathbf{a}_4\sigma\lambda_0/\lambda_1). \tag{10.30}$$

Since the left-hand side of eq. (10.29) represents the normal component of $\overline{\mathbf{q}_2\mathbf{p}_2}$ and the right-hand side that of $\overline{\mathbf{q}_2\mathbf{p}_4}$, they are naturally equal. Equation (10.30) now shows that they can be modified to

$$\begin{aligned}\overline{\mathbf{q}_2\mathbf{p}_2} &= \mathbf{a}_1(\sigma\lambda_1/\lambda_0)\\ \overline{\mathbf{q}_2\mathbf{p}_4} &= \mathbf{a}_4(\sigma\lambda_0/\lambda_1).\end{aligned}$$

This means that locations of the control points on the edges $\mathbf{b}_1$ and $\mathbf{b}_2$ can be moved along the respective edges. If $\sigma = 1$, the connection is $C^{(2)}$, but if σ is an arbitrary positive constant, $C^{(2)}$ reduces to $G^{(2)}$.

To indicate that σ is used at the junction of two B polygons B_0 and B_1 connecting in $G^{(2)}$, we attach a subscript to the strtch factor σ such as σ_0. For connection of B_i and B_{i+1} we use σ_i . Then the dividing ratio list of an edge $\mathbf{a}_i$ is modified from $L_i^{(3)}$ for $C^{(2)}$ to $L_i'^{(3)}$ for $G^{(2)}$:

$$L_i'^{(3)} = (\sigma_{i-1}\lambda_{i-2},\ \lambda_{i-1},\ \lambda_i\sigma_i). \tag{10.31}$$

When applying $G^{(2)}$ connection, notice that $L_i^{(2)} = (\lambda_{i-1}, \lambda_i)$ does not change. We can consider that a stretch factor σ_i belongs to a vertex $\mathbf{q}_{i+1}$ of an S polygon of degree three. We can determine the locations of B polygons connecting in $G^{(2)}$ using $L_i'^{(3)}$ and $L_{i+1}'^{(3)}$ for the control points on $\mathbf{b}_i$ and on $\mathbf{b}_{i+1}$ instead of $L_i^{(3)}$ and $L_{i+1}^{(3)}$.

For example, consider the effects of changing values of σ on the shape of connected curve segments. In Fig. 10.7(a) where all $\sigma = 1$, the variation of radius of curvature (relative values) is shown along the curve. Its profile shows moderate shape change. In (b) where all $\sigma = 0.3$, the curve shape approaches that of the S polygon. The radius of curvature becomes small at the junctions. In (c) where all $\sigma = 1.3$, the curve segments approach a flat shape at their junctions. The effect of σ is to stretch the shape of connecting curves around their junction it belongs to.

In the literature [11][22], parameters called tension (β_2) and bias (β_1) are included in coefficients of the polynomial expressing a curve segment which connects in $G^{(2)}$. The curves including these parameters are called Beta splines. In deriving the expressions for Beta splines, automatic algebraic symbol manipulation by a computer was used. The bias corresponds to λ in this section, but the tension is a complicated function of geometric entities and is taken as a constant at all junctions. When it varies at each junction, the expression of the curve segment becomes a complex rational representation of high degree. Compared

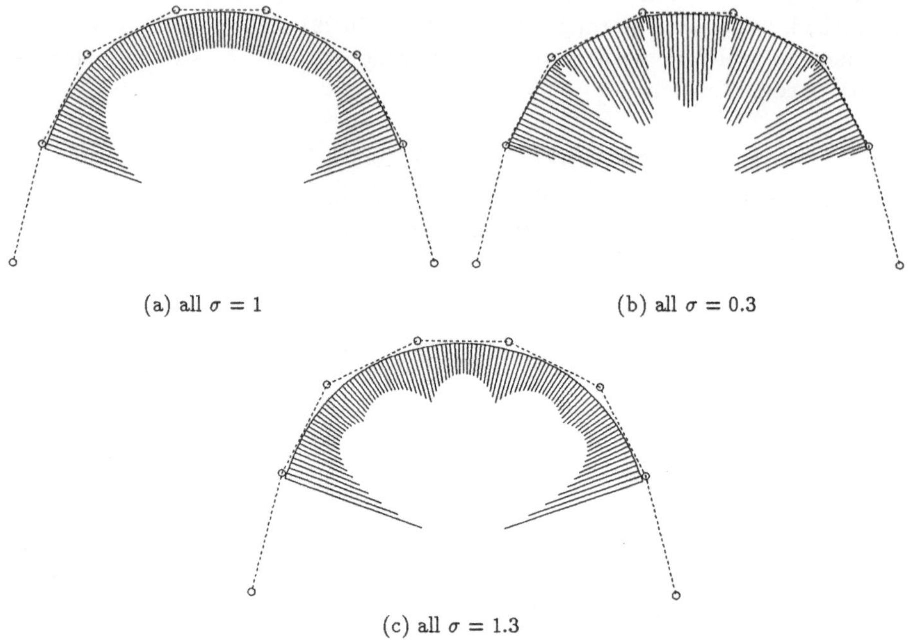

Figure 10.7 Effects of values of σ on shapes of connected curves. Patterns of radius of curvature are also shown

with the Beta spline, our expression of curves with $G^{(2)}$ connection is simple, that is, they are connected Bézier polygons of degree three, whose control point locations are easily determined from the S polygon with the modified dividing ratio lists of degree three. Effects of values of the stretch factor σ on the location of the control points are easily determined even without a computer.

10.5 Curvature Profile Problem in Design

10.5.1 Geometric Interpretation of Dividing Ratios

We explain in this section the meaning of the dividing ratio list and a practical method to determine the values in the list. We defined ratio of scale factors λ in Sect. 10.2 as

$$k_0 = \frac{\lambda_{-1}}{\lambda_0} = \frac{(dt)_0}{(dt)_{-1}} = \frac{\dot{\mathbf{r}}_{-1}(0;n)}{\dot{\mathbf{r}}_0(1;n)}. \tag{10.32}$$

This is the ratio of scales of the parameters of connecting curves at their junction or the ratio of absolute values of the tangent vectors there. If the value of this scale ratio is not equal to 1, the magnitudes of the tangent vectors of the connecting curves at the junction are not equal, also their derivatives of higher degree with respect to the parameter are not the same, though the geometrical quantities such as unit tangent directions, curvatures and torsions are continuous there.

When we want to construct shapes by connecting curve segments, it is difficult to determine their ratios exactly in advance. Fortunately their exact ratio values are not needed. We explain two methods of estimating k_0 at junctions.

1. When a curve segment $\mathbf{r}(t; n)$ or its B polygon is divided at $t = t^*$ into two curve segments $\mathbf{r}_{-1}$ and $\mathbf{r}_0$ or B_{-1} and B_0, eq. (10.32) holds exactly. The value of k_0 is equal to λ_{-1}/λ_0 . Accordingly, if the shape of two connecting curve segments can roughly be simulated by one curve segment, and its division at $t = t^*$ gives also an approximation to the desired connecting curve shape, then $t^*/(1-t^*)$ is a good estimate for the desired scale ratio.

2. The other method is to take k_0 equal to the ratio of their chord lengths. In this case the list of λ is proportional to the list of chord lengths of connecting curve segments. This is effective when the shapes of connected segments (more precisely their B polygons) are expected to be almost similar.

When we design a curve shape using an S polygon, we determine the λ_i to be all equal or proportional to expected chord lengths of the segments. The former is simple, but when there are considerable differences in the chord lengths of the generated curve segments, the distribution of radius of curvature is affected. Figure 10.8 shows them for the same S polygon: (a) profile of radius of curvature when the list of scale factors λ is (1,1,1,1,1,1,1), (b) that for λ list (1,1,1,0.3,1,1,1), (c) that for λ list (1,1,1,2,1,1,1). The region of small radius of curvature at the central part of the curve seems proportional to the value of the central element of the λ list. As we explain in the next section, this profile is important for evaluation of the shape quality of curves.

The shape of a curve generated from an S polygon is predictable, because the polygon represents its approximate shape, but the reverse does not always give predictable results. The former process is to get points $\mathbf{p}_i$ from points $\mathbf{q}_i$, the latter is its reverse. In the latter case, simultaneous equations for the vertices of an S polygon have to be solved. Curves generated from the S polygon show sometimes quite different shapes from those expected due to inappropriate selection of the dividing ratios. This is the same problem with the scale ratios discussed in Sect. 8.5. The equations to be solved and examples are given there.

We can have more freedom of shape design by applying the $G^{(2)}$ connection treated in Sect. 10.6 or the rational expressions explained in Chap. 12, but we do not yet have good research results on controlling the curvature profile along the curve.

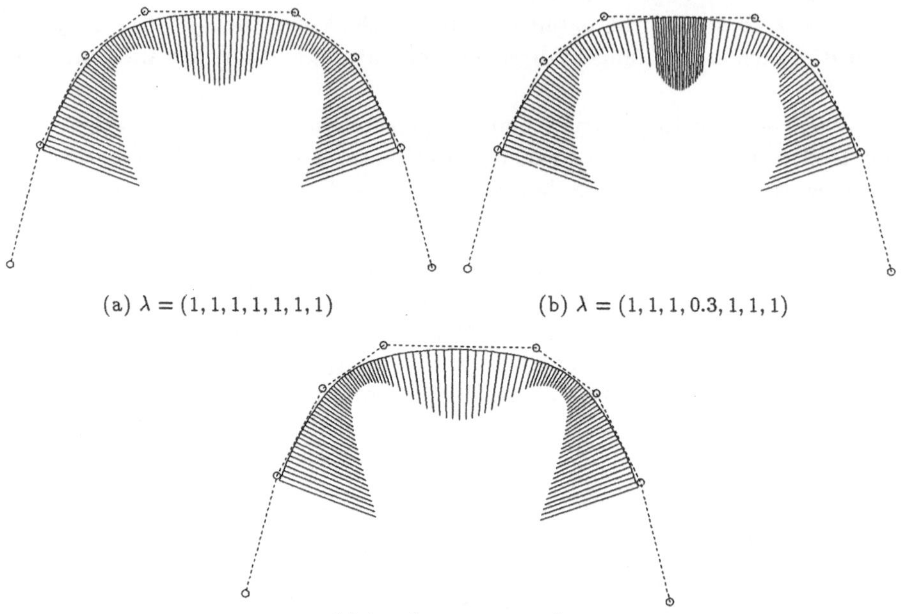

(a) $\lambda = (1, 1, 1, 1, 1, 1, 1)$

(b) $\lambda = (1, 1, 1, 0.3, 1, 1, 1)$

(c) $\lambda = (1, 1, 1, 2, 1, 1, 1)$

Figure 10.8 Effects of values of scale ratios on shapes of connected curves. Patterns of radius of curvature are also shown

10.5.2 Control of Curvature Distribution of Connected Curves

When connecting two curve segments in $C^{(2)}$ or in $G^{(2)}$, usually the shape of the distribution of curvature along the curve has not been treated. But when quality of curve shape has to be considered, the shape of the curvature profile becomes important. It is very sensitive to locations of control points. For example, Fig. 10.9(a) and (b) show the radius of curvature distribution for lists of λ ratios $(1, 1, 1, 1, 1, 1)$ and $(1, 1, 1, 0.5, 0.5, 1)$ for the same S polygon. The profiles are not smooth in either case, though they are continuous. As shown in Fig. 10.9(c), a slight displacement of one vertex changes the profile remarkably.

When we connect curve segments to synthesize a larger segment, frequently its profile of curvature distribution takes shapes as shown in Fig. 10.8 or Fig. 10.9. This type of connection is not always accepted by car style designers though the curvature is continuous. So we have to devise a new method to synthesize a curve shape with a good curvature profile by connecting two segments. At first we make a segment of degree five to adjust its control points to have a good curvature profile. Then we divide the segment at t^* to approximate each part by a segment of degree three keeping the scale ratio of the approximated segments to be nearly equal to $t^*/(1-t^*)$. When this process is performed well, we have

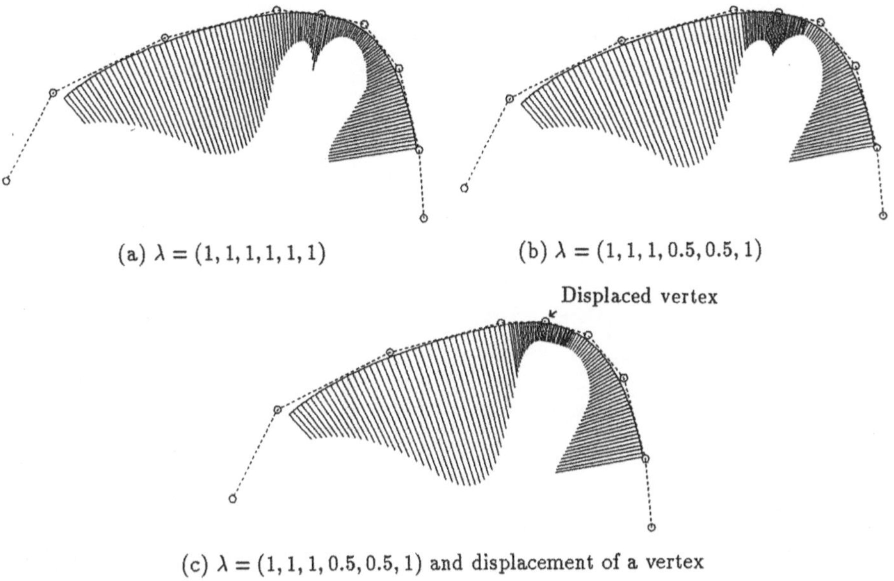

Figure 10.9 Profiles of radius curvature. Effects of variation of λ, and of displacement of one vertex

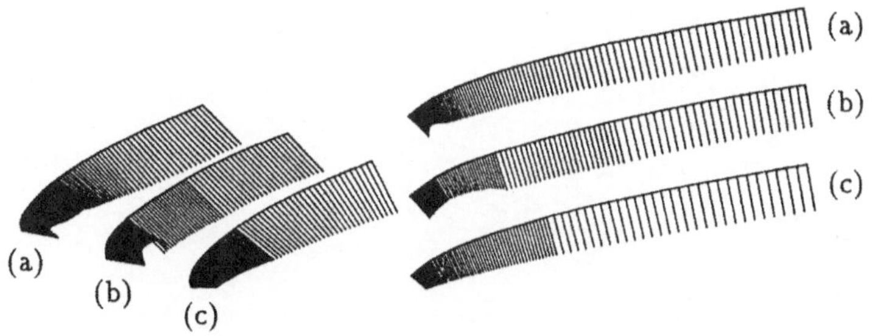

Figure 10.10 Selection of an optimum scale ratio at a junction. Profiles of radius of curvature(logarithmic values) are shown: (a) scale ratio, 1 : 1, (b) scale ratio = chord ratio, (c) scale ratio obtained from division of higher degree curve

a synthesized shape with a good profile of curvature.

Examples are shown in Fig. 10.10. In (a), the scale ratio of two segments is taken as 1, in (b) it is equal to the ratio of the chord lengths of the segments. The scale ratio in (c) is determined from the division of an approximated curve represented by a Bézier curve segment of degree five. As for their shapes, cases of (a) and (b) are not considered good, but (c) is evaluated as good by car body stylists. Evaluation of surface quality is more difficult; its methods are described in Chap. 7 and in Chap. 17.

11 Connection of Bézier Patches and Geometry of Spline Polygons and Nets

11.1 Introduction

In Sect. 11.2 of this chapter, we treat surfaces which consist of Bézier patches connecting in $C^{(n-1)}$ in all directions. Techniques for obtaining B polygons from an S polygon are extended to those for generating nets of Bézier control points from a given spline net. For simplicity of mathematical manipulation and programming, we adopt a tensor product technique for producing Bézier nets. Connection in $G^{(2)}$ and insertion of vertices of an S net are also included in the first section. In the next three sections, we explain the theoretical aspects of the S polygon, which are the bases of the practical methods in the previous chapter.

Properties of S polygons are discussed using Menelaus' theorem in the primary geometry. Then we deduce two general algorithms to determine B polygons from an S polygon and a method of inserting a vertex in an S polygon. We introduce a general formula for locating B points from a given S polygon of degree n. When n is equal to or lower than 3, methods of locating B points from an S polygon are well known, but in higher degree such methods are rarely found. So we explain our method, which can be applied for arbitrary n. In this way we treat the theory of connection and control in curves and surfaces using our knowledge of Bézier curves and patches as much as possible. Finally in Sect. 11.7, we explain an orthodox approach to the B spline curve, using the B spline basis functions.

11.2 Spline Nets and Connected Bézier Nets

11.2.1 Tensor Product Surfaces

Just as a row of B polygons of degree n connected in $C^{(n-1)}$ can be generated from an S polygon with its dividing ratio lists $L_i^{(n)}$ of degree n, an array of $C^{(n-1)}$ connected B nets of degree $n \times n$ can be generated from an S net with its dividing ratio lists in u and v directions $Lu_i^{(n)}$ and $Lv_j^{(n)}$ of degree n. See Fig. 11.1.

Let mesh points or vertices of the S net be $\mathbf{q}_{ij}$. Their first subscript i belongs to the u direction and the second subscript j to the v direction. In figures, the u direction is taken vertically and the v direction horizontally. This coincides with the order of subscripts of elements in a matrix. When we treat a matrix of vertices of a unit S net or the corresponding matrix of B points of a complete B net derived from the unit S net, we call them an S matrix or a B matrix for simplicity.

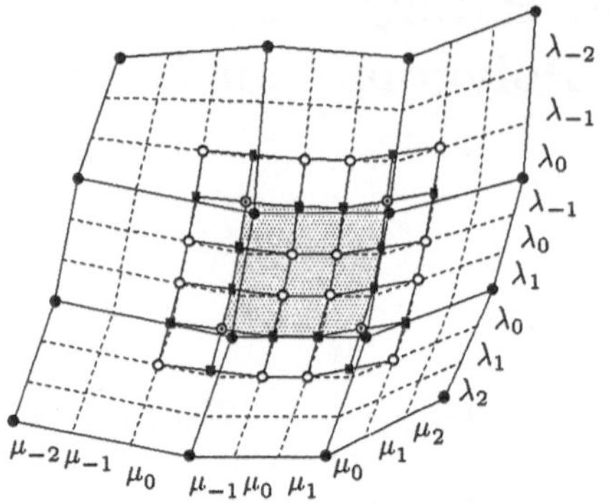

Figure 11.1 A unit S net and its derived B net

In an S polygon, its basic element is an edge with its dividing ratio list, while in an S net, its basic element is a ruled surface of degree 1×1 with its u and v dividing ratio lists. The ruled surface is defined by four vertices of a mesh. Dividing points of edges in an S polygon correspond to dividing lines in the v and u directions, which coincide with generating lines of the ruled surface in the v and u directions. The dotted lines in Fig. 11.1 are the dividing lines. Dividing ratio lists $Lu_i^{(n)}$ of the dividing lines in v direction are common to all the ruled surfaces in the v direction (row), and $Lv_j^{(n)}$ of the dividing lines in the u direction are common to all the ruled surfaces in the u direction (column). Accordingly, the dividing lines in the v direction divide the dividing lines in the u direction in the same dividing ratios as given by $Lu_i^{(n)}$ and vice versa.

Now, we explain a tensor product method for determining locations of elements of a B matrix from a given unit S matrix.

1. Convert each column of a unit S matrix with the common dividing ratio list $Lu_i^{(n)}$ into the corresponding column of the B polygons, then the S matrix is changed to an intermediate matrix. If we make a surface from the control points defined by this matrix, it consists of Bézier patches of degree $n \times 1$ connected in the u direction in $C^{(n-1)}$ and in the v direction in $C^{(0)}$. The $u = const$ lines of this composite patches make polygons in the row direction and can be considered S polygons with the common dividing ratio list $Lv_j^{(n)}$.
2. Convert each row of the intermediate matrix with the common $Lv_j^{(n)}$ into the corresponding complete B polygons, then we have a matrix of vertices of a complete B net, from which we can produce a B patch. This B patch connects with its adjacent B patches of the same kind in $C^{(n-1)}$.

Example. Let a unit S matrix of $\mathbf{s}_{00}(u, v; 3, 3)$ be

$$\begin{bmatrix} \mathbf{q}_{00} & \mathbf{q}_{01} & \mathbf{q}_{02} & \mathbf{q}_{03} \\ \mathbf{q}_{10} & \mathbf{q}_{11} & \mathbf{q}_{12} & \mathbf{q}_{13} \\ \mathbf{q}_{20} & \mathbf{q}_{21} & \mathbf{q}_{22} & \mathbf{q}_{23} \\ \mathbf{q}_{30} & \mathbf{q}_{31} & \mathbf{q}_{32} & \mathbf{q}_{33} \end{bmatrix} . \tag{11.1}$$

Converting each column of the above array with the dividing ratio lists

$$Lu_i^{(3)} = (\ \lambda_{-2+i},\ \ \lambda_{-1+i},\ \ \lambda_i\ ), \qquad i = 0, 1, 2, \tag{11.2}$$

into a B polygon of degree three, we obtain the intermediate matrix. Let it be

$$\begin{bmatrix} \mathbf{q}'_{00} & \mathbf{q}'_{01} & \mathbf{q}'_{02} & \mathbf{q}'_{03} \\ \mathbf{q}'_{10} & \mathbf{q}'_{11} & \mathbf{q}'_{12} & \mathbf{q}'_{13} \\ \mathbf{q}'_{20} & \mathbf{q}'_{21} & \mathbf{q}'_{22} & \mathbf{q}'_{23} \\ \mathbf{q}'_{30} & \mathbf{q}'_{31} & \mathbf{q}'_{32} & \mathbf{q}'_{33} \end{bmatrix}, \tag{11.3}$$

where the elements of each column of the above matrix are obtained according to the location symbols used in the item (3) of Sect. 10.3. as follows,

$$\begin{aligned} \mathbf{q}'_{0j} &= (\mathbf{q}_{0.j})_{.2.1}, \\ \mathbf{q}'_{1j} &= (\mathbf{q}_{1.j})_{.1}, \\ \mathbf{q}'_{2j} &= (\mathbf{q}_{1.j})_{.2}, \\ \mathbf{q}'_{3j} &= (\mathbf{q}_{1.j})_{.2.1}, \quad j = 0, \ldots 3. \end{aligned}$$

Then, considering each row of the above intermediate matrix as a unit S polygon with the dividing ratio lists

$$Lv_j^{(3)} = (\ \mu_{-2+j},\ \ \mu_{-1+j},\ \ \mu_j\ ), \qquad j = 0, 1, 2,$$

and converting it into a B polygon of degree three, we obtain a B point matrix

$$\begin{bmatrix} \mathbf{p}_{00} & \mathbf{p}_{01} & \mathbf{p}_{02} & \mathbf{p}_{03} \\ \mathbf{p}_{10} & \mathbf{p}_{11} & \mathbf{p}_{12} & \mathbf{p}_{13} \\ \mathbf{p}_{20} & \mathbf{p}_{21} & \mathbf{p}_{22} & \mathbf{p}_{23} \\ \mathbf{p}_{30} & \mathbf{p}_{31} & \mathbf{p}_{32} & \mathbf{p}_{33} \end{bmatrix},$$

where locations of the B polygon for each row of the above matrix are given by

$$\begin{aligned} \mathbf{p}_{i0} &= (\mathbf{q}'_{i.0})_{.2.1}, \\ \mathbf{p}_{i1} &= (\mathbf{q}'_{i,1})_{.1}, \\ \mathbf{p}_{i2} &= (\mathbf{q}'_{i.1})_{.2}, \\ \mathbf{p}_{i3} &= (\mathbf{q}'_{i.1})_{.2.1}, \quad i = 0, \ldots 3. \end{aligned}$$

The scale of the B net thus determined is $\lambda_i \times \mu_j$. Figure 11.1 shows a unit S net of degree three and its derived B nets.

For a connection of $G^{(2)}$ the dividing ratio lists $L'u_i^{(3)}$ and $L'v_j^{(3)}$ should be used instead of $Lu_i^{(3)}$ and $Lv_j^{(3)}$ (refer to eq. (10.31)) where $L'u_i^{(3)}$ and $L'v_j^{(3)}$ are the dividing ratio lists in the u and v directions for $G^{(2)}$ connection.

For a higher-degree S net, the conversion process is similar to the above. The order of the conversion procedures, for columns or rows, does not affect the resulting surface.

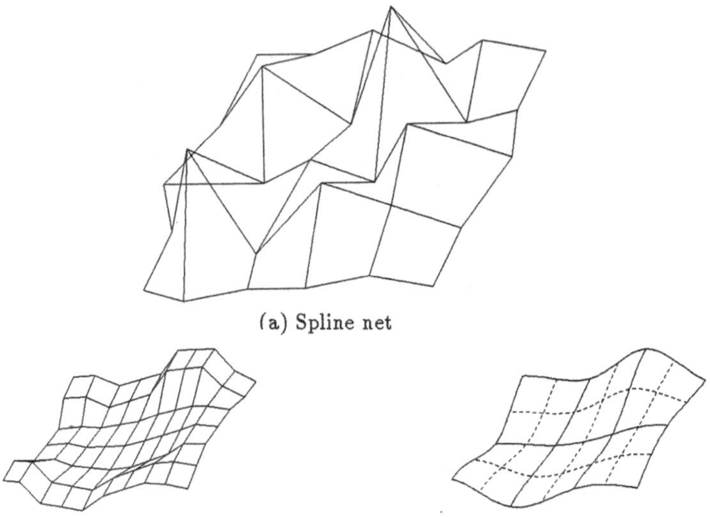

(a) Spline net

(b) Derived B nets

(c) B patches

Figure 11.2 Shape of an S net and its derived B nets and six generated patches connected in $C^{(2)}$

In Fig. 11.2, an S net, its generating array of B nets and the shape of its surface are shown in the same scale. For the S net, the shape of the generated surface is globally similar, but its detail is quite different, whereas for the array of B nets, the shapes of both the generated surface and the B nets are visually almost the same, and their true shapes and geometric properties can easily be calculated. In the S net, shape modification by displacements of their vertices is possible, but in the B nets it is not allowed. Accordingly, in the design of surface shape, representations and displays of both the S net and the B net should be used interchangeably.

11.2.2 Division of an S Net

To divide an S net or to increase its vertices, we use a method similar to one used in derived B nets from a S net.

For each column of an S net, we make new vertices in all sub-edges corresponding to a certain scale factor, for instance λ_0, by dividing them in the ratio $(u : 1-u)$. We replace the columns of the old vertices by the new ones. This procedure corresponds to the division of an S polygon. The number of rows increases by one. For each row of this new net, we make new vertices in all sub-edges corresponding to a certain scale factor, for instance μ_0, by dividing them in the ratio $(v : 1-v)$ and replace the old vertices by the new ones.

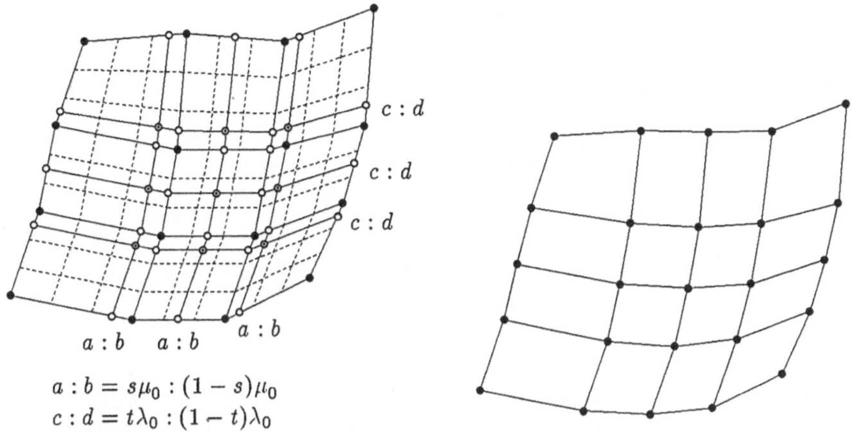

(a) Locations of new vertices

(b) New S net with increased vertices

Figure 11.3 Division of S net

This is the increase of vertices of an S net. A B net corresponding to $\lambda_0 \times \mu_0$ is divided into four B nets corresponding to $\lambda_0 u \times \mu_0 v$, $\lambda_0(1-u) \times \mu_0 v$, $\lambda_0 u \times \mu_0(1-v)$ and $\lambda_0(1-u) \times \mu_0(1-v)$ connected in $C^{(n-1)}$.

An example of insertion of control points in an S net is shown in Fig. 11.3. The initial 4×4 vertices of an S net of degree 3×3 are increased to 5×5 vertices. In Fig. 11.3(a), the original unit S net is shown together with the dividing lines in the u and v directions and locations of the new vertices. Figure 11.3(b) shows the resulting S net with increased vertices.

11.3 Geometric Structure of S Polygons

In the upper part of Fig. 11.4, an S polygon with its vertices $\mathbf{q}_0$ $\mathbf{q}_1$ $\mathbf{q}_2$ $\mathbf{q}_3$ is schematically shown in a straight line with the vertices indicated by small blank circles. In the following explanation, any number n is used, but if n is taken equal to three, the schematic diagrams in Fig. 11.4 become helpful for understanding. In Fig. 11.5 the dividing ratios and sub-edges are shown for n=4.

First, we assume that each of n edge vectors of a B polygon B_0 of degree n be constructed by a linear combination of consecutive n edge vectors of an S polygon and the similar B polygon B_1 be determined from the one-edge right-shifted n consecutive edges of the same S polygon, and that B_0 and B_1 connect in $C^{(n-1)}$, which is guaranteed by the Connection Defining Polygons (CDPs) explained in Sect. 10.2.4. Since we assume the degree of B polygons B_0 and B_1 is n, and that of their constraining CDP $\tilde{B}_{0,1}$ is $n-1$, the degree of freedom of

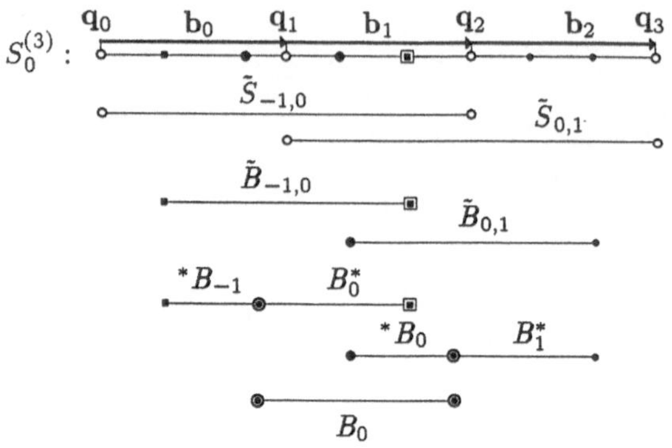

Figure 11.4 Schematic diagram for relations between S polygon and its various derived polygons: a unit S polygon, truncated S polygons, CDPs, truncated B polygons and a B polygon

the edge vectors of the connected B_0 and B_1 is $2n - (n-1) = n+1$.

Accordingly, we take consecutive $n+1$ edges of the S polygon, from which $C^{(n-1)}$ connected B polygons B_0 and B_1, can be constructed. We consider that in these consecutive $n+1$ edges of the S polygon, the first n edges produce B_0 with scale λ_0 and the last n edges produce B_1 with scale λ_1. There are $n-1$ overlapped edges. Then it is appropriate to assume that a CDP $\tilde{B}_{0,1}$ with scale $(\lambda_0 + \lambda_1)$ is generated from their common overlapped $n-1$ edges. Since the S polygon extends in both directions indefinitely, any consecutive $n-1$ edges of the S polygon can generate one CDP.

Let edges of the S polygon be $\cdots, \mathbf{b}_0, \mathbf{b}_1, \mathbf{b}_2, \cdots$, and a B polygon B_0 be generated from consecutive n edges $(\mathbf{b}_0 \cdots \mathbf{b}_{n-1})$ which begin at a vertex $\mathbf{q}_0$. Then, the S polygon must have the following properties:

1. A list of scale factors $(\lambda_{i-n+1}, \lambda_{i-n+2}, \cdots, \lambda_i)$ of n elements belongs to each edge $\mathbf{b}_i$ of the S polygon. The index of the last element of the list is the same as that of the edge to which it belongs. We call this list the *dividing ratio list* of the edge $\mathbf{b}_i$ and denote it by

$$L_i^{(n)} = (\lambda_{-n+i+1}, \cdots, \lambda_i). \tag{11.4}$$

To assign an edge $\mathbf{b}_i$ the data of its dividing ratio list, we divide the edge into n sub-edges in the ratio proportional to each element of $L_i^{(n)}$ given by (11.4). For simplicity, we call this process "divide $\mathbf{b}_i$ by $L_i^{(n)}$" or "dividing ratio of $\mathbf{b}_i$ is $L_i^{(n)}$". A "sub-edge of λ_j" is a sub-edge which is a λ_j part of an edge divided by $L_i^{(n)}$.

The eq.(11.4) is derived as follows. Referring to Fig. 11.5, since a polygon B_0 whose scale is λ_0 is constructed from n consecutive edges $(\mathbf{b}_0 \cdots \mathbf{b}_{n-1})$

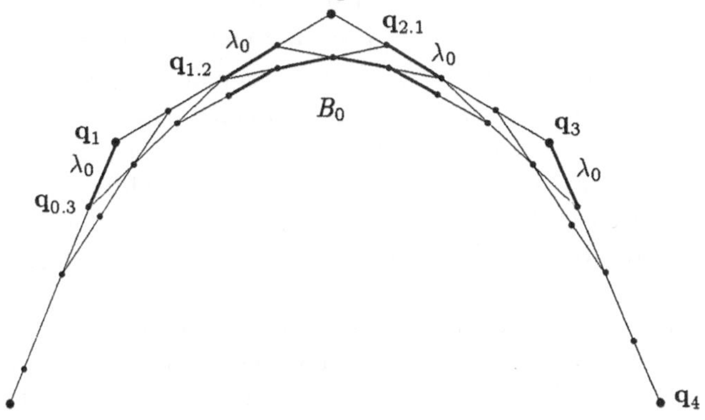

Figure 11.5 Dividing ratios and sub-edges

of the S polygon, we consider that the polygon B_0 is constructed from all the λ_0 sub-edges, each of which belongs to each of the above consecutive n edges of the S polygon, but not from the other edges.
Its adjacent polygon B_{-1} is produced by λ_{-1} sub-edges from each of n edges $(\mathbf{b}_{-1} \cdots \mathbf{b}_{n-2})$, and similar relations continue until $B_{-(n-1)}$ is constructed by $\lambda_{-(n-1)}$ sub-edges from each of n edges $(\mathbf{b}_{-(n-1)} \cdots \mathbf{b}_0)$. Therefore, the edge $\mathbf{b}_0$ influences the n B polygons $(B_{-(n-1)} \cdots B_0)$, that is, the edge $\mathbf{b}_0$ has the dividing ratio list $L_0^{(n)}$, which contains only n elements of λ from λ_{-n+1} to λ_0.
We call a set of n consecutive edges $(\mathbf{b}_i \cdots \mathbf{b}_{i+n-1})$ beginning at a vertex $\mathbf{q}_i$ a *unit S polygon* $S_i^{(n)}$ and its scale λ_i *the main scale* of $S_i^{(n)}$, which is the common element of the $L_j^{(n)}$ of all its n edges for $j = i, \cdots (i+n-1)$. The index of the main scale is the same as that of the unit S polygon.

2. The common part of adjacent unit S polygons $S_{i-1}^{(n)}$ and $S_i^{(n)}$ has $n-1$ consecutive edges and common scales λ_{i-1} and λ_i. So the common part can be considered a unit S polygon of degree $n-1$ with the main scale $(\lambda_{i-1} + \lambda_i)$. We use a notation $\tilde{S}_{i-1,i}$ for it and call it the reduced S polygon of second kind. It produces a *CDP* $\tilde{B}_{i-1,i}$. The reduced S polygon of the first kind appears in the farther section.
3. Referring to Fig. 11.4 again, a unit S polygon $S_i^{(n)}$ generates three B polygons ${}^*B_{i-1}, B_i, B_{i+1}^*$ of degree n, the central one complete and the other two incomplete. They connect in $C^{(n-1)}$ and their scales are λ_{i-1}, λ_i and λ_{i+1}.
This is proved as follows. A unit S polygon $S_i^{(n)}$ is looked upon as union of two common parts: the common edges of $\{S_{i-1}^{(n)}$ and $S_i^{(n)}\}$, which make $\tilde{S}_{i-1,i}$, and the common edges of $\{S_i^{(n)}$ and $S_{i+1}^{(n)}\}$, which form $\tilde{S}_{i,i+1}$. The reduced S polygon $\tilde{S}_{i-1,i}$ generates a *CDP* $\tilde{B}_{i-1,i}$, whose division in the ratio $(\lambda_{i-1} : \lambda_i)$

gives incomplete B polygons $\{{}^*B_{i-1}, B_i^*\}$ and similarly from $\tilde{S}_{i,i+1}$, a *CDP* $\tilde{B}_{i,i+1}$ is obtained, whose division by ratio $(\lambda_0 : \lambda_1)$ produces $\{{}^*B_i, B_{i+1}^*\}$. Their union is ${}^*B_{i-1}, B_i, B_{i+1}^*$. So the statement is correct.

4. The common edges $(\mathbf{b}_1 \cdots \mathbf{b}_{n-2})$ of three unit S polygons $S_{-1}^{(n)}, S_0^{(n)}$ and $S_1^{(n)}$ constitute also a unit S polygon of degree $(n-2)$ with the main scale $(\lambda_{-1} + \lambda_0 + \lambda_1)$. This is also the common part of $\tilde{S}_{-1,0}$ and $\tilde{S}_{0,1}$, which can generate two adjacent *CDPs* $\tilde{B}_{-1,0}$ and $\tilde{B}_{0,1}$. Their common part is the both-end truncated incomplete B polygon ${}^*B_0^*$. So we use a notation ${}^*S_0^*$ for this truncated S polygon, whose main scale is $(\lambda_{-1} + \lambda_0 + \lambda_1)$ and the degree is $(n-2)$.
If we can deduce from this ${}^*S_0^*$ a unit S polygon of degree $(n-2)$ with the main scale λ_0, we can easily determine the locations of the B points from an S polygon of degree n, because the degree of the given S polygon can be reduced rapidly with the same main scale, and a unit S polygon of degree one or zero has the same vertices or vertex as a B polygon of the same degree. Examples of the method are explained in Sect. 11.4.3.

11.4 Menelaus Edges and Their Dividing Points

To determine B polygons connected in $C^{(n-1)}$ from an S polygon, we apply Menelaus' theorem (refer to Sect. 3.3) to the geometric structure of the S polygon, which we explain in this section.

11.4.1 Dividing Points and Sub-edges

The dividing ratios lists of all edges of the unit S polygon $S_0^{(n)}$ are $\{L_0^{(n)}, \cdots, L_{n-1}^{(n)}\}$, whose common element is λ_0. As an edge of the polygon advances to the right, the relative position of a sub-edge of λ_0 in each edge shifts by one position leftwards from the right-end position of the first edge to the left-end position of the last edge. See Fig. 11.5. Except for these consecutive n edges there are no sub-edges of λ_0 in the S polygon.

We divide an edge $\mathbf{b}_0$ by the ratio $L_0^{(n)}$ and the next edge $\mathbf{b}_1$ by $L_1^{(n)}$ each into n sub-edges and let the dividing points of $\mathbf{b}_0$ be $(\mathbf{q}_{0.1}\ \mathbf{q}_{0.2} \cdots \mathbf{q}_{0.n-1})$ and those of $\mathbf{b}_1$ be $(\mathbf{q}_{1.1}\ \mathbf{q}_{1.2} \cdots \mathbf{q}_{1.n-1})$.

11.4.2 Relations Among Dividing Points and Menelaus Edges

We define Menelaus edges which are used in the process of locating B points from an S polygon and explain their properties. Owing to the division of edges into sub-edges by the scale lists and the shifting of the dividing ratios in the consecutive edges, we can apply Menelaus' theorem (Sect. 3.3) to each pair of adjacent edges.

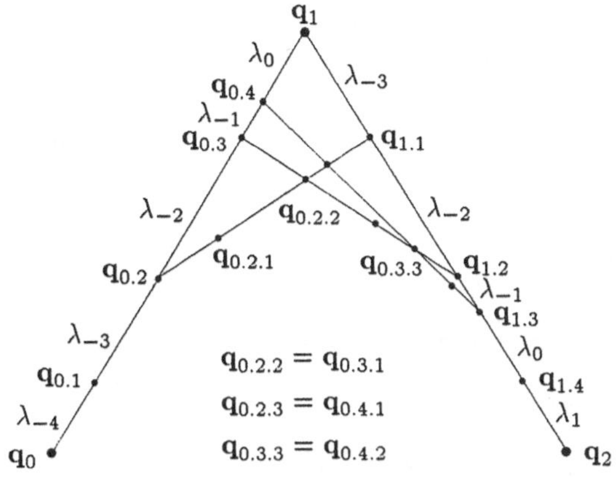

Figure 11.6 Dividing points and Menelaus edge

Refer to Fig. 11.6. We select one dividing point $\mathbf{q}_{0.i}$ on the edge $\mathbf{b}_0 = \mathbf{q}_0\mathbf{q}_1$ and the other point $\mathbf{q}_{1.i-1}$ on the adjacent edge $\mathbf{b}_1 = \mathbf{q}_1\mathbf{q}_2$. The subscripts of $\mathbf{q}_{i.i-1}$ are determined by increasing the first index and decreasing the second index of $\mathbf{q}_{0.i}$ each by one. We call this indexing operation *skew shifting.* We divide a line segment joining $\mathbf{q}_{0.i}$ and its skew shifted $\mathbf{q}_{1.i-1}$, in the ratio $L_0^{(n-1)}$ into $n-1$ sub-edges at the dividing points $(\mathbf{q}_{0.i.1}\ \mathbf{q}_{0.i.2} \cdots \mathbf{q}_{0.i.n-2})$. In each symbol of the dividing points of this array, the first two index numbers indicate the index of the starting point of this line segment which starts at $\mathbf{q}_{0.i}$ and ends at $\mathbf{q}_{1.i-1}$ and the third one indicates the sequence number of the dividing points.

The following relations hold for the dividing points.

(i) Three points $\mathbf{q}_{0.j}, \mathbf{q}_{0.i.j-1}$ and $\mathbf{q}_{1.j-1}$ are collinear. When two of the three point are determined first, a line joining these two points or its extension intersects a line, on which the third point is to be placed. This intersecting point gives the location of the third point.

(ii) Two points, whose location symbols are $\mathbf{q}_{0.i.j}$ and $\mathbf{q}_{0.i+k.j-k}$ are the same point if $j \neq i-1$ and $k < n-i$.

These two relations are easily proved by application of Menelaus' theorem and its inverse (refer to Sect. 3.3).

A line segment, which joins the points $\mathbf{q}_{0.j}$ and $\mathbf{q}_{1.j-1}$ each on the adjacent edges and have $n-2$ dividing points by the ratio $L_0^{(n-1)}$, appears frequently. For convenience, we call this type of line segment the *Menelaus edge*, which subtends an angle at the vertex $\mathbf{q}_1$.

Any one of the Menelaus edges which subtends an angle made by the two edges $\mathbf{b}_i$ and $\mathbf{b}_{i+1}$ can replace the parts of these two edges cut off by the Menelaus edge on their vertex side, because it preserves geometric properties of the cut-off parts by the properties stated above in (i) and (ii).

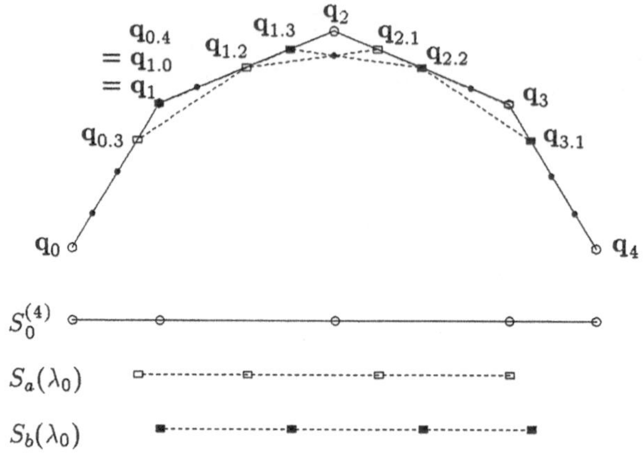

Figure 11.7 Relation between a unit S polygon $S_0^{(4)}$ and its reduced S polygons $S_a(\lambda_0), S_b(\lambda_0)$

11.5 Derivation of B Polygons from an S Polygon

We explain procedures or formulas which can determine the locations of B points from a given S polygon with its scale list. First we introduce auxiliary polygons, which we call reduced S polygons, derived from the given S polygon. With the help of these polygons, we can locate the B points as well as insert a new vertex in the given S polygon.

11.5.1 Reduced S Polygons

A unit S polygon of degree n is resolved into unit S polygons of degree $n-1$, which can generate all the B points belonging to the original S polygon. These resolved S polygons are called the *reduced* S polygons of a unit S polygon $S_0^{(n)}$, for convenience. There are two kinds of the reduced S polygon. One is used for determining the locations of B polygons of degree n from the unit S polygons of degree $n-1$. The other is directly related to a CDP as stated in item 3 of Sect. 11.3 and also used for insertion of new vertices to increase freedom of the shape design.

1. Refer to Fig. 11.7. On each edge of $S_0^{(n)}$ there exists a sub-edge corresponding to the main scale λ_0 . On the first edge $\mathbf{b}_0$, two dividing points $\mathbf{q}_{0.n-1}$ and $\mathbf{q}_{0.n}$ $\{= \mathbf{q}_1\}$ are the end points of this sub-edges. We construct two polygons S_a and S_b of n vertices by repeated skew shifting from $\mathbf{q}_{0.n-1}$ and $\mathbf{q}_{0.n}$, which are the starting points.

$$\left.\begin{array}{llllll} S_a: & \mathbf{q}_{0.n-1} & \mathbf{q}_{1.n-2} & \cdots & \mathbf{q}_{n-2.1} & \mathbf{q}_{n-1.0} \\ S_b: & \mathbf{q}_{0.n} & \mathbf{q}_{1.n-1} & \cdots & \mathbf{q}_{n-2.2} & \mathbf{q}_{n-1.1} \end{array}\right\} \tag{11.5}$$

where $\mathbf{q}_{n-1.0} = \mathbf{q}_{n-1}$ and $\mathbf{q}_{0.n} = \mathbf{q}_1$.
The degree of S_a and S_b is $n-1$ and their corresponding edges intersect owing to their construction. The intersection of these two polygons determines a both-end-truncated B polygon. Proof of this and use of this polygon are described in Sect. 11.5.3.

2. We divide each of $n-1$ edges of S_a into $n-1$ sub-edges by the dividing ratios $L_0^{(n-1)}, \cdots, L_{n-1}^{(n-1)}$. For S_b we apply the same procedure. Then the two polygons are the unit S polygons of degree $n-1$ with the main scale λ_0, so notations $S_a(\lambda_0)$ and $S_b(\lambda_0)$ are used for them.
3. Each edge of $S_a(\lambda_0)$ and $S_b(\lambda_0)$ becomes a Menelaus edge which subtends a vertex of $S_0^{(n)}$. Hence, edges either of $S_a(\lambda_0)$ or of $S_b(\lambda_0)$ can be used as substitutes of parts of the edges of $S_0^{(n)}$ which are cut off by these Menelaus edges.
   Since $S_a(\lambda_0)$ and $S_b(\lambda_0)$ hold all the geometric data of the sub-edges corresponding to λ_0, they can generate $B_0^{(n)}$ which is originally to be generated from $S_0^{(n)}$. Since the last edge of $S_0^{(n)}$ does not belong to edges of $S_a(\lambda_0)$ and the first edge of $S_0^{(n)}$ is also not included in $S_b(\lambda_0)$, we can conclude:
   $S_a(\lambda_0)$ can generate an incomplete B polygon $B_0^{*(n)}$ and $S_b(\lambda_0)$ can generate ${}^*B_0^{(n)}$.
   It follows that if we can determine locations of a complete B polygon of degree $n-1$ from a unit S polygons of degree $n-1$, we can determine a complete B polygon of degree n from the two incomplete B polygons of degree n. Those incomplete B polygons are derived from the two reduced S polygons of degree $n-1$.
   Since for degree 1, a unit S polygon and the complete B polygon are the same, in the case of $n = 1$ the above statement holds, so it holds by induction for general n.
4. In $S_0^{(n)}$, there are still two unit S polygons of degree $n-1$ other than those stated above. One is a polygon with vertices whose first index values are one less than those given by S_b in (11.5). The locations of its vertices are:
$$\mathbf{q}_0 \quad \mathbf{q}_{0.n-1} \quad \cdots \quad \mathbf{q}_{n-2.1} \tag{11.6}$$
   Since the main scale of this unit S polygon (11.6) of degree $n-1$ is λ_{-1}, we can use the notation $S_b(\lambda_{-1})$ for this polygon, which can generate $B_{-1}^{*(n)}$.
5. Similarly, a unit S polygon of degree $n-1$ with vertices whose first index values are one more than those given by S_a in (11.5) is:
$$\mathbf{q}_{1.n-1} \quad \cdots \quad \mathbf{q}_{n-1.1} \quad \mathbf{q}_{n.0}. \tag{11.7}$$
   It has the main scale λ_1 with the right-end point coinciding with that of $S_0^{(n)}$. This polygon is denoted by $S_a(\lambda_1)$ and generates ${}^*B_1(n)$.
   Thus $S_b(\lambda_{-1}), S_a(\lambda_0), S_b(\lambda_0), S_a(\lambda_1)$ are the reduced S polygons of the first kind, and from them four incomplete B polygons ${}^*B_{-1}, B_0^*, {}^*B_0, B_1^*$ can be produced. The union of $B_0^*, {}^*B_0$ is the complete B polygon $B_0^{(n)}$.

6. Besides the reduced S polygons stated above, from item (3) of Sect. 11.3.1 there are another two reduced S polygons of the second kind: $\tilde{S}_{-1,0}$ and $\tilde{S}_{0,1}$ of degree $n-1$, which are deduced from $S_0^{(n)}$. They generate two *CDPs* $\tilde{B}_{-1,0}$ and $\tilde{B}_{0,1}$.
7. Division of $\tilde{B}_{-1,0}$ gives ${}^*B_{-1}$ and B_0^*, which are also obtained from $S_b(\lambda_{-1})$ and $S_a(\lambda_0)$. Similarly, division of $\tilde{B}_{0,1}$ gives ${}^*B_0$ and B_1^* which are also generated from $S_b(\lambda_0)$ and $S_a(\lambda_1)$. Accordingly, $\tilde{S}_{-1,0}$ and $\tilde{S}_{0,1}$ are the reduced S polygons of $S_0^{(n)}$. They are of the second kind. The B polygons B_{-1}, B_0 and B_1 are connected in $C^{(n-1)}$. From (6) and (7), we can prove that a procedure of division of an S polygon, or insertion of a vertex in an S polygon, does not change the shape of the original curve. This is explained in Sect. 11.5.4.

Thus, all the conditions for S polygons given in Sect. 11.3.1 are satisfied without contradiction.

11.5.2 Examples

We apply the rule in item (3) of Sect. 11.5.1 for locating B polygons from a unit S polygon.
For n=1, it is evident that $B_0^{(1)} = (\mathbf{p}_0\ \mathbf{p}_1) = (\mathbf{q}_0\ \mathbf{q}_1)$ is the same as $S_0^{(1)} = (\mathbf{q}_0\ \mathbf{q}_1)$.
For n=2, a unit S polygon of degree two is $S_0^{(2)} = (\mathbf{q}_0\ \mathbf{q}_1\ \mathbf{q}_2)$ whose dividing ratios of two edges are $(\lambda_{-1} : \lambda_0)$ and $(\lambda_0 : \lambda_1)$. Its reduced S polygons are given by

$$S_a^{(1)}(\lambda_0) = (\mathbf{q}_{0.1}\ \mathbf{q}_1), \qquad S_b^{(1)}(\lambda_0) = (\mathbf{q}_1\ \mathbf{q}_{1.1}). \tag{11.8}$$

The former gives the location of $B_0^{*(2)}$ and the latter determines that of ${}^*B_0^{(2)}$, because they are actually of degree one and the result of the case of n=1 is directly applied. Accordingly we have

$$B_0^{(2)} = (\mathbf{p}_0\ \mathbf{p}_1\ \mathbf{p}_2) = (\mathbf{q}_{0.1}\ \mathbf{q}_1\ \mathbf{q}_{1.1}). \tag{11.9}$$

Similarly we obtain from the reduced S polygons:

$$ {}^*B_{-1}^{(2)} = S_b^{(1)}(\lambda_{-1}) = (\mathbf{q}_0\ \mathbf{q}_{0.1}) \tag{11.10}$$

$$B_1^{*(2)} = S_a^{(1)}(\lambda_1) = (\mathbf{q}_{1.1}\ \mathbf{q}_2). \tag{11.11}$$

By incrementing the first index of the dividing points of $B_0^{(2)}$ and $S_0^{(2)}$, we have the right-hand adjacent polygon $B_1^{(2)}$:

$$B_1^{(2)} = (\mathbf{q}_{1.1}\ \mathbf{q}_2\ \mathbf{q}_{2.1}), \qquad S_1^{(2)} = (\mathbf{q}_1\ \mathbf{q}_2\ \mathbf{q}_3) \tag{11.12}$$

This step is more thoroughly treated in Sect. 11.6.

11.5.3 Locating B Polygons from Reduced-Truncated S Polygons

If a both-end-incomplete B polygon ${}^*B_i^*$ is simply obtained from a truncated polygon ${}^*S_i^*$, it is very useful for determining B points of higher degree from the corresponding S polygon, because polygons ${}^*B_i^*$ are sufficient for obtaining all the locations of B points from the given S polygon. Though ${}^*S_i^*$ is of degree $n-2$, its main scale is not λ_i. Accordingly, it cannot easily be used for repeated reduction of S polygons of higher degree to those of lower degree of the same main scale.

The objective of this section is to deduce an S polygon of lower degree with the same main scale from a given S polygon of higher degree. Without losing generality, we set $i = 0$ and use a notation $\check{S}_0^{(n-2)}$ and $\check{L}_0^{(n-2)}$ for such an S polygon and its dividing ratio of the first edge. We call this polygon $\check{S}_0^{(n-2)}$ a reduced-truncated S polygon. This polygon was already used in Sect. 10.4.1.

Since $S_a(\lambda_0)$ and $S_b(\lambda_0)$ generate B_0^* and ${}^*B_0$, if we can produce an equivalent S polygon with the common part of $S_a(\lambda_0), S_b(\lambda_0)$, we can obtain the common part of B_0^* and ${}^*B_0$, which is ${}^*B_0^*$. Since the requirements of such an S polygon have been given, at first we explain a method of locating vertices of this reduced-truncated S polygon.

Refer to Fig. 11.7. There are $n-1$ intersections between $S_a(\lambda_0)$ and $S_b(\lambda_0)$, because each pair of their corresponding edges intersect. Let the intersecting points be $\mathbf{r}_0, \mathbf{r}_1, \cdots \mathbf{r}_{n-3}, \mathbf{r}_{n-2}$. They make a polygon S of degree $n-2$. Its first vertex $\mathbf{r}_0$ is at $\mathbf{q}_{1.(n-2)}$ and the last one $\mathbf{r}_{n-2}$ is at $\mathbf{q}_{(n-2).2}$ by definition.

Each of the intermediate vertices is determined at an intersecting point of the two Menelaus edges, which are edges of $S_a(\lambda_0)$ and $S_b(\lambda_0)$ and subtend a common angle at a vertex of $S_0^{(n)}$. Their locations are determined by using Menelaus' theorem. Thus, all the locations of the vertices of $\check{S}_0^{(n-2)}$ are given by

$$\left.\begin{array}{rcll} \mathbf{r}_0 & = & \mathbf{q}_{1.(n-2)} & \\ \mathbf{r}_i & = & \mathbf{q}_{i.(n-i-1).(n-i-1)} = \mathbf{q}_{i.(n-i).(n-i-2)}, & i = 1 \cdots n-3 \\ \mathbf{r}_{n-2} & = & \mathbf{q}_{(n-2).2} & \end{array}\right\}. \tag{11.13}$$

(In Fig. 11.7, since n=4, the degree of the reduced-truncated S polygon is two, so there exist $\mathbf{r}_0 = \mathbf{q}_{1.2}$, $\mathbf{r}_1 = \mathbf{q}_{1.2.2}$ and $\mathbf{r}_2 = \mathbf{q}_{2.2}$ only.)

We prove that the polygon thus determined has the desired dividing ratio lists and the main scale.

1. The first edge of $\check{S}_0^{(n-2)}$ is the same as the second edge of $S_a(\lambda_0)$ except for the rightmost sub-edge which corresponds to λ_1. Since the dividing ratio list of the second edge of $S_a(\lambda_0)$ is $L_1^{(n-1)} = (\lambda_{-n+3}, \cdots, \lambda_0, \lambda_1)$, the dividing ratio list of the first edge of $\check{S}_0^{(n-2)}$ becomes $L_0^{(n-2)} = (\lambda_{-n+3}, \cdots, \lambda_0)$, which we can denote as $\check{L}_0^{(n-2)}$.
2. The last edge of $\check{S}_0^{(n-2)}$ is the $(n-2)$-th edge of $S_b(\lambda_0)$ except for the leftmost sub-edge which corresponds to λ_{-1}. Since the $(n-2)$-th edge of $S(\lambda_0)$ has

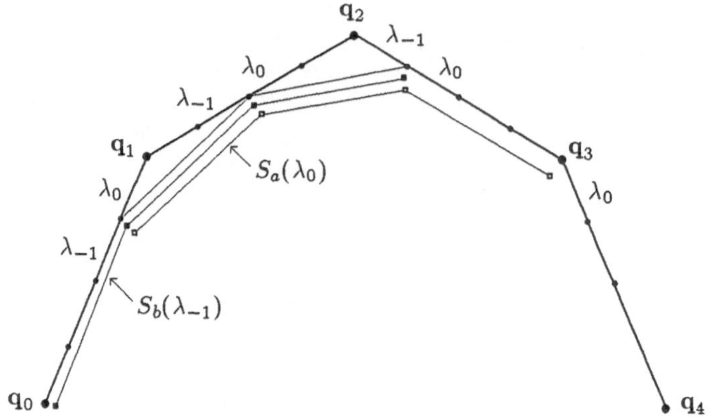

Figure 11.8 Resolution of $\tilde{S}_{-1,0}$ into $S_b(\lambda_{-1})$ and $S_a(\lambda_0)$

the dividing ratio list $L_{n-2}^{(n-1)} = (\lambda_{-1}, \lambda_0, \cdots, \lambda_{n-3})$, the dividing ratio list of the last edge of $\check{S}_0^{(n-2)}$ becomes $\check{L}_{n-2}^{(n-2)} = (\lambda_0, \cdots, \lambda_{n-3})$.

3. The intermediate edges of $\check{S}$ are Menelaus edges of level one with respect to the vertices of $S_a(\lambda_0)$ and also to those of $S_b(\lambda_0)$. For either case the dividing ratio lists of $\check{S}_0^{(n-2)}$ are $\check{L}_1^{(n-2)} \cdots \check{L}_{n-3}^{(n-2)}$. From the properties stated above, the polygon $\check{S}_0^{(n-2)}$ is a unit S polygon of degree $n-2$ with the main scale λ_0, and its degree is $n-2$. Hence, $\check{S}_0^{(n-2)}$ is the S polygon which we desired, that is, the common part of $S_a(\lambda_0)$ and $S_b(\lambda_0)$.

Application examples have already appeared in the previous chapter in Sect. 3.4.

11.5.4 Division of an S Polygon and Insertion of a Vertex

Refer to Fig. 11.8. Two unit polygons $S_b(\lambda_{-1})$ and $S_a(\lambda_0)$ are the divided S polygons of $\tilde{S}_{-1,0}$ in the ratio $(\lambda_{-1} : \lambda_0)$. As stated in Sect. 11.5.1(6) and (7), $S_b(\lambda_{-1})$ and $S_a(\lambda_0)$ produce two B polygons ${}^*B_{-1}$ and B_0^*, while $\tilde{S}_{-1,0}$ produces B polygon $\tilde{B}_{-1,0}$, division of which in the ratio $(\lambda_{-1} : \lambda_0)$ produces the same B polygons as those produced from $S_b(\lambda_{-1})$ and $S_a(\lambda_0)$.

But the number of vertices of $\tilde{S}_{-1,0}$ is n and that of $S_b(\lambda_{-1})$ and $S_a(\lambda_0)$, which have $n-1$ common edges, is $n+1$ except for the duplicate ones. The division of S polygon $\tilde{S}_{-1,0}$ into $S_b(\lambda_1)$ and $S_a(\lambda_0)$ is equivalent to increase or insertion of a vertex of the S polygon without change of shape of the generated curve segments.

In this case we considered division of a unit S polygon of degree $n-1$ with main scale $(\lambda_{-1} + \lambda_0)$ into two unit S polygons of the same degree with the main scales λ_{-1} and λ_0 respectively. If we are to divide a unit S polygon of degree n

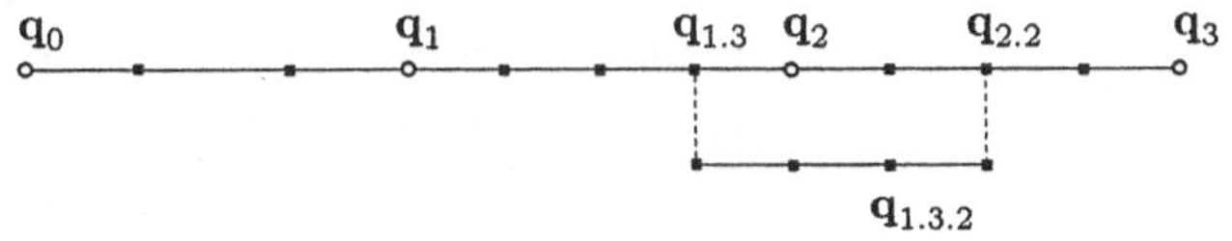

Figure 11.9 Rules on expressions of locations of B points

with the main scale λ_i into $(t : 1-t)$, we make n dividing points, each of which divides each sub-edge of λ_i in the ratio $(t : 1-t)$. There are n such dividing points which replace the $n-1$ intermediate vertices of the old unit S polygon. This is the increase of the vertices of an S polygon. An example has already appeared in Sects. 10.5 and 11.2.2.

11.6 General Formulas for Locations of B Points

We give formulas of locations of B polygons connected in $C^{(n-1)}$ from a given S polygon with its list of dividing ratios. In lower-degree cases, for example $n = 2$ and $n = 3$, methods of determining locations of B points are well known. And the method using reduced-truncated S polygons, which is given in Sect. 11.5.3 is effective for higher-degree polygons. But in this section, we present a general formula to give locations of B points from an S polygon of arbitrary degree n.

11.6.1 Rules of Location Symbols of B Points and Their Properties

The location of each dividing point on a Menelaus edge is determined by (i) the sequence number from the left starting point together with its dividing ratio list which determines division of the edge concerned, and (ii) the location of the left starting point of the edge, which is also a dividing point of a Menelaus edge of one level lower.

Referring to Fig. 11.9, for instance, $\mathbf{q}_{1.3.2}$ is the second dividing point of an edge starting from $\mathbf{q}_{1.3}$ and ending at $\mathbf{q}_{2.2}$. The left end point $\mathbf{q}_{1.3}$ is the third dividing point of an edge joining $\mathbf{q}_1$ and $\mathbf{q}_2$. The location of the right end of a Menelaus edge is not explicitly shown, but it is the point whose index is skew shifted from that of the left end point of the same edge.

11.6.1.1 Level of Menelaus Edges and Dividing Points

As for the level of an edge and a dividing point, the lowest level is defined as level zero. Edges and vertices of the level zero are those of the original S polygon. A dividing point of level one is a point on an edge of level zero, and a Menelaus edge of level one starts from a dividing point of level one. Dividing points of level two are on a Menelaus edge of level one. The dividing ratio list is inherited as follows. For instance, if the level zero edge joining $\mathbf{q}_1$ and $\mathbf{q}_2$ has dividing ratio list $L_1^{(4)} = (\lambda_{-2}, \lambda_{-1}, \lambda_0, \lambda_1)$, the dividing ratio list of the level one edge joining $\mathbf{q}_{1.3}$ and $\mathbf{q}_{2.2}$ is $L_1^{(3)} = (\lambda_{-1}, \lambda_0, \lambda_1)$. That is, the first element of $L_1^{(4)}$ is removed and the rest is the same. The dividing ratio list of level two edge joining $\mathbf{q}_{1.3.2}$ and $\mathbf{q}_{2.2.1}$ has the dividing ratio list $L_1^{(2)} = (\lambda_0, \lambda_1)$.

11.6.1.2 Symbols for Location of Control Points

Now we generalize the above notation rule for dividing points. Let θ_i be an index variable indicating a sequence number of a dividing point of level i. The location of a dividing point of level i is given in the form:

$$\mathbf{q}\theta_0.\theta_1.\theta_2.\cdots\theta_i. \tag{11.14}$$

The characteristics of this expression are :

- The location of this dividing point is on a Menelaus edge of level $i-1$ whose left end is $\mathbf{q}\theta_0.\theta_1.\theta_2.\cdots\theta_{i-1}$. Its dividing ratio list is given by $L\theta_0^{(n-i+1)}$, because the lowest level is zero by definition and its superscript of the dividing ratio list eq.(11.4) is (n). Accordingly for the level $i-1$, the superscript becomes $(n-(i-1))$.
- The subscript of the dividing ratio list is coincident with values of the first index θ_0 of the location symbol, because the last element of the dividing ratio of a Menelaus edge is inherited from the last element of the dividing ratio of the Menelaus edge of one level lower. Accordingly it is the same as that of the lowest level: it is the scale factor of the rightmost sub-edge of an edge of the original polygon.
- A vertex of the polygon is represented by $\mathbf{q}\theta_0$ and a dividing point on its edge is $\mathbf{q}\theta_0.\theta_1$. The value of θ_0 can be any number because it represents vertices of the original S polygon, but for the other θ_k, $k > 0$, its maximum value is $n-k$.
- The operation of skew shifting increases value of θ_0 by one and decreases each values of θ_k by one for $k > 0$. If the values of indices of two dividing points are the same to a certain level and the values of the following two indices are (i, j) and the others are $(i+k, j-k)$ and if $j \neq i-1$ and $k < n-i$, the two dividing points are the same point. Refer to Sect. 11.4.2. For example, three dividing points with location symbols

  $$\text{(i) } \mathbf{q}_{0.3.3.2} \quad \text{(ii) } \mathbf{q}_{0.4.2.2} \quad \text{(iii) } \mathbf{q}_{0.4.3.1}$$

  represent the same location, because in the expressions (i) and (ii), $\mathbf{q}_0$ is common and $i = 3$, $j = 3$, $k = 1$. In (ii) and (iii), $\mathbf{q}_{0.4}$ is common and $i = 2$, $j = 2$, $k = 1$. They satisfy the conditions given above.

Table 11.1 Values of indices for B point locations

n	p_0	p_1	p_2	p_3	p_4	p_5
1	0	1				
2	0.1	1	1.1			
3	0.2.1	1.1	1.2	1.2.1		
4	0.3.2.1	1.2.1	1.2.2	1.3.2	1.3.2.1	
5	0.4.3.2.1	1.3.2.1	1.3.2.2	1.3.3.2	1.4.3.2	1.4.3.2.1

11.6.2 General Expressions of B Point Locations

11.6.2.1 Application of Location Symbols

From location symbols of vertices of a B polygon of degree n, which are derived from a unit S polygon of the same degree, we can deduce vertex locations of a B polygon of degree $n+1$ derived from a unit S polygon of degree $n+1$. For $n = 1$ and $n = 2$, we have shown the deduction process in Sect. 11.4.2. The method is similar for general n. In Table 11.1 the locations of B points for $n = 1, \cdots 5$ are shown. We give rules for deducing ${}^*B_0^{(i+1)}$ and $B_0^{*(i+1)}$ from $B_0^{(i)}$:

- To convert location symbols for B points of $B_0^{(i)}$, which is assumed known, into those of $B_0^{*(i+1)}$, which is one degree higher and misses the rightmost point, replace the first index position of the location symbol in $B_0^{(i)}$ by two index numbers: if the first one is 0, it is replaced by $0.i$ and if the first one is 1, it is replaced by $1.(i-1)$. Do not change the subscripts of the control points $\mathbf{p}_j$ of B polygons.
  For example, let us assume for $i = 3$ we know $\mathbf{p}_0^{(3)} = \mathbf{q}_{0.2.1}$, and $\mathbf{p}_1^{(3)} = \mathbf{q}_{1.1}$. Using the above rule we obtain $\mathbf{p}_0^{(4)} = \mathbf{q}_{0.3.2.1}$ and $\mathbf{p}_1^{(4)} = \mathbf{p}_{1.2.1}$.
- To convert location symbols for $B_0^{(i)}$ into those of ${}^*B_0^{(i+1)}$, replace the first index position of the symbol by two index numbers: if the first one is 0, it is replaced by 1 and if the first one is 1, it is replaced by $1.i$. Increase values of the subscript j of the control points $\mathbf{p}_j$ by one.
  For example, from $\mathbf{p}_0^{(4)} = \mathbf{q}_{0.3.2.1}$, we get $\mathbf{p}_1^{(5)} = \mathbf{q}_{1.3.2.1}$ and from $\mathbf{p}_1^{(4)} = \mathbf{q}_{1.2.1}$, we obtain $\mathbf{p}_2^{(5)} = \mathbf{q}_{1.4.2.1}$, which is equivalent to $\mathbf{q}_{1.3.3.1}$ and $\mathbf{q}_{1.3.2.2}$. The last one is written in the Table 11.1.

The number of B points in $B_0^{(n+1)}$ is $n+2$, but the total number of B points obtained by the above method is $2n+2$, so there are n duplicates. For convenience, the first $n-1$ points are taken from those of $B^*{}_0^{(n+1)}$ and the last two points are from those of ${}^*B_0^{(n+1)}$, which are shown in Table 11.1.

11.6.2.2 Formulas for Location Symbols

The general formulas for location symbols of B points of $B_0^{(n)}$ are given in the form (11.14):

- the first B point $\mathbf{p}_0$:

$$\begin{cases} \theta_0 = 0 \\ \theta_k = n - k \quad : \text{for } k = 1 \ldots n-1 \end{cases}$$

- the second B point $\mathbf{p}_1$:

$$\begin{cases} \theta_0 = 1 \\ \theta_k = n - 1 - k \quad : \text{for } k = 1 \ldots n-2 \end{cases}$$

- the intermediate B points $\mathbf{p}_j \quad (j = 2 \ldots n-2)$:

$$\begin{cases} \theta_0 = 1 \\ \theta_k = n - 1 - k \quad : \text{for } k = 1 \ldots n-j-1 \\ \theta_k = n - k \quad : \text{for } k = n-j \ldots n-2 \end{cases} \tag{11.15}$$

- the penultimate point $\mathbf{p}_{n-1}$:

$$\begin{cases} \theta_0 = 1 \\ \theta_k = n - k \quad : \text{for } k = 1 \ldots n-2 \end{cases}$$

- the last point $\mathbf{p}_n$:

$$\begin{cases} \theta_0 = 1 \\ \theta_k = n - k \quad : \text{for } k = 1 \ldots n-1 \end{cases}$$

To determine the real location of a B point from its symbolic notation, we just translate it according to the definition.

For example, we take $\mathbf{p}_1$ and $\mathbf{p}_2$ of $B_0^{(5)}$. According to the above formulas (11.15) for $n = 5$, $j = 1$ and $j = 2$ we have $\mathbf{p}_1 = \mathbf{q}_{1.3.2.1}$ and $\mathbf{p}_2 = \mathbf{q}_{1.3.2.2}$. They are the first and the second dividing points of the level two Menelaus edge divided in the ratio $L_1^{(3)}$, because $n = 5$, $i = 3$ and $\theta_0 = 1$ and $L\theta_0^{(n-i+1)}$ according to the first item of Sect. 11.6.1.2. The end points of this Menelaus edge are $\mathbf{q}_{1.3.2}$ and its skew shifted point $\mathbf{q}_{2.2.1}$.

The point $\mathbf{q}_{1.3.2}$ is the second dividing point of the Menelaus edge of level one, which joins $\mathbf{q}_{1.3}$ and$\mathbf{q}_{2.2}$, divided in $L_1^{(4)}$. The point $\mathbf{q}_{1.3}$ is the third dividing point of the edge of level zero between $\mathbf{q}_1$ and $\mathbf{q}_2$ divided in $L_1^{(5)}$. The point $\mathbf{q}_{2.2.1}$ is the first dividing point of the edge of level two between $\mathbf{q}_{2.2}$ and $\mathbf{q}_{3.1}$, divided in $L_2^{(4)}$. The points $\mathbf{q}_{2.2}$ and $\mathbf{q}_{3.1}$ are the second and the first dividing points of the edges $(\mathbf{q}_2\mathbf{q}_3)$ and $(\mathbf{q}_3\mathbf{q}_4)$ of level zero divided by $L_2^{(5)}$ and $L_3^{(5)}$ respectively.

11.6.2.3 Level of Edges of a B Polygon

The relation between the levels of the Menelaus edges and edges of a B polygon is:

- Both end edges of the $B_i^{(n)}$ polygon deduced from an S polygon of degree n belong to the sub-edges of the Menelaus edges of level $(n-2)$.
- The other intermediate $(n-2)$ edges belong to the Menelaus edges of level $(n-3)$.
- Each of the edges of a $B_i^{(n)}$ polygon, whose scale factor is λ_i, is the sub-edge of λ_i of the respective Menelaus edge of level $(n-2)$ or the level $(n-3)$.

For example, the first and the last edges of $B_0^{(5)}$ are on level 3 and the other three are on level 2, and their sub-edge belong to the λ_0 part of the respective Menelaus edges.

Exercise. When the scale factors λ_i are all equal, prove that the locations of B points are as shown below.

For all n,

$$\mathbf{p}_0^{(n)} = (1/2)(\mathbf{p}_1 + \mathbf{p}_{-1})^{(n)}, \qquad \mathbf{p}_n^{(n)} = (1/2)(\mathbf{p}_{n+1} + \mathbf{p}_{n-1})^{(n)}. \tag{11.16}$$

For $n = 2$,

$$\mathbf{p}_1^{(2)} = \mathbf{q}_1.$$

For $n = 3$,

$$(\mathbf{p}_1\ \mathbf{p}_2) = (1/3)(\mathbf{q}_1\ \mathbf{q}_2)\begin{bmatrix} 2 & 1 \\ 1 & 2 \end{bmatrix}.$$

For $n = 4$,

$$(\mathbf{p}_1\ \mathbf{p}_2\ \mathbf{p}_3) = (1/12)(\mathbf{q}_1\ \mathbf{q}_2\ \mathbf{q}_3)\begin{bmatrix} 4 & 2 & 1 \\ 7 & 8 & 7 \\ 1 & 2 & 4 \end{bmatrix}.$$

For $n = 5$,

$$(\mathbf{p}_1\ \mathbf{p}_2\ \mathbf{p}_3\ \mathbf{p}_4) = (\frac{1}{120})(\mathbf{q}_1\ \mathbf{q}_2\ \mathbf{q}_3\ \mathbf{q}_4)\begin{bmatrix} 8 & 4 & 2 & 1 \\ 33 & 30 & 24 & 18 \\ 18 & 24 & 30 & 33 \\ 1 & 2 & 4 & 8 \end{bmatrix}.$$

11.7 Appendix. Orthodox Approach to a B Spline Curve

In order for basis functions or weighting functions for control points not to have a global influence on the shape they express, they must have the property that each one has value zero outside a restricted range called a support. Each of the basis functions of this property together with derivative continuity conditions can be represented not by a single polynomial, but by the connection of polynomials defined between successive points called knots. The B spline curve is a linear combination of the B spline basis functions, each of which is multiplied by a control point vector $\mathbf{q}_i$, such as

$$\mathbf{r}(x) = \sum_{i=0}^{n} \mathbf{q}_i N_i^{(n)}(x). \tag{11.17}$$

The control points $\mathbf{q}_i$ make a B spline control polygon. The normalized B spline basis functions $N_i^{(n)}(x)$ of degree n are piecewise polynomials of degree n which are defined over the knots $\cdots < x_0 < x_1 < x_2 < \cdots$ and have the following properties:

$$\begin{aligned}
\sum_{i=0}^{n} N^{(n)}{}_i(x) &= 1, \\
N^{(n)}{}_i(x) &\geq 0, \\
N^{(n)}{}_i(x) &= 0 \quad \text{if } x < x_i \text{ and } x > x_{i+n+1}\ , \\
N^{(n)}{}_i(x) &\ \text{is} \ (n-1) \text{ times continuously differentiable.}
\end{aligned}$$

The recurrence formula by deBoor is

$$N_i^{(j)}(x) = (x - x_i)\frac{N_i^{(j-1)}(x)}{(x_{i+j} - x_i)} + (x_{i+j+1} - x)\frac{N_{i+1}^{(j-1)}(x)}{(x_{j+i+1} - x_{i+1})}. \tag{11.18}$$

where

$$N_i^{(0)}(x) = \begin{cases} 1 & \text{if } x_i < x < x_{i+1} \\ 0 & \text{otherwise.} \end{cases}$$

The vertices $\mathbf{q}_i$ and the knots $\cdots x_0, x_1, x_2, \cdots$ having been supplied, an orthodox method of calculation of a curve expression (11.17) at given x is to obtain values of the spline basis functions by using the recurrence formula (11.18) given above.

Compared with the method of determining Bézier points from the vertices of the B spline control polygon with the given scale ratio list, the orthodox method of B spline curves is not easy for understanding the geometric properties of the curve or for its manipulation. Moreover, when we need to calculate various characteristic curves on a spline surfaces or to determine interference curves between spline surfaces, their surface expressions derived from the tensor product of eq.(11.17) are not adequate because calculation of a large number of surface points is required. For further detail on splines consult the references [22][33].

12 Rational Bézier and Spline Expressions

12.1 Introduction

Bézier expressions for curves and surfaces are easy to understand and have favorable features for use in CAD and CAM. But they cannot exactly express conics and quadrics such as circles and spheres. By extending the expressions from polynomials to rational forms, we can provide them not only with the ability to express conics and quadrics, but also with possibly increased freedom for design [28][29][32].

A Bézier curve is constructed by repeated linear interpolation between two points, that is, division of a line joining two points in the ratio $t : (1-t)$. Whereas its counterpart, a rational Bézier, uses repeated rational interpolation which divides a line joining two end points in the ratio $w_1 t : w_0(1-t)$, where w_0 and w_1 are weights of the respective end points, and the weight of the intermediate dividing point is $w_0(1-t) + w_1 t$. The end points and their weights may be functions of t. For the same end points and the same value of the parameter t, $0 \leq t \leq 1$, the linear and the rational interpolations represent different points. Similarly a ruled surface is expressed by a tensor product of two line segments in linear or rational interpolation forms. For the same values of parameters u and v, their generators are different. But the rational and the linear interpolations have close relations. For instance, in a perspective projection (see Fig. 12.1), the control polygon $(\mathbf{p}_0\ \mathbf{p}_1\ \mathbf{p}_2\ \mathbf{p}_3)$ of a rational Bézier and the corresponding control polygon $(\hat{\mathbf{p}}_0\ \hat{\mathbf{p}}_1\ \hat{\mathbf{p}}_2\ \hat{\mathbf{p}}_3)$ of a non-rational Bézier give the same projection seen from a view point $\hat{\mathbf{o}}$. A ratio of distances from $\hat{\mathbf{o}}$ to the corresponding control points, $\hat{\mathbf{o}}\hat{\mathbf{p}}_i : \hat{\mathbf{o}}\mathbf{p}_i$, is the weight w_i belonging to $\mathbf{p}_i$.

Analogous to this correspondence between the control points, a point on a rational Bézier curve can be mapped to a point on the usual Bézier curve or on the corresponding control polygon. Many simple and elegant properties which characterize Bézier and spline polygons are inherited by their rational equivalents unless the projection violates the original relations. An S polygon of degree n is mapped to a rational S polygon of the same degree. An S polygon of degree three with $G^{(2)}$ connection is also mapped to a rational S polygon of degree three with $G^{(2)}$ connection. The corresponding locations of B points are easily determined. This approach makes their mathematical analysis and geometric understanding much easier than other methods which use 4D space (wx, wy, wz, w) or only rational forms in 3D.

Currently, the merits in use of the rational curves and surfaces are not always recognized except for representation of conics. This is because appropriate selection of weight values used in the rational forms is not always clear and their

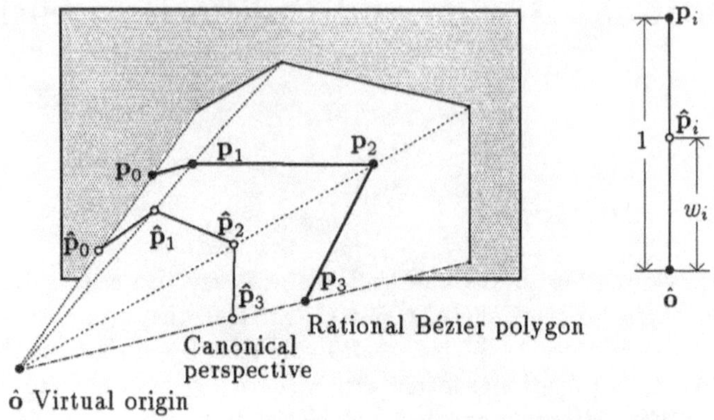

Figure 12.1 Perspective relation between control polygons of a rational and a non-rational Bézier

shape calculation is more complicated than non-rational ones [32]. But owing to their generalized features, their use has begun to be considered recently. Accordingly, we explain in this chapter characteristics and uses of the rational Bézier and spline curves and surfaces, applying a new geometric interpretation, though much space in the last section is allocated to rational Bézier curves of degree two, which are practically useful. An optimum curve fitting method with conics is explained, which uses rational expressions of degree two with positive as well as negative weights.

12.2 Rational Bézier Curves

12.2.1 Rational Division Between Two Points and Its Perspective Map

In a rational polygon each point has its weight together with its coordinate values. When a point on an edge of the polygon is determined by division of its two end points, let the linear dividing ratio between the two end points be$a : b$, then the rational dividing ratio is defined as $w_1a : w_0b$ and the weight of the corresponding rational dividing point has a value determined by the linear dividing ratio of the weights w_0 and w_1 of the end points of the edge.

We explain a graphical method of locating a dividing point between two points $\mathbf{p}_0$ and $\mathbf{p}_1$ in the ratio $w_1t : w_0(1-t)$, which we call a rational division with weights w_0 and w_1. Referring to Fig. 12.2(a), we define points with their symbols in the figure.

1. Let ô be an arbitrary point which we call a virtual origin. The following procedure takes a point $\hat{\mathbf{p}}_0$ on a line joining ô and $\mathbf{p}_0$ or its extension to

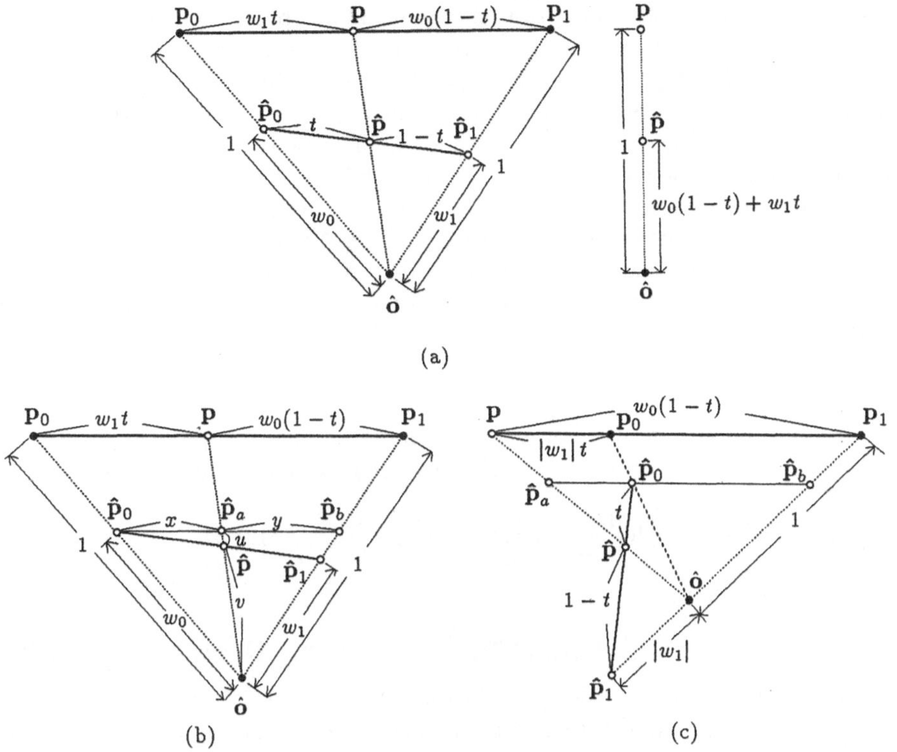

Figure 12.2 (a) Rational division of a line segment, (b) figure for the proof (when both weights are positive), (c) figure for the proof (when one weight is negative)

make a ratio $\{\hat{\mathbf{o}}\mathbf{p}_0 : \hat{\mathbf{o}}\hat{\mathbf{p}}_0 = 1 : w_0\}$, similarly takes a point $\hat{\mathbf{p}}_1$ on $\hat{\mathbf{o}}\mathbf{p}_1$ so that $\{\hat{\mathbf{o}}\mathbf{p}_1 : \hat{\mathbf{o}}\hat{\mathbf{p}}_1 = 1 : w_1\}$.

2. Let $\hat{\mathbf{p}}$ be a dividing point on $\hat{\mathbf{p}}_0\hat{\mathbf{p}}_1$ in the ratio $(t : 1-t)$, which we call a linear division, and let $\mathbf{p}$ be an intersecting point of the line $\mathbf{p}_0\mathbf{p}_1$ and a line joining $\hat{\mathbf{o}}$ and $\hat{\mathbf{p}}$ or its extension.

3. Then we can prove that the point $\mathbf{p}$ divides the line $\mathbf{p}_0\mathbf{p}_1$ in the ratio $\{w_1 t : w_0(1-t)\}$, which is a rational division, and the point $\mathbf{p}$ has a weight $(1-t)w_0 + tw_1$, that is, a ratio $[\hat{\mathbf{o}}\mathbf{p} : \hat{\mathbf{o}}\hat{\mathbf{p}} = 1 : \{(1-t)w_0 + tw_1\}$ holds.

Line segments, for instance $\mathbf{p}'_0\mathbf{p}'_1$, each of whose end points are on $\hat{\mathbf{o}}\mathbf{p}_0$ and $\hat{\mathbf{o}}\mathbf{p}_1$ or their extensions, are perspectives of the original line $\mathbf{p}_0\mathbf{p}_1$ with respect to the point $\hat{\mathbf{o}}$. Among the perspectives the line $\hat{\mathbf{p}}_0\hat{\mathbf{p}}_1$ is special, because locations of its end points are directly associated with the end points of the original segment with weights w_0 and w_1. Accordingly, let us call this line $\hat{\mathbf{p}}_0\hat{\mathbf{p}}_1$ the canonical perspective of the original line $\mathbf{p}_0\mathbf{p}_1$ with respect to the point $\hat{\mathbf{o}}$. Any points $\mathbf{p}$ between $\mathbf{p}_0$ and $\mathbf{p}_1$ correspond to their canonical perspective point $\hat{\mathbf{p}}$, which is a dividing point between $\hat{\mathbf{p}}_0$ and $\hat{\mathbf{p}}_1$ in the ratio $t : 1-t$. We can prove that the corresponding point $\mathbf{p}$ is the dividing point of the original line $\mathbf{p}_0\mathbf{p}_1$ in the ratio $tw_1 : (1-t)w_0$. The point $\mathbf{p}$ has the weight $tw_1 + (1-t)w_0$ just as both end points $\mathbf{p}_0$ and $\mathbf{p}_1$ have the weights w_0 and w_1. The canonical perspective has an important role in the construction of rational Bézier or spline curves.

We give a proof for the above stated relations between the original line and its canonical perspective line. Referring to Fig. 12.2(b), draw a line from $\hat{\mathbf{p}}_0$ parallel to $\mathbf{p}_0\mathbf{p}_1$ and let its intersection with $\hat{\mathbf{o}}\mathbf{p}$ and $\hat{\mathbf{o}}\mathbf{p}_1$ be $\hat{\mathbf{p}}_a$ and $\hat{\mathbf{p}}_b$, which divide $\hat{\mathbf{o}}\mathbf{p}$ and $\hat{\mathbf{o}}\mathbf{p}_1$ in the same ratio $w_0 : 1-w_0$. Then there is a relation of length ratios:

$$|\hat{\mathbf{o}}\hat{\mathbf{p}}_1| : |\hat{\mathbf{p}}_1\hat{\mathbf{p}}_b| : |\hat{\mathbf{p}}_b\mathbf{p}_1| = w_1 : w_0 - w_1 : 1 - w_0. \tag{12.1}$$

Let $\hat{\mathbf{o}}\mathbf{p}$ divide $\hat{\mathbf{p}}_0\hat{\mathbf{p}}_b$ in the ratio $x : y$ and $\hat{\mathbf{p}}_0\hat{\mathbf{p}}_1$ divide $\hat{\mathbf{o}}\hat{\mathbf{p}}_a$ in the ratio $u : v$. Applying Menelaus' theorem to a triangle $\Delta\hat{\mathbf{p}}_0\hat{\mathbf{p}}_b\hat{\mathbf{p}}_1$ and a line $\hat{\mathbf{o}}\mathbf{p}$, we have

$$\left(\frac{|\hat{\mathbf{p}}_0\hat{\mathbf{p}}_a|}{|\hat{\mathbf{p}}_a\hat{\mathbf{p}}_b|}\right)\left(\frac{|\hat{\mathbf{p}}_b\hat{\mathbf{o}}|}{|\hat{\mathbf{o}}\hat{\mathbf{p}}_1|}\right)\left(\frac{|\hat{\mathbf{p}}_1\hat{\mathbf{p}}|}{|\hat{\mathbf{p}}\hat{\mathbf{p}}_0|}\right) = \left(\frac{x}{y}\right)\left(\frac{w_0}{w_1}\right)\left(\frac{1-t}{t}\right) = 1.$$

From this we obtain

$$|\hat{\mathbf{p}}_0\hat{\mathbf{p}}_a| : |\hat{\mathbf{p}}_a\hat{\mathbf{p}}_b| = |\mathbf{p}_0\mathbf{p}| : |\mathbf{p}\mathbf{p}_1| = x : y = tw_1 : (1-t)w_0. \tag{12.2}$$

Therefore, the point $\mathbf{p}$ divides $\mathbf{p}_0\mathbf{p}_1$ in the ratio $tw_1 : (1-t)w_0$.

Then applying the theorem to a triangle $\Delta\hat{\mathbf{p}}_a\hat{\mathbf{o}}\hat{\mathbf{p}}_b$ and a line $\hat{\mathbf{p}}_0\hat{\mathbf{p}}_1$, we have

$$\left(\frac{|\hat{\mathbf{o}}\hat{\mathbf{p}}|}{|\hat{\mathbf{p}}\hat{\mathbf{p}}_a|}\right)\left(\frac{|\hat{\mathbf{p}}_a\hat{\mathbf{p}}_0|}{|\hat{\mathbf{p}}_0\hat{\mathbf{p}}_b|}\right)\left(\frac{|\hat{\mathbf{p}}_b\hat{\mathbf{p}}_1|}{|\hat{\mathbf{p}}_1\hat{\mathbf{o}}|}\right) = \left(\frac{u}{v}\right)\left(\frac{x}{x+y}\right)\left(\frac{w_0-w_1}{w_1}\right) = 1. \tag{12.3}$$

From these geometrical relations (12.2) and (12.3) another relation

$$|\hat{\mathbf{o}}\mathbf{p}| : |\hat{\mathbf{o}}\hat{\mathbf{p}}| = (u+v)/w_0 : u = 1 : \{(1-t)w_0 + tw_1\}, \tag{12.4}$$

is obtained. This indicates that a point $\mathbf{p}$ has a weight $\{(1-t)w_0 + tw_1\}$. At $t = 0$, the weight of the point $\mathbf{p}_0$ is w_0 and at $t = 1$, the weight of the point $\mathbf{p}_1$ is w_1. The above stated relations can be proved even if one or both of the weights has or have negative value(s). See Fig. 12.2(c). When one of the weights is negative, the line segment starting from $\mathbf{p}_0$ and reaching at $\mathbf{p}_1$ or vice versa should be considered to pass an infinite point. The application is found in Sect. 12.7.3.

12.2.2 Rational Bézier Curves and Their Canonical Perspectives

Since the expression for a dividing point between two end points represents an equation of a line connecting them, a line $\hat{\mathbf{p}}_0\hat{\mathbf{p}}_1$ is considered to be a usual Bézier curve of degree one given by

$$\hat{\mathbf{r}}_0(t;1) = (1-t)\hat{\mathbf{p}}_0 + t\hat{\mathbf{p}}_1 = (1-t+\mathsf{E}t)\hat{\mathbf{p}}_0. \tag{12.5}$$

We use the shift operator E for concise description.

Similarly, from the definition of the rational dividing point, a line $\mathbf{p}_0\mathbf{p}_1$ is represented by an equation

$$\mathbf{r}_0(t;1) = \frac{(1-t)w_0\mathbf{p}_0 + tw_1\mathbf{p}_1}{(1-t)w_0 + tw_1}, \tag{12.6}$$

which we call a rational Bézier curve of degree one. We use the notation $\mathbf{r}_0(t;1)$ for it. Since there are relations $\hat{\mathbf{p}}_0 = w_0\mathbf{p}_0$ and $\hat{\mathbf{p}}_1 = w_1\mathbf{p}_1$ together with equations (12.5) and (12.6), we have

$$\hat{\mathbf{r}}_0(t;1) = \{(1-t)w_0 + tw_1\}\mathbf{r}_0(t;1).$$

We use the notation $w_0(t;1)$ for the weight of a point $\mathbf{r}_0(t;1)$:

$$w_0(t;1) = (1-t)w_0 + tw_1 = (1-t+\mathsf{E}t)w_0. \tag{12.7}$$

Substituting the above equation into eq. (12.6), we have

$$\mathbf{r}_0(t;1) = \frac{(1-t+\mathsf{E}t)(w_0\mathbf{p}_0)}{w_0(t;1)}. \tag{12.8}$$

Equation (12.8) is a rational Bézier of degree one constructed from rational Béziers of degree zero, $\mathbf{p}_0 = \mathbf{r}_0(t;0)$ and $\mathbf{p}_1 = \mathbf{r}_1(t;0)$, together with their weights. In Fig. 12.3, we show a construction process of a rational Bézier of degree three from four control points $(\mathbf{p}_0, \mathbf{p}_1, \mathbf{p}_2, \mathbf{p}_3)$ with their weights $(w_0\ w_1\ w_2\ w_3)$. Two rational Béziers of degree two are produced during the process.

1. From the control polygon with its weights and a virtual origin $\hat{\mathrm{o}}$, make its canonical perspective $(\hat{\mathbf{p}}_0\ \hat{\mathbf{p}}_1\ \hat{\mathbf{p}}_2\ \hat{\mathbf{p}}_3)$. See Fig. 12.3(a).
2. Determine a point on a Bézier curve $\hat{\mathbf{p}}$ from the canonical control polygon as shown in Fig. 12.3(b). Three points $\hat{\mathbf{p}}_a, \hat{\mathbf{p}}_b$ and $\hat{\mathbf{p}}_c$ are linear interpolated points on the edges $(\hat{\mathbf{p}}_0\hat{\mathbf{p}}_1), (\hat{\mathbf{p}}_1\hat{\mathbf{p}}_2)$ and $(\hat{\mathbf{p}}_2\hat{\mathbf{p}}_3)$. These points are $\hat{\mathbf{r}}_0(t;1), \hat{\mathbf{r}}_1(t;1)$ and $\hat{\mathbf{r}}_2(t;1)$. Two dividing points $\hat{\mathbf{p}}_d$ and $\hat{\mathbf{p}}_e$, which are $\hat{\mathbf{r}}_0(t;2)$ and $\hat{\mathbf{r}}_1(t;2)$, are linear interpolated points on $(\hat{\mathbf{p}}_a\hat{\mathbf{p}}_b)$ and $(\hat{\mathbf{p}}_b\hat{\mathbf{p}}_c)$. Then a point $\hat{\mathbf{p}}$, which is $\hat{\mathbf{r}}_0(t;3)$, is a linear interpolated point of $(\hat{\mathbf{p}}_d\hat{\mathbf{p}}_e)$.
3. Referring to Fig. 12.3(c), take the corresponding points $\mathbf{p}_a, \mathbf{p}_b, \mathbf{p}_c, \mathbf{p}_d, \mathbf{p}_e$ and $\mathbf{p}$ on the edges of the rational control polygon and its derived edges. Then $\mathbf{p}$ is a point on a rational Bézier curve $\mathbf{r}_0(t;3)$, and $\mathbf{p}_d$, $\mathbf{p}_e$ are $\mathbf{r}_0(t;2)$ and $\mathbf{r}_1(t;2)$,and $\mathbf{p}_a$, $\mathbf{p}_b$, $\mathbf{p}_c$ are $\mathbf{r}_0(t;1)$, $\mathbf{r}_1(t;1)$, $\mathbf{r}_2(t;2)$.

Since the weights of the points $\mathbf{p}_a, \mathbf{p}_b$ and $\mathbf{p}_c$ are $w_0(t;1), w_1(t;1)$ and $w_2(t;1)$, the weights of the points $\mathbf{p}_d$ and $\mathbf{p}_e$ are $w_0(t;2)$ and $w_1(t;2)$, whichare the linear interpolations of $w_0(t;1)$ and $w_1(t;1)$, and $w_1(t;1)$ and $w_2(t;1)$ respectively.

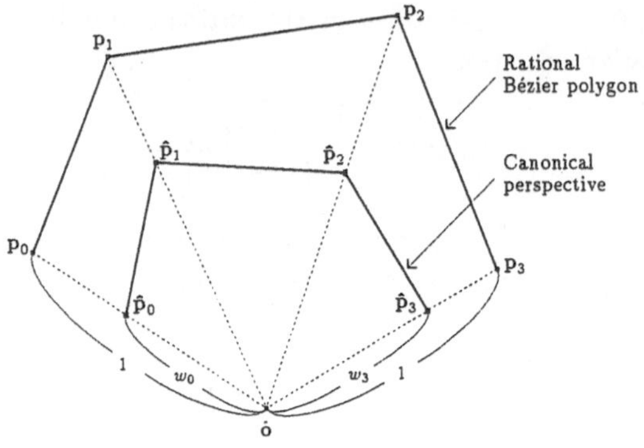

(a)

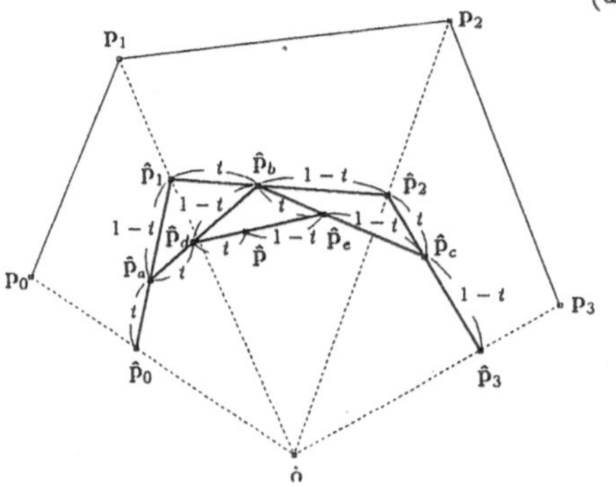

(b)

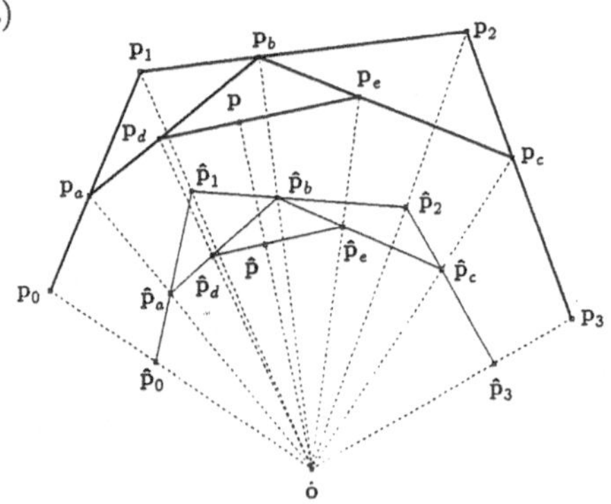

(c)

Figure 12.3 (a) Rational polygon and its canonical perspective, (b) determining a point from Bézier polygon of degree three, (c)corresponding point of rational polygon

Similarly that of $\mathbf{p}$ is $w_0(t;3)$, which is the linear interpolation of $w_0(t;2)$ and $w_1(t;2)$. So we can use the expressions:

$$w_0(t;2) = (1-t)w_0(t;1) + tw_1(t;1) = (1-t+\mathsf{E}t)w_0(t;1),$$
$$w_0(t;3) = (1-t)w_0(t;2) + tw_1(t;2) = (1-t+\mathsf{E}t)w_0(t;2).$$

Hence, we have

$$\begin{aligned} \mathbf{r}_0(t;2) &= \frac{(1-t)w_0(t;1)\mathbf{r}_0(t;1) + tw_1(t;1)\mathbf{r}_1(t;1)}{(1-t)w_0(t;1)+tw_1(t;1)} \\ &= \frac{(1-t+\mathsf{E}t)(w_0(t;1)\mathbf{r}_0(t;1))}{w_0(t;2)} \\ &= \frac{(1-t+\mathsf{E}t)^2(w_0\mathbf{p}_0)}{(1-t+\mathsf{E}t)^2 w_0}, \end{aligned} \tag{12.9}$$

$$\begin{aligned}\mathbf{r}_0(t;3) &= \frac{(1-t+\mathsf{E}t)(w_0(t;2)\mathbf{r}_0(t;2))}{w_0(t;3)} \\ &= \frac{(1-t+\mathsf{E}t)^3(w_0\mathbf{p}_0)}{(1-t+\mathsf{E}t)^3 w_0}.\end{aligned} \tag{12.10}$$

With a similar procedure, we can obtain a rational Bézier curve of degree n from two rational Bézier curves of degree $n-1$:

$$\begin{aligned}\mathbf{r}_0(t;n) &= \frac{(1-t+\mathsf{E}t)(w_0(t;n-1)\mathbf{r}_0(t;n-1))}{(1-t+\mathsf{E}t)w_0(t;n-1)} \\ &= \frac{(1-t+\mathsf{E}t)^n(w_0\mathbf{p}_0)}{w_0(t;n)},\end{aligned} \tag{12.11}$$

where

$$w_0(t;n) = (1-t+\mathsf{E}t)^n w_0. \tag{12.12}$$

Since the graphical construction of a rational Bézier curve corresponds to that of a linear Bézier curve, we can easily understand the effects of weights of vertices of the rational polygon using its canonical perspective.

12.2.3 Effects of Weights

A rational Bézier curve is a linear combination of the weighting functions multiplied by their respective control points. The weighting function of the j-th control point $\mathbf{p}_j$ is the coefficient of $\mathbf{p}_j$ in eq. (12.11) which is given by

$$C_j^n\, t^j(1-t)^{n-j} w_j / w_0(t;n)$$

and the summation of all the weighting functions for $j = 0, \cdots n$ is 1. The denominator $w_0(t;n)$ is considered a scalar Bézier curve, whose control points are determined by values $(w_0\ ,\ w_1\ ,\ \cdots\ ,\ w_n)$ given at $t = 0\ ,\ (1/n),\ \cdots\ ,\ ((n-1)/n)\ ,\ 1$. See Fig. 12.4. If the weights are equal, the denominator becomes a constant. Then eq. (12.11) becomes a usual Bézier curve.

Since all the weights are assumed positive, each of the weighting functions for the control points has positive values less than one, but it is not always of single peak type, though the summation of the functions is equal to one. Anyway a rational Bézier curve is in a convex hull of its control points by the same reason as that for non-rational one. If a certain w_j is much larger than the others, then the denominator becomes large at $t = (j/n)$ and the curve $\mathbf{r}(t;n)$ approaches $\mathbf{p}_j$ whose weight is w_j. This is more precisely explained in the following. If we increase w_j to $w_j + \Delta w_j$ and let $\mathbf{r}_0(t;n)$ become $\mathbf{r}'_0(t;n)$, then eq. (12.11) is transformed to

$$\mathbf{r}'_0(t;n) = \frac{\mathbf{r}_0(t;n) + \gamma\Delta w_j \mathbf{p}_j}{1+\gamma\Delta w_j}, \tag{12.13}$$

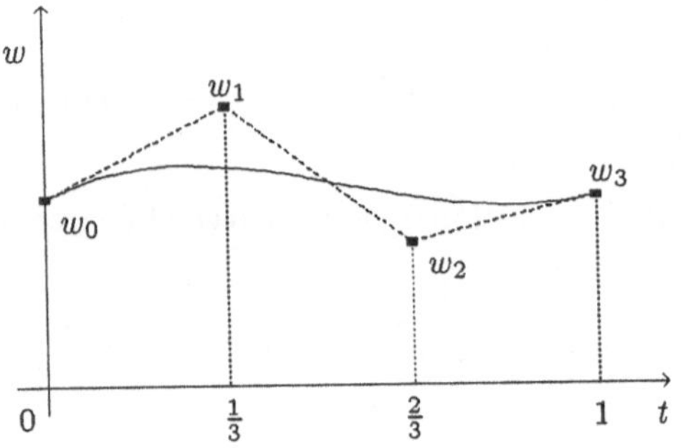

Figure 12.4 Behavior of a weight function of degree three

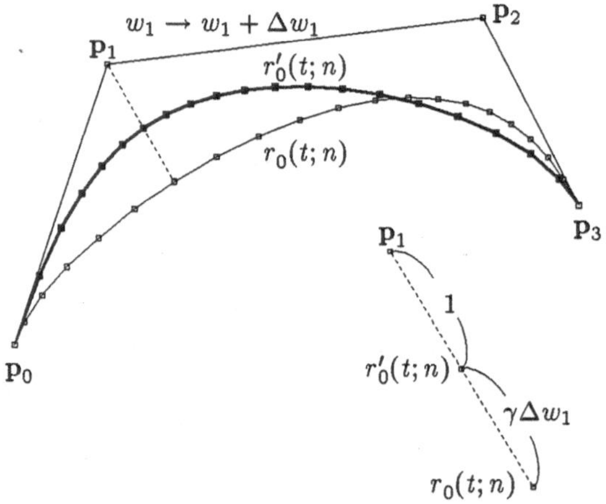

Figure 12.5 Variation of shape of a rational Bézier curve with weight

where $\gamma = C_j^n\ t^j(1-t)^{n-j}/w_0(t;n)$. The right-hand side of this equation can be interpreted as a rational interpolation between the original $\mathbf{r}_0(t;n)$ and a control point $\mathbf{p}_j$, that is, division in the ratio $(\gamma\Delta w_j : 1)$. See Fig. 12.5.

Accordingly, all the points of the original curve move towards $\mathbf{p}_j$. Whereas if a control point $\mathbf{p}_j$ moves by $\Delta\mathbf{p}_j$, all the points move in the same direction as that of $\Delta\mathbf{p}_j$.

So far, we have assumed that the weights are greater than zero, but here we treat a case of a weight of zero. If $w_i = 0$, then the value of the denominator becomes small at $t = (i/n)$, but always positive. Though the control point $\mathbf{p}_i$ has no influence on the shape of the curve, the summation of the weighting functions of the other control points is equal to 1. So the generated curve $\acute{\mathbf{r}}(t,w)$ is contained in a convex hull made by the control points without the vertex $\mathbf{p}_i$. Now we consider a case in which a control point $\mathbf{p}_i$ is moved to infinity in a direction $\Delta\mathbf{p}_i$ as the weight w_i approaches zero, while a product $w_i\mathbf{p}_i$ converges to a constant relative vector $\Delta\mathbf{p}_i$. Then a relative vector $\Delta\mathbf{r}(t)$ which is equal to $\Delta\mathbf{p}_i$ multiplied by its weighting function affects the shape of the curve $\mathbf{r}$ in the following way,

$$\mathbf{r}(t,w) = \acute{\mathbf{r}}(t,w) + \Delta\mathbf{r}(t), \tag{12.14}$$

where $\Delta\mathbf{r}(t) = \Delta\mathbf{p}_i(C_i^n)(1-t)^i t^{n-i}/w_0(t;n)$. An example of the use of zero weight is given in Sect. 12.6.2.

For curves of degree two, the weights classify the curves into an ellipse, a parabola and a hyperbola. This is discussed in more detail in Sect. 12.6.

12.2.4 Division and Degree Elevation

Division and degree elevation of a rational Bézier correspond to those of its linear one, which are easily performed. When $\hat{\mathbf{r}}_0(t;n)$ is divided at $t=\tau$ into two Bézier curves $\hat{\mathbf{r}}_0^a(t;n)$ and $\hat{\mathbf{r}}_0^b(t;n)$, the locations of their control points are given by, from equations (9.24) and (9.25),

$$\begin{aligned}
\hat{\mathbf{p}}_i^a &= \mathbf{r}_0(\tau;i) = (1-\tau+\mathsf{E}\tau)^i\hat{\mathbf{p}}_0,\\
\hat{\mathbf{p}}_{n-i}^b &= \grave{\mathbf{r}}_n(\tau;i) = \{(1-\tau)\mathsf{E}^{-1}+\tau\}^i\hat{\mathbf{p}}_n,\\
i &= 0,\dots n.
\end{aligned} \tag{12.15}$$

A rational Bézier curve $\mathbf{r}_0(t;n)$ is divided at $t=\tau$ into two rational Béziers $\mathbf{r}_0^a(t;n)$ and $\mathbf{r}_0^b(t;n)$, the locations of whose control points are easily determined from eq. (12.15) or by the graphical method.

$$\begin{aligned}
\mathbf{p}_i^a &= \mathbf{r}_0(\tau;i) = \frac{(1-\tau+\mathsf{E}\tau)^i w_0\mathbf{p}_0}{(1-\tau+\mathsf{E}\tau)^i w_0}, \qquad (12.16)\\
\mathbf{p}_{n-i}^b &= \grave{\mathbf{r}}_n(\tau;i) = \frac{\{(1-\tau)\mathsf{E}^{-1}+\tau\}^i w_n\mathbf{p}_n}{\{(1-\tau)\mathsf{E}^{-1}+\tau\}^i w_n}.
\end{aligned}$$

Their weights are

$$\begin{aligned}
w_i^a(\tau;i) &= (1-\tau+\mathsf{E}\tau)^i w_0,\\
w_{n-i}^b(\tau;i) &= \{(1-\tau)\mathsf{E}^{-1}+\tau\}^i w_n.
\end{aligned}$$

For n=3, the locations of the control points of divided rational Bézier polygons are shown in Fig. 12.3. They are $(\mathbf{p}_0, \mathbf{p}_a, \mathbf{p}_d, \mathbf{p})$ and $(\mathbf{p}, \mathbf{p}_e, \mathbf{p}_c, \mathbf{p}_3)$, corresponding to $(\hat{\mathbf{p}}_0, \hat{\mathbf{p}}_a, \hat{\mathbf{p}}_d, \hat{\mathbf{p}})$ and $(\hat{\mathbf{p}}, \hat{\mathbf{p}}_e, \hat{\mathbf{p}}_c, \hat{\mathbf{p}}_3)$.

For elevation of degree of a rational Bézier, since new control points $\hat{\mathbf{p}}_i'$ of the corresponding linear Bézier are given from eq.(9.50) by

$$\hat{\mathbf{p}}_i' = \frac{1}{n+1}\{i\hat{\mathbf{p}}_{i-1} + (n+1-i)\hat{\mathbf{p}}_i\}, \qquad i = 0,\dots n+1,$$

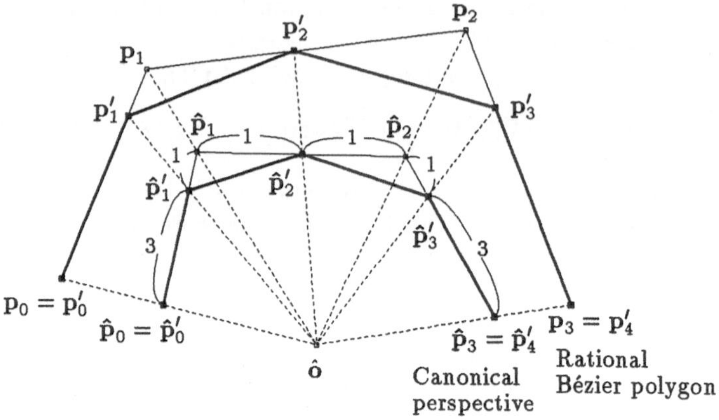

Figure 12.6 Degree elevation of a rational Bézier polygon

their rational equivalents $\mathbf{p}'_i$ become

$$\begin{aligned}
\mathbf{p}'_i &= \frac{iw_{i-1}\mathbf{p}_{i-1} + (n+1-i)w_i\mathbf{p}_i}{iw_{i-1} + (n+1-i)w_i}, \qquad (12.17)\\
w'_i &= \frac{iw_{i-1} + (n+1-i)w_i}{n+1},\\
i &= 0, \ldots n+1.
\end{aligned}$$

Alternatively, they are easily determined by the graphical method. See Fig. 12.6.

12.2.5 Derivatives at Ends of a Segment

The i-th derivatives at an end point of a linear Bézier curve are expressed by $i+1$ control points from the end. In a rational Bézier, the same statement holds, but its derivatives cannot be expressed by simple projection of those of the linear Bézier curve.

To avoid differentiation in a rational form, we change eq. (12.11) into the form:

$$(1 - t + \mathsf{E}t)^n w_0 \mathbf{r}_0(t; n) = (1 - t + \mathsf{E}t)^n w_0 \mathbf{p}_0. \qquad (12.18)$$

Differentiating eq. (12.18) i times by t and setting $t = 0$, we have

$$w_0\mathbf{r}_0^{(i)} + \sum_{j=1}^{i} C_j^i \frac{n!}{(n-j)!}(\mathsf{E}-1)^j w_0 \mathbf{r}_0^{i-j} = \frac{n!}{(n-i)!}(\mathsf{E}-1)^i w_0 \mathbf{p}_0.$$

From the above relation, $\mathbf{r}_0, \mathbf{r}_0^{(1)}, \mathbf{r}_0^{(2)}, \cdots$ can be obtained successively.

$$\mathbf{r}_0 = \mathbf{p}_0,$$

$$
\begin{aligned}
&w_0\mathbf{r}_0^{(1)} + n(w_1 - w_0)\mathbf{p}_0 = n(w_1\mathbf{p}_1 - w_0\mathbf{p}_0),\\
&w_0\mathbf{r}_0^{(2)} + 2n^2(w_1 - w_0)w_1/w_0(\mathbf{p}_1 - \mathbf{p}_0) + n(n-1)(w_2 - 2w_1 + w_0)\mathbf{p}_0\\
&\quad = n(n-1)(w_2\mathbf{p}_2 - 2w_1\mathbf{p}_1 + w_0\mathbf{p}_0).
\end{aligned}
$$

Then the derivatives at one end $t = 0$ are:

$$
\begin{aligned}
\mathbf{r}_0 &= \mathbf{p}_0,\\
\mathbf{r}_0^{(1)} &= n\frac{w_1}{w_0}(\mathbf{p}_1 - \mathbf{p}_0) && (12.19)\\
\mathbf{r}_0^{(2)} &= n(n-1)\frac{w_2}{w_0}(\mathbf{p}_2 - \mathbf{p}_1),\\
&+n\frac{\{2nw_1^2 - (n-1)w_0w_1 - 2w_0w_1\}}{w_0^2}(\mathbf{p}_1 - \mathbf{p}_0). && (12.20)
\end{aligned}
$$

With a similar manipulation, the derivatives at $t = 1$ are given by

$$
\begin{aligned}
\mathbf{r}_0 &= \mathbf{p}_n,\\
\mathbf{r}_0^{(1)} &= n\frac{w_{n-1}}{w_n}(\mathbf{p}_n - \mathbf{p}_{n-1}) && (12.21)\\
\mathbf{r}_0^{(2)} &= -n(n-1)\frac{w_{n-2}}{w_n}(\mathbf{p}_{n-2} - \mathbf{p}_{n-1}),\\
&+n\frac{\{2nw_{n-1}^2 - (n-1)w_n - 2w_n - 2w_{n-1}w_n\}}{w_n^2}(\mathbf{p}_n - \mathbf{p}_{n-1}). && (12.22)
\end{aligned}
$$

The tangent direction at an end of a rational Bézier expression is that of the first control edge of its polygon. The direction of the second derivative is determined by the difference of the second and the first edges, each multiplied by a different factor. Hence, its direction is not the same as that of the non-rational Bézier expression. Its magnitude of curvature at the end point is equal to that of a non-rational Bézier multiplied by a factor (w_0w_2/w_1^2).

12.3 Rational Bézier Patches

Since a linear Bézier patch of degree one is given by

$$\hat{\mathbf{s}}_{00}(u, v; 1, 1) = (1 - u + \mathsf{E}u)(1 - v + \mathsf{F}v)\hat{\mathbf{p}}_{00}, \tag{12.23}$$

its rational equivalent is

$$\mathbf{s}_{00}(u, v; 1, 1) = \frac{(1 - u + \mathsf{E}u)(1 - v + \mathsf{F}v)w_{00}\mathbf{p}_{00}}{(1 - u + \mathsf{E}u)(1 - v + \mathsf{F}v)w_{00}}, \tag{12.24}$$

where $\{\mathbf{p}_{ij}\}$ for $i, j = 0, 1$ is its control point matrix and $\{w_{ij}\}$ is its corresponding weight matrix. See Fig. 12.7. Equation (12.23) is a canonical perspective of eq. (12.24). Locations of the control points $\hat{\mathbf{p}}_{ij}$ of eq. (12.23) are determined on the line joining a virtual origin $\hat{\mathbf{o}}$ and $\mathbf{p}_{ij}$ so as to make $\hat{\mathbf{o}}\mathbf{p}_{ij} : \hat{\mathbf{o}}\hat{\mathbf{p}}_{ij} = 1 : w_{ij}$.

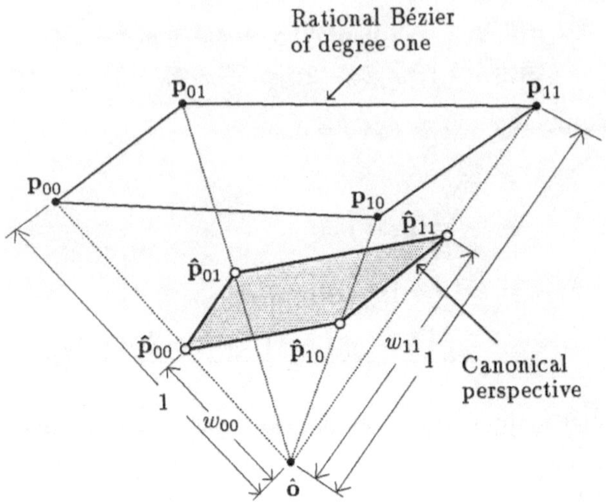

Figure 12.7 Rational Bézier patch of degree one

Equation (12.24) is derived also in the following way. A rational line segment $\mathbf{s}_{00}(0, v)$ joining $\mathbf{p}_{00}$ and $\mathbf{p}_{01}$ is given by

$$\mathbf{s}_{00}(0, v) = \frac{(1 - v + \mathsf{F}v)w_{00}\mathbf{p}_{00}}{(1 - v + \mathsf{F}v)w_{00}}, \tag{12.25}$$

and another rational line segment $\mathbf{s}_{00}(1, v)$ joining $\mathbf{p}_{10}$ and $\mathbf{p}_{11}$ is

$$\mathbf{s}_{00}(1, v) = \frac{(1 - v + \mathsf{F}v)w_{10}\mathbf{p}_{10}}{(1 - v + \mathsf{F}v)w_{10}}. \tag{12.26}$$

For $v = const$ a rational line segment joining $\mathbf{s}_{00}(0, v)$ and $\mathbf{s}_{00}(1, v)$, whose weights are $(1-v+\mathsf{F}v)w_{00}$ and $(1-v+\mathsf{F}v)w_{10}$, is given, referring to eq. (12.8) formallyby

$$\mathbf{s}_{00}(u, v) = \frac{(1 - u + \mathsf{E}u)(w'_0(v)p'_0(v))}{(1 - u + \mathsf{E}u)w'_0(v)} \tag{12.27}$$

where

$$\begin{aligned} p'_i(v) &= \frac{(1 - v + \mathsf{F}v)(w_{i0}\mathbf{p}_{i0})}{(1 - v + \mathsf{F}v)w_{i0}}, \\ w'_i(v) &= (1 - v + \mathsf{F}v)w_{i0} \ (i = 0, 1). \end{aligned}$$

Substituting the above equations into (12.27), we obtain eq. (12.24).

For $n = 2$, we make a rational Bézier patch of degree one whose four vertices are given by $\mathbf{s}_{00}(u, v; 1, 1), \mathbf{s}_{10}(u, v; 1, 1), \mathbf{s}_{01}(u, v; 1, 1)$ and $\mathbf{s}_{11}(u, v; 1, 1)$, that is, each point is on one of four adjacent rational Bézier patches of degree one, with weights $w_{ij}(u,v;1,1)$, $i,j=0,1$. This makes a rational Bézier patch of degree two:

$$\mathbf{s}_{00}(u,v;2,2)=\frac{(1-u+\mathsf{E}u)(1-v+\mathsf{F}v)(w_{00}(u,v;1,1)\mathbf{s}_{00}(u,v;1,1))}{(1-u+\mathsf{E}u)(1-v+\mathsf{F}v)w_{00}(u,v;1,1)} \tag{12.28}$$

where $w_{00}(u,v;1,1)=(1-u+\mathsf{E}u)(1-v+\mathsf{F}v)w_{00}$. The same expression can be obtained from the tensor product of two rational Bézier curve of degree two.

We make a general form of a rational Bézier patch of degree n when its control point matrix $\{\mathbf{p}_{ij}\}$ and weight matrix $\{w_{ij}\}$ are given. Since a canonical perspective of a Bézier surface patch is

$$\hat{\mathbf{s}}_{00}(u,v;n,n)=(1-v+\mathsf{F}v)^n(1-u+\mathsf{E}u)^n\hat{\mathbf{p}}_{00},$$

the corresponding rational Bézier surface patch is given by

$$\mathbf{s}_{00}(u,v;n,n)=\frac{(1-v+\mathsf{F}v)^n(1-u+\mathsf{E}u)^n(w_{00}\mathbf{p}_{00})}{(1-v+\mathsf{F}v)^n(1-u+\mathsf{E}u)^n w_{00}}. \tag{12.29}$$

Figure 12.8 shows examples of the shape of rational Bézier patches of degree three. The contour curves are shown to indicate variation of shapes according to the variation of values of weights at the different vertices. Effects of the weights on its shape are similar to those in the rational Bézier curve: each point on the patch is moved in the direction of a control point with increased weight.

12.4 Rational Splines

12.4.1 Rational *B* Polygons from a Rational *S* Polygon

From a spline polygon of degree n with its scale list, we can produce a series of Bézier polygons of degree n connecting in $C^{(n-1)}$. To distinguish a rational S polygon from a usual spline polygon, vertices of the latter are denoted by $\hat{\mathbf{q}}$ instead of $\mathbf{q}$. As explained in Chap. 11, locations of vertices of these B polygons are determined from those of the S polygon$(\cdots\hat{\mathbf{q}}_{-1},\hat{\mathbf{q}}_0,\hat{\mathbf{q}}_1,\hat{\mathbf{q}}_2,\cdots)$. Now attaching a weight k_i (to distinguish the weights of vertices of a rational S polygon from the weights w_i of vertices of Bézier polygons, different symbols k_i are used) to each vertex $\mathbf{q}_i$ of the rational S polygon we can deduce a series of rational Bézier polygons of degree n connecting in $C^{(n-1)}$.

Though we can use a canonical perspective of a rational spline polygon as we did in a rational Bézier, we utilize directly the result of the rational division for locating the rational B polygons from the rational S polygon.

As explained in Sect. 12.3 and here, locations of Bézier control points can be determined by repeated linear division of Menelaus edges of the given S polygon. Similarly, locations of rational Bézier control points are determined by repeated rational division of rational Menelaus edges of the rational S polygon. The

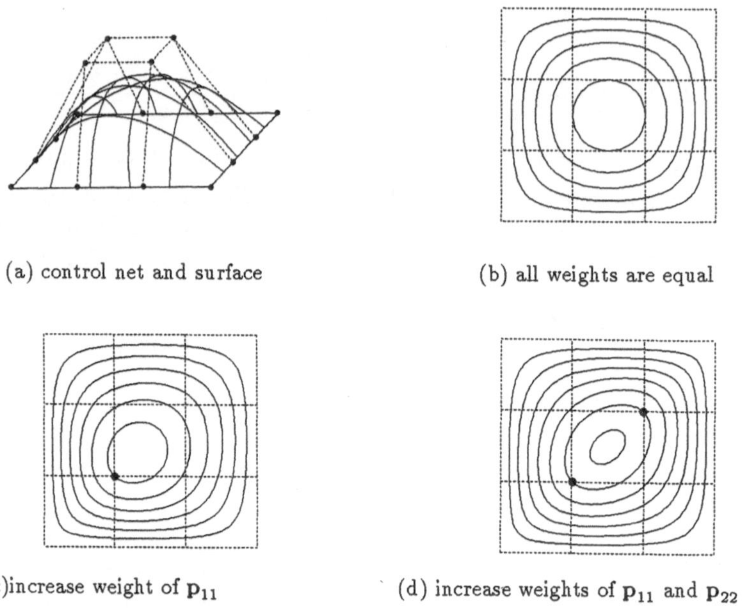
(a) control net and surface
(b) all weights are equal
(c)increase weight of $\mathbf{p}_{11}$
(d) increase weights of $\mathbf{p}_{11}$ and $\mathbf{p}_{22}$

Figure 12.8 Effects of weights on shape of a rational patch of degree three. Contour curves and the control point(s) of increased weight(s) indicated by dots are shown

rational Menelaus edge is similar to the Menelaus edge and its end points are rational dividing points of the rational Menelaus edge of one level lower. The rational dividing point has a weight which is determined by linear division of the weights of the end points of the edge. The location of a rational dividing point is at the dividing point in the ratio $(w'a : wb)$ of the edge, when its corresponding linear dividing point divides the edge in the ratio $(a : b)$ and has the weight 1. For instance, the rational dividing points $q_{0.1}$ and $q_{0.2}$ of an edge of a rational S polygon of degree three are located at the dividing points of ratios $(k_1\lambda_{-2} : k_0(\lambda_{-1} + \lambda_0))$ and $(k_1(\lambda_{-2} + \lambda_{-1}) : k_0\lambda_0)$ of the edge (q_0q_1). Their weights are the dividing value in the ratios $(\lambda_{-2} : (\lambda_{-1} + \lambda_0))$ and $((\lambda_{-2} + \lambda_{-1}) : \lambda_0)$ of k_0 and k_1, which are the weights of the vertices q_0 and q_1.

Accordingly, formulas of weights of the rational B points have the same form as those of the B points derived from the linear S polygon, but the symbols of vertices are replaced by its weights of the vertices. Formulas for the rational B points are similar to those of the B points from the linear S polygon, in which the linear dividing ratios are replaced by the rational dividing ratios. We give an example for $n = 2$.

Referring to the formula given in Chap. 11 and Fig. 12.9, vertices of a Bézier

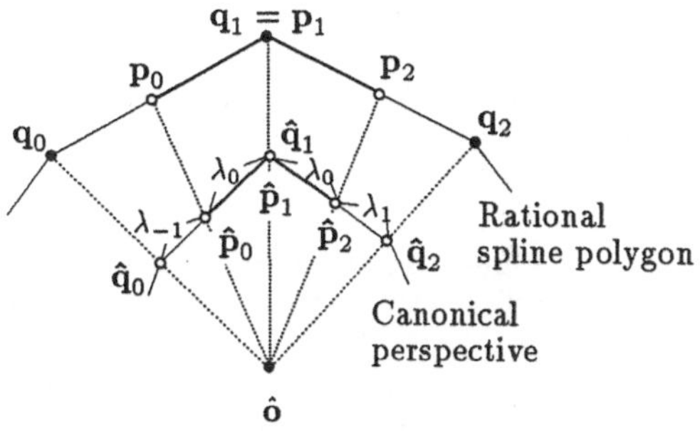

Figure 12.9 Connected rational Bézier polygons and curves of degree two from a rational spline polygon

polygon of degree two are determined from the unit S polygon of degree two:

$$\begin{aligned} \mathbf{p}_0 &= \frac{\lambda_0 \mathbf{q}_0 + \lambda_{-1} \mathbf{q}_1}{\lambda_0 + \lambda_{-1}}, \\ \mathbf{p}_1 &= \mathbf{q}_1, \\ \mathbf{p}_2 &= \frac{\lambda_1 \mathbf{q}_1 + \lambda_0 \mathbf{q}_2}{\lambda_1 + \lambda_0}. \end{aligned} \tag{12.30}$$

Replacing $\mathbf{p}_i$ by k_i, we obtain

$$\begin{aligned} w_0 &= \frac{\lambda_0 k_0 + \lambda_{-1} k_1}{\lambda_{-1} + \lambda_0}, \\ w_1 &= k_1, \\ w_2 &= \frac{\lambda_1 k_1 + \lambda_0 k_2}{\lambda_0 + \lambda_1}. \end{aligned} \tag{12.31}$$

We obtain the locations of the rational Bézier control points from eq. (12.30) by replacing the linear dividing ratios with the rational ones.

$$\begin{aligned} \mathbf{p}_0 &= \frac{\lambda_0 k_0 \mathbf{q}_0 + \lambda_{-1} k_1 \mathbf{q}_1}{\lambda_0 k_0 + \lambda_{-1} k_1}, \\ \mathbf{p}_1 &= \mathbf{q}_1, \\ \mathbf{p}_2 &= \frac{\lambda_1 k_1 \mathbf{q}_1 + \lambda_0 k_2 \mathbf{q}_2}{\lambda_1 k_1 + \lambda_0 k_2}. \end{aligned} \tag{12.32}$$

If a weight k_1 of the vertex $\mathbf{q}_1$ increases, Bézier points $\mathbf{p}_0$,$\mathbf{p}_2$ approach $\mathbf{q}_1$ and their weights w_0, w_2 increase. This makes the curve $\mathbf{r}_0(t; 2)$ shrink and its adjacent $\mathbf{r}_{-1}$ and $\mathbf{r}_1$ grow. An example is shown in Fig. 12.10.

Since the relation between a unit S polygon of degree n and its derived B polygon is formulated in Chap. 11, it is easy to write down its rational equivalents such as eq. (12.31) and eq. (12.32) for any degree n. Application of the graphical method for determining points on a rational spline curve is also easy.

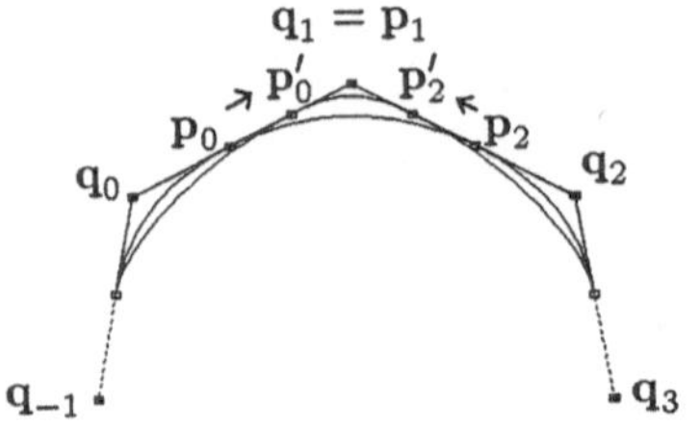

Figure 12.10 Shape variation of a spline curve with an increasing weight k_1

Rational Bézier curves derived from a rational S polygon of degree n connect in $C^{(n-1)}$. This is proved as follows. In the canonical perspective, $C^{(n-1)}$ connection of Bézier polygons of degree n is guaranteed by the S polygon of degree n. There is a connection defining polygon (CDP, refer to Sect. 10.2.4) of degree $n-1$ at each junction of the Bézier polygons, which can be constructed from the S polygon. Division of the CDP determines the Bézier points of the $C^{(n-1)}$ connecting B polygons. These B points together with the S polygon are mapped to the corresponding rational B points and the rational S polygon. Accordingly, the rational CDP guarantees $C^{(n-1)}$ connection of rational B polygons derived from the rational S polygon. Of course, the condition of $C^{(n-1)}$ connection is also proved by manipulation of the expressions of rational Béziers deduced from the rational S polygon.

12.4.2 $G^{(2)}$ Connection of Curves from a Rational S Polygon

If two rational Bézier polygons of degree three connect in $G^{(2)}$, the condition of curvature continuity at the junction of B_0 and B_1 gives the relation from the derivatives given in Sect. 12.2.5:

$$-\frac{w_3 w_5}{w_4^2}\frac{(\mathbf{n}\cdot\mathbf{a}_4)}{\mathbf{a}_3^2} = \frac{w_3 w_1}{w_2^2}\frac{(\mathbf{n}\cdot\mathbf{a}_1)}{\mathbf{a}_2^2}.$$

This equation is similar to eq. (10.27) of Sect. 10.4. Therefore, we obtain an equation similar to eq. (10.30) of Sect. 10.4. From this relation we can deduce an arbitrary constant σ by which locations of the control points can be adjusted.

If we change the linear dividing ratios from $(\lambda_{-2} : \lambda_{-1} : \lambda_0)$ and $(\lambda_{-1} : \lambda_0 : \lambda_1)$ to $(\sigma_0\lambda_{-2} : \lambda_{-1} : \sigma_1\lambda_0)$ and $(\sigma_1\lambda_{-1} : \lambda_0 : \sigma_2\lambda_1)$, where σ_i are the stretch factors, then by the rational division of the edges of level zero using the above revised ratios we determine the rational dividing points of level one and their weights. And we determine the rational B points of level two considering that their corresponding linear dividing ratios are $(\lambda_{-1} : \lambda_0)$ and $(\lambda_0 : \lambda_1)$.

Effects on derived shapes of displacing control points by changing values of σ_i are similar to the non-rational case. See Fig. 12.11.

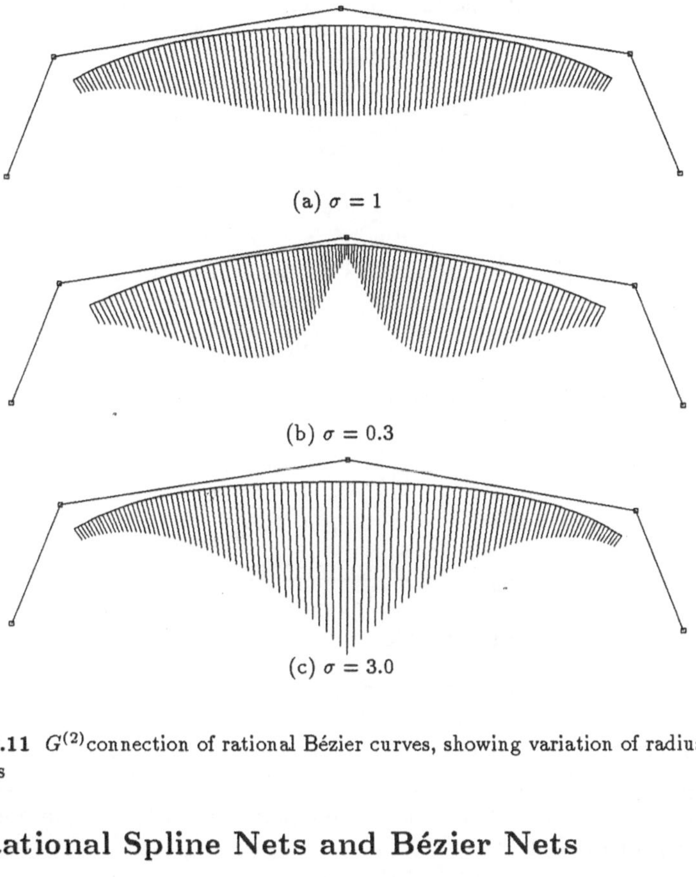

Figure 12.11 $G^{(2)}$connection of rational Bézier curves, showing variation of radius of curvature profiles

12.5 Rational Spline Nets and Bézier Nets

From a rational spline net we can determine its canonical perspective from which we obtain Bézier nets. Accordingly, we can easily determine their rational equivalents. Or we can use the tensor product method for the derivation of a rational spline surface. Anyway we can determine rational Bézier nets connecting in $C^{(n-1)}$ with their adjacent rational Bézier nets.

An example is shown in Fig. 12.12. Given a vertex matrix $\{\mathbf{q}_{ij}\}$ and a weight matrix $\{k_{ij}\}$ of a rational spline net of degree 2×2, we deduce those of a rational Bézier net of degree 2×2 by a tensor product scheme. From each column of $\{\mathbf{q}_{ij}\}$ and $\{k_{ij}\}$, for $j = 0, 1, 2$, we obtain intermediate Bézier points $\mathbf{p}'_{ij}$ and their weights k'_{ij}.

$$\begin{aligned} \mathbf{p}'_{0j} &= \frac{\lambda_0 k_{0j} \mathbf{q}_{0j} + \lambda_{-1} k_{1j} \mathbf{q}_{1j}}{\lambda_0 k_{0j} + \lambda_{-1} k_{1j}}, \\ \mathbf{p}'_{1j} &= \mathbf{q}_{1j}, \end{aligned}$$

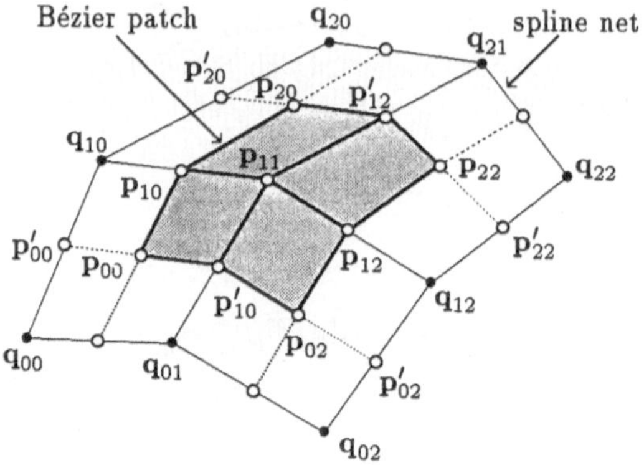

Figure 12.12 A rational Bézier net of degree 2 × 2 from a rational spline net of the same degree

$$
\begin{aligned}
\mathbf{p}'_{2j} &= \frac{\lambda_1 k_{1j}\mathbf{q}_{1j} + \lambda_0 k_{2j}\mathbf{q}_{2j}}{\lambda_1 k_{1j} + \lambda_0 k_{2j}}, \\
k'_{0j} &= \frac{\lambda_0 k_{0j} + \lambda_{-1} k_{1j}}{\lambda_{-1} + \lambda_0}, \\
k'_{1j} &= w_{1j}, \\
k'_{2j} &= \frac{\lambda_1 k_{1j} + \lambda_0 k_{2j}}{\lambda_0 + \lambda_1}.
\end{aligned}
$$

Bézier points $\{\mathbf{p}_{ij}\}$ of the patch and their weights $\{w_{ij}\}$ are determined from each row of the array of intermediate points and their weights obtained above, which we consider as rational spline polygons with the common scale list (μ_{-1}, μ_0, μ_1) for $i = 0, 1, 2$:

$$
\begin{aligned}
\mathbf{p}_{i0} &= \frac{\mu_0 k'_{i0}\mathbf{p}'_{i0} + \mu_{-1} k'_{i1}\mathbf{p}'_{i1}}{\mu_0 k'_{i0} + \mu_{-1} k'_{i1}}, \\
\mathbf{p}_{i1} &= \mathbf{p}'_{i1} \\
\mathbf{p}_{i2} &= \frac{\mu_1 k'_{i1}\mathbf{p}'_{i1} + \mu_0 k'_{i2}\mathbf{p}'_{i2}}{\mu_1 k'_{i1} + \mu_0 k'_{i2}}, \\
w_{i0} &= \frac{\mu_0 k'_{i0} + \mu_{-1} k'_{i1}}{\mu_0 + \mu_{-1}}, \\
w_{i1} &= k'_{i1}, \\
w_{i2} &= \frac{\mu_1 k'_{i1} + \mu_0 k'_{i2}}{\mu_1 + \mu_0}.
\end{aligned}
$$

Using the above data, the rational Bézier patch of degree 2 × 2 is given by

$$
\mathbf{s}_{00}(u, v; 2, 2) = \frac{(1 - u + \mathsf{E}u)^2(1 - v + \mathsf{F}v)^2 w_{00}\mathbf{p}_{00}}{(1 - u + \mathsf{E}u)^2(1 - v + \mathsf{F}v)^2 w_{00}} \tag{12.33}
$$

This patch connects with the adjacent patches $\mathbf{s}_{2i,2j}(u,v)$, for $i,j = -1,0,1$, in $C^{(1)}$. A technique similar to the above is applicable for the derivation of a rational Bézier patch of higher degree from a rational spline net of the same degree. Figure 12.13 shows a rational spline net and its generating rational Bézier patches connected in $C^{(1)}$. Projections of boundary curves together with parametric curves of the Bézier patches are also shown. Though they are very distorted by values of the weights, the contour curves of the surfaces are not so distorted and the shape variation is smooth. This shows that we cannot recognize the shape of surface only from its parametric curves.

Figure 12.14 shows a composite surface consisting of a surface of revolution smoothly connecting with a surface of elliptical cross section. This is an example of ability of the rational spline surfaces.

12.6 Expressions for Conics

12.6.1 Conversion to an Implicit Form

Now we examine the case of a rational Bézier expression of degree two, which is given by

$$\mathbf{r}(t;2) = \frac{(1-t)^2 w_0 \mathbf{p}_0 + 2t(1-t)w_1\mathbf{p}_1 + t^2 w_2 \mathbf{p}_2}{(1-t)^2 w_0 + 2t(1-t)w_1 + t^2 w_2}.$$

To observe effects of weights, transforming the above equation by

$$t = \frac{\tau}{(w_1/w_0)(1-\tau)+\tau}, \qquad w = \frac{w_1}{\sqrt{w_0 w_2}} \tag{12.34}$$

and then replacing τ by t in the above equation , we have

$$\mathbf{r}(t,w) = \frac{(1-t)^2\mathbf{p}_0 + 2t(1-t)w\mathbf{p}_1 + t^2\mathbf{p}_2}{(1-t)^2 + 2t(1-t)w + t^2}. \tag{12.35}$$

We change the notation from $\mathbf{r}(t;2)$ to $\mathbf{r}(t,w)$, because the single weight w is the important parameter and the degree two is already assumed.

The three weights are reduced to one. Now eliminating a parameter t, we make an implicit expression of eq. (12.35) in terms of x and y. The general method is explained in Sect. 5.6. For convenience, we introduce a variable u which replaces t:

$$u = \frac{t}{1-t}. \tag{12.36}$$

Then eq. (12.35) is changed to

$$\mathbf{r} = \frac{\mathbf{p}_0 + 2uw\mathbf{p}_1 + u^2\mathbf{p}_2}{1 + 2uw + u^2}. \tag{12.37}$$

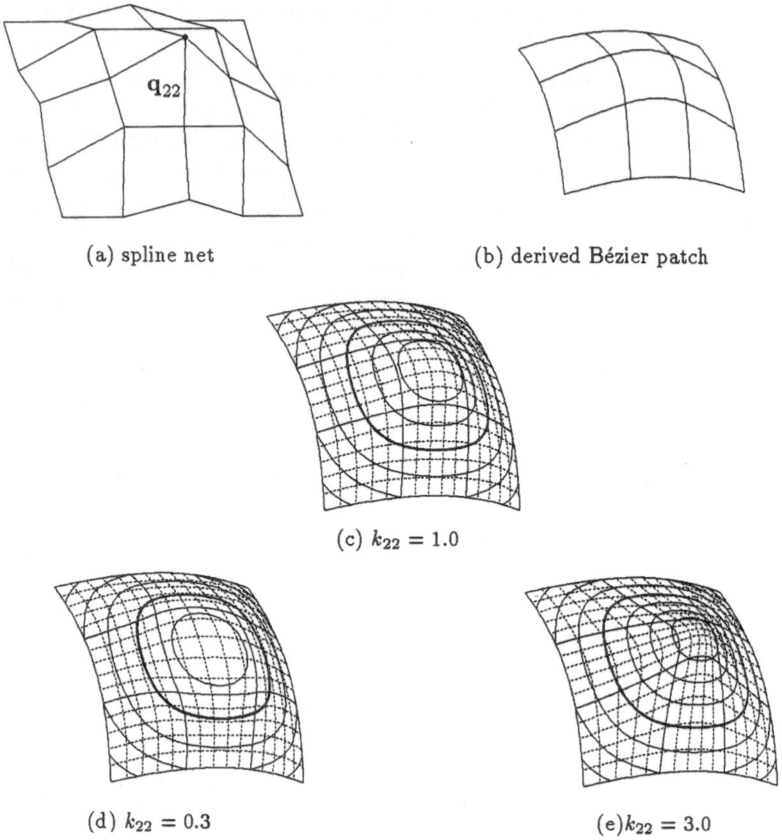

Figure 12.13 Effects of a weight k_{22} on the shape of a patch. Dotted lines are parameter lines, solid lines are boundary curves of a patch. Shapes of the patches may be grasped from their contour curves, but not from their parametric curves

Let coordinates of the points in eq. (12.37) be

$$\mathbf{o} = (0,0),\ \mathbf{p}_0 = (-c,0),\ \mathbf{p}_1 = (b,h),\ \mathbf{p}_2 = (c,0),\ \mathbf{r} = (x,y). \tag{12.38}$$

Resolving eq. (12.37) into its components and rearranging them, we obtain

$$\begin{aligned} u^2(c-x) + 2wu(b-x) - (c+x) &= 0, \\ u^2(-y) + 2wu(h-y) - y &= 0. \end{aligned} \tag{12.39}$$

Considering u^2 and u as unknowns, we get the solution of eq. (12.40):

$$u^2 = 2w\{(c+x)h - y(b+c)\}/D, \tag{12.40}$$

(a) Oblique view

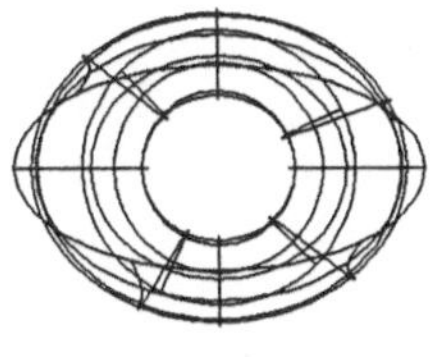

(b) Top view

Figure 12.14 Smooth connection of a surface of revolution with another surface

$$\begin{aligned} u &= 2cy/D, \\ \text{where } D &= 2w\{(c-x)h + y(b-c)\}. \end{aligned} \tag{12.41}$$

Eliminating u from the above equation, finally we obtain

$$h^2x^2 - 2hbxy + \{c^2/w^2 + (b^2 - c^2)\}y^2 + 2c^2hy - c^2h^2 = 0. \tag{12.42}$$

This represents a general conic equation which is described in Chap. 4.

12.6.2 Classification by Weight

Since the discriminant of this quadratic equation is given by $c^2h^2(1/w^2 - 1)$, the conic it expresses is an ellipse for $w < 1$, a parabola for $w = 1$ and a hyperbola for $w > 1$. By differentiating eq. (12.42) with respect to x and setting it to zero, we obtain a relation $hx = by$. This means the extremum value of y is on the median $\mathbf{op}_1$ and its extension. See Fig. 12.15. Its value y_0 and the corresponding value t_0 are obtained by using equations (12.42) and (12.41).

$$y_0 = \frac{wh}{1+w}, \qquad t_0 = \frac{1}{2}. \tag{12.43}$$

In eq. (12.42), w is not necessarily limited to a positive value. Equation (12.35) represents a curve segment for $w > 0$. If we want to extend the segment, we can do it by changing w to $-w$. The extended curve segment is also a part of the conic expressed by eq. (12.42). We can express this part together with the original segment in the parametric form of eq. (12.35) with new control points $\mathbf{p}_0, \mathbf{p}_1$ and $\mathbf{p}_2$. This technique is treated later in Sect. 4.4.2.

We examine the effects of the weight w on the shape of curve of eq. (12.42). Refer to Fig. 12.16.

- $w > 1$ or $w < -1$: the coefficient of y^2 is negative, so the curve is a hyperbola, and the y coordinate values y_0 of its intersecting points with the median are,

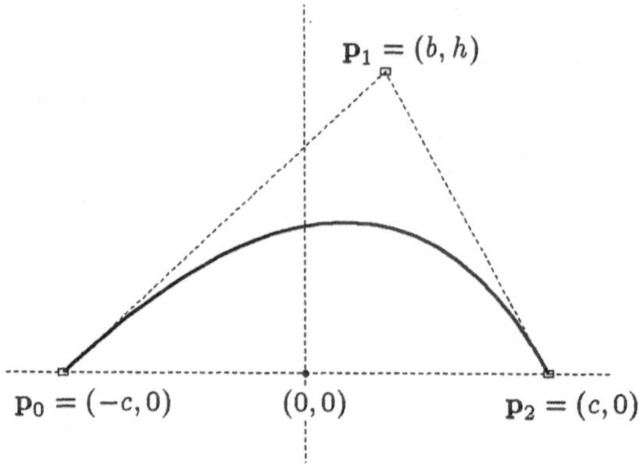

Figure 12.15 A rational Bézier control polygon of degree two and a conic segment

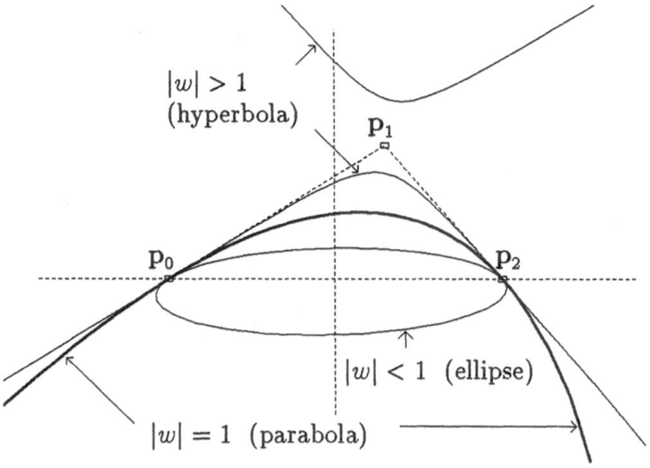

Figure 12.16 Conics and their weight

if $\infty > w > 1$, between h and ∞, and between h and $h/2$,
if $-1 > w > -\infty$, between $h/2$ and h, and between ∞ and h.

- $w = \pm 1$: the curve is a parabola and crosses the median at $y_0 = h/2$.
- $-1 < w < 1$: the curve is an ellipse, its intersecting points with the median are $y_0 = wh/(1 + w)$.
- When an absolute value of w is ∞ , the curve expressed by eq. (12.42) becomes

$$\{h(x - c) - y(b - c)\}\{h(x + c) + y(b + c)\} = 0,$$

which is two straight lines through $(\mathbf{p}_0, \mathbf{p}_1)$ and $(\mathbf{p}_1, \mathbf{p}_2)$.

The conditions that the ellipse given by eq. (12.42) is a circle are that the

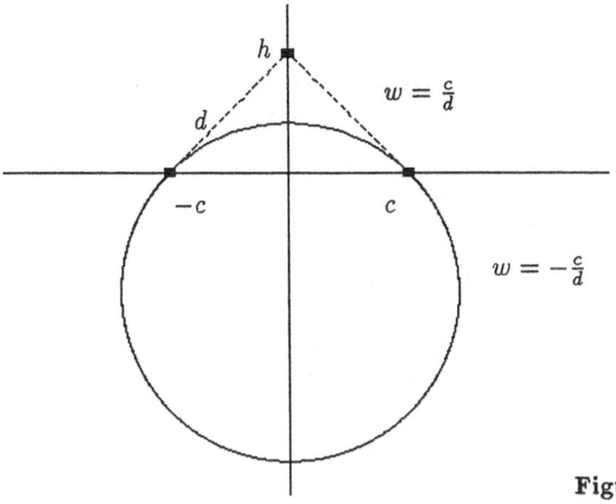

Figure 12.17 Representation of circular arcs

coefficients of x^2 and y^2 are equal and that of xy is zero. Hence we obtain $h^2 = c^2(1/w^2 - 1)$ and $b = 0$. Let d be distance between $\mathbf{p}_0$ and $\mathbf{p}_1$, then we have

$$w = \pm\frac{c}{\sqrt{h^2+c^2}} = \pm\frac{c}{d} \tag{12.44}$$

$$y_0 = \frac{ch}{c+d} \quad \text{or} \quad \frac{-ch}{d-c}. \tag{12.45}$$

If $w = c/d$, eq. (12.35) expresses the upper part of a circle above the chord $\mathbf{p}_0\mathbf{p}_2$, and if $w = -c/d$, it represents the lower part of the circle. See Fig. 12.17.

The case of $w = 0$ is treated in another way from the definition of the curve. Eq.(12.42) with $b = 0$ is

$$x^2 + c^2(1/h^2w^2 - 1/h^2)y^2 + 2c^2y/h - c^2 = 0, \tag{12.46}$$

Keeping the product of w and h equal to a constant c, if we increase h to infinity, then the above equation becomes: $x^2 + y^2 = c^2$, which is the equation of a circle of radius c. Thus, we can also make a rational Bézier expression (12.37) express a circle of radius c with its center at the origin by taking $\mathbf{p}_1$ to infinity while $w\mathbf{p}_1$ converges to a vector $\pm c\mathbf{j}$ and setting $\mathbf{p}_0$ and $\mathbf{p}_2$ to $-c\mathbf{i}$ and $c\mathbf{i}$:

$$\begin{aligned}\mathbf{r} &= \frac{(1-t)^2\mathbf{p}_0 + 2t(1-t)w\mathbf{p}_1 + t^2\mathbf{p}_2}{(1-t)^2 + 2t(1-t)w + t^2} \\ &= \frac{c\{(2t-1)\mathbf{i} \pm 2t(1-t)\mathbf{j}\}}{(1-t)^2 + t^2}.\end{aligned} \tag{12.47}$$

Exercise 1. Prove that eq. (12.47) represent an exact circle of radius c.
Exercise 2. Prove that in eq. (12.47), if $\mathbf{j}$ is replaced by a vector parallel to $\mathbf{op}_1$,

the resulting equation represents an ellipse with $\mathbf{p}_0\mathbf{p}_2$ and $\hat{\mathbf{o}}\mathbf{p}_1$ as its conjugate axes.

Exercise 3. Prove that a conic is uniquely determined if two points with directions of tangents there and any other point on the curve are given.

12.6.3 Sphere and Surface of Revolution

A surface of revolution can be expressed by control points of its meridian curve moving along circular orbits of latitudes. As a simple example, we give an equation of a sphere [17].

In the xy plane a half-circle of radius r with its center at the origin and its diameter on the y axis is expressed fromeq. (12.47) by

$$\mathbf{r} = \frac{(1-u)^2\mathbf{p}_0 + 2u(1-u)\mathbf{p}_1 + u^2\mathbf{p}_2}{(1-u)^2+u^2}, \tag{12.48}$$

where $\mathbf{p}_0 = (0, -r, 0)$, $\mathbf{p}_1 = (r, 0, 0)$ and $\mathbf{p}_2 = (0, r, 0)$.

If the control point $\mathbf{p}_1$ of the above equation rotates around the y axis with a parameter v, then $\mathbf{r}$ is a function of u and v and represents a sphere with its center at the origin. The locus of $\mathbf{p}_1$ is given by

$$\frac{(1-v)^2\mathbf{p}_{1,0} \pm 2v(1-v)\mathbf{p}_{1,1} + v^2\mathbf{p}_{1,2}}{(1-v)^2+v^2}, \tag{12.49}$$

where $\mathbf{p}_{1,0} = (r, 0, 0)$, $\mathbf{p}_{1,1} = (0, 0, r)$, $\mathbf{p}_{1,2} = (-r, 0, 0)$.

Substituting the above expression into $\mathbf{p}_1$ of eq. (12.48), we obtain a spherical surface:

$$\begin{aligned} \mathbf{s}(u,v) &= \frac{(1-u)^2\mathbf{p}_0 + u^2\mathbf{p}_2}{(1-u)^2+u^2} + \\ &+ \frac{2u(1-u)(1-v)^2\mathbf{p}_{1,0} \pm 2v(1-v)\mathbf{p}_{1,1} + v^2\mathbf{p}_{1,2}}{\{(1-u)^2+u^2\}\{(1-v)^2+v^2\}}. \end{aligned} \tag{12.50}$$

To check its spherity, we form the squared sum of the x, y and z components of eq. (12.50) and obtain r^2 which is equal to the squared value of the radius of the sphere.

If three points $\mathbf{p}_0, \mathbf{p}_1$ and $\mathbf{p}_2$ of eq. (12.48) rotate around a certain axis, a torus surface is constructed. If the control points of a plane curve are rotated around an arbitrary axis, then a surface of revolution is easily produced.

12.7 Interpolation and Extrapolation with Conics

12.7.1 Weight of a Control Point and Parameter Values

To make the geometric structure of $\mathbf{r}(t, w)$ clear we change the right-hand side of eq. (12.35) into the following form. Refer to Fig. 12.18.

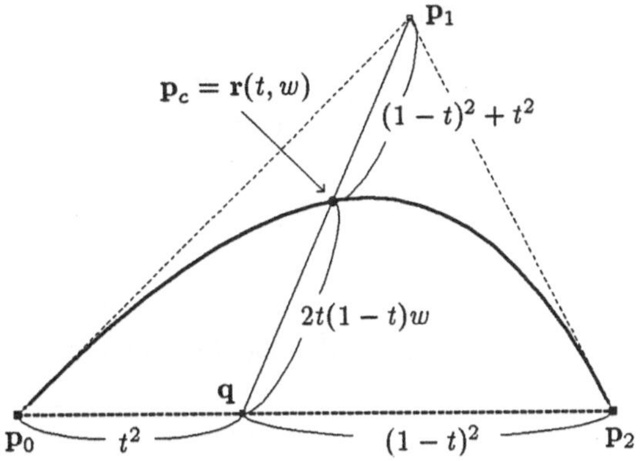

Figure 12.18 Structure of a conic segment

$$\mathbf{r}(t, w) = \frac{(1-t)^2 + t^2}{(1-t)^2 + 2wt(1-t) + t^2} \frac{(1-t)^2\mathbf{p}_0 + t^2\mathbf{p}_2}{(1-t)^2 + t^2} + \frac{2t(1-t)w\mathbf{p}_1}{(1-t)^2 + 2wt(1-t) + t^2}. \qquad (12.51)$$

Let $\mathbf{p}_c$ be a point on $\mathbf{r}(t, w)$ and $\mathbf{q}$ a dividing point of the chord $\mathbf{p}_0\mathbf{p}_2$ in the ratio $t^2 : (1-t)^2$. The second factor of the first term of the above equation is equal to $\mathbf{q}$. Accordingly, the right-hand side of eq. (12.51) gives a dividing point $\mathbf{p}_c$ between $\mathbf{p}_1$ and $\mathbf{q}$ in the ratio $(1-t)^2 + t^2 : 2t(1-t)w$.

We can determine a value of a parameter t of a point $\mathbf{p}_c$ on a curve and the weight of the curve from length ratios $\overline{\mathbf{p}_0\mathbf{q}} : \overline{\mathbf{q}\mathbf{p}_2}$ $(= m : n)$ and $\overline{\mathbf{p}_1\mathbf{p}_c} : \overline{\mathbf{p}_c\mathbf{q}}$ $(= m' : n')$:

$$t = \frac{\sqrt{m'}}{\sqrt{m'} + \sqrt{n'}}, \qquad (12.52)$$

$$w = (\frac{n}{2m})\{\sqrt{\frac{m'}{n'}} + \sqrt{\frac{n'}{m'}}\}. \qquad (12.53)$$

If $\mathbf{q}$ is the mid-point of $\mathbf{p}_0\mathbf{p}_2$, a vector $\mathbf{p}_c\mathbf{q}$ is a camber vector of the curve with the chord $\mathbf{p}_0\mathbf{p}_2$. This gives the extremum y value.

12.7.2 Division of a Rational Polygon

We divide the rational Bézier curve of eq. (12.35) at $t = \tau$ into two parts $\mathbf{r}^a$ and $\mathbf{r}^b$ with their control polygons $(\mathbf{p}_0^a\ \mathbf{p}_1^a\ \mathbf{p}_2^a)$ and $(\mathbf{p}_0^b\ \mathbf{p}_1^b\ \mathbf{p}_2^b)$. Their locations and weights given from eq. (12.16) are

$$\mathbf{p}_0^a = \mathbf{r}_0(\tau; 0) = \mathbf{p}_0,$$

$$
\begin{aligned}
\mathbf{p}_1^a &= \mathbf{r}_0(\tau;1) = \frac{(1-\tau+\mathsf{E}\tau)w_0\mathbf{p}_0}{(1-\tau+\mathsf{E}\tau)w_0}, \\
\mathbf{p}_2^a &= \mathbf{r}_0(\tau;2) = \frac{(1-\tau+\mathsf{E}\tau)^2 w_0\mathbf{p}_0}{(1-\tau+\mathsf{E}\tau)^2 w_0},
\end{aligned}
\tag{12.54}
$$

whose weights are

$$
\begin{aligned}
w_0^a &= 1, \\
w_1^a &= (1-\tau) + w\tau, \\
w_2^a &= (1-\tau)^2 + 2w\tau(1-\tau) + \tau^2,
\end{aligned}
\tag{12.55}
$$

and

$$
\begin{aligned}
\mathbf{p}_0^b &= \mathbf{r}_2(\tau;2) = \frac{\{(1-\tau)\mathsf{E}^{-1}+\tau\}^2 w_2\mathbf{p}_2}{\{(1-\tau)\mathsf{E}^{-1}+\tau\}^2 w_2}, \\
\mathbf{p}_1^b &= \mathbf{r}_2(\tau;1) = \frac{\{(1-\tau)\mathsf{E}^{-1}+\tau\} w_2\mathbf{p}_2}{\{(1-\tau)\mathsf{E}^{-1}+\tau\} w_2}, \\
\mathbf{p}_2^b &= \mathbf{r}_2(\tau;0) = \mathbf{p}_2,
\end{aligned}
\tag{12.56}
$$

whose weights are

$$
\begin{aligned}
w_0^b &= (1-\tau)^2 + 2w\tau(1-\tau) + \tau^2, \\
w_1^b &= (1-\tau)w + \tau, \\
w_2^b &= 1.
\end{aligned}
\tag{12.57}
$$

In this section the weight of the first and third control points of the rational Bézier polygon of degree two must have the value 1. Accordingly, we have to make a correction for the expression of the weights given by equations (12.55) and (12.57). By the relation (12.34), the weights (12.55) and (12.57) are changed to

$$
w_0^a = 1,\ w_1^a = \frac{w\tau + (1-\tau)}{\sqrt{(1-\tau)^2 + 2w\tau(1-\tau) + \tau^2}},\ w_2^a = 1, \tag{12.58}
$$

$$
w_0^b = 1,\ w_1^b = \frac{\tau + w(1-\tau)}{\sqrt{(1-\tau)^2 + 2w\tau(1-\tau) + \tau^2}},\ w_2^b = 1. \tag{12.59}
$$

12.7.3 Extension of a Curve Segment

When we try to fit a given sequence of points by connected curve segments of conics, we make a segment as large as possible within a given tolerance to reduce the number of conics, whose information has to be stored [34][35].

Accordingly, we consider extension of a curve segment which has been determined. Refer to Fig. 12.19. The segment $\mathbf{r}(t, w)$ starts from $\mathbf{p}_0$ and ends at $\mathbf{p}_2$ in

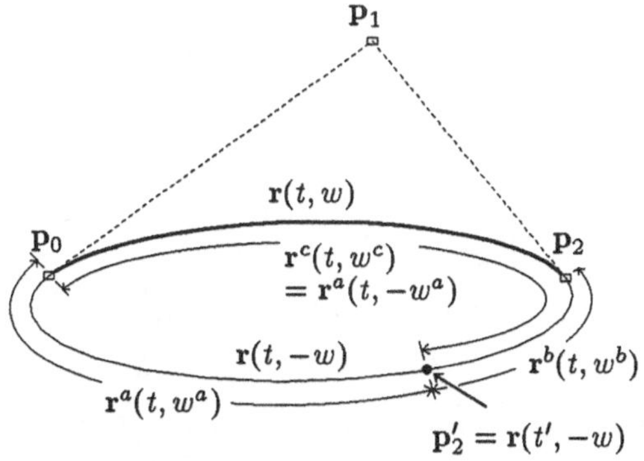

Figure 12.19 Extension of conic segment

the convex hull ($\mathbf{p}_0$-$\mathbf{p}_1$ $\mathbf{p}_2$). If we change the sign of the weight, the curve $\mathbf{r}(t,-w)$ starts from $\mathbf{p}_0$ and ends at $\mathbf{p}_2$, but it is outside of the convex hull. Hence, if we want to extend the curve segment $\mathbf{r}(t,w)$ beyond $\mathbf{p}_2$ in the same conic, we use $\mathbf{r}(t,-w)$ from $t=1$ to t' which corresponds to a segment from $\mathbf{p}_2$ to $\mathbf{p}_2'$. We want to combine the original curve $\mathbf{r}(t,w)$ from $t=0$ to $t=1$ and the extended segment $\mathbf{r}(t,-w)$ from $t=1$ to $t=t'$ into one segment $\mathbf{r}^c(t,w^c)$ with its control points ($\mathbf{p}_0$ $\mathbf{p}_1'$ $\mathbf{p}_2'$) and the weight w^c of the second control point $\mathbf{p}_1'$. Suppose that value of t' has been determined, then a procedure to determine $\mathbf{r}^c(t,w^c)$ or $\mathbf{p}_1'$ and w^c is given in the following:

1. The three control points ($\mathbf{p}_0$ $\mathbf{p}_1'$ $\mathbf{p}_2'$) are obtained by the division of $\mathbf{r}(t,-w)$ at $t=t'$. They are given by eq. (12.54) with $\tau=t'$ and $w_1=-w$.
2. The weight w^c is equal to $-w_1^a$, which is given by (12.58), with $\tau=t'$ and the sign of w reversed.

$$w^c = -\frac{-wt' + (1-t')}{\sqrt{(1-t')^2 - 2wt'(1-t') + t'^2}}. \tag{12.60}$$

The reason is as follows. At $t=t'$ we divide $\mathbf{r}(t,-w)$ into two parts, $\mathbf{r}^a(t,w^a)$ starting at $\mathbf{p}_0$ and ending at $\mathbf{p}_2'$, and $\mathbf{r}^b(t,w^b)$, which starts at $\mathbf{p}_2'$ and ends at $\mathbf{p}_2$ and is the extended part adjacent to $\mathbf{r}(t,w)$. Therefore, a curve segment whose weight is equal to $-1\times$ weight of $\mathbf{r}^a(t,w^a)$, that is, $\mathbf{r}^a(t,-w^a)$, starts at $\mathbf{p}_0$ and passes through $\mathbf{p}_2$ and ends at $\mathbf{p}_2'$. This is the curve segment$\mathbf{r}^c(t,w^c)$ which we want.

To understand this dividing processes we can use the canonical perspectives of these rational Béziers. Refer to Fig. 12.20(a). We set the virtual origin $\hat{\mathbf{o}}$ at the mid-point between $\mathbf{p}_0$ and $\mathbf{p}_2$. We construct a $(1/w)$-times magnified canonical Bézier polygon ($\hat{\mathbf{p}}_0$ $\hat{\mathbf{p}}_1$ $\hat{\mathbf{p}}_2$) of the rational Bézier polygon ($\mathbf{p}_0$ $\mathbf{p}_1$ $\mathbf{p}_2$) with respect to the virtual origin $\hat{\mathbf{o}}$. Then we have

$$\begin{aligned}\hat{\mathbf{o}}\hat{\mathbf{p}}_0 &= (1/w)\hat{\mathbf{o}}\mathbf{p}_0, \quad \hat{\mathbf{o}}\hat{\mathbf{p}}_2 = (1/w)\hat{\mathbf{o}}\mathbf{p}_2,\\ \hat{\mathbf{o}}\hat{\mathbf{p}}_1 &= (1/w)w\hat{\mathbf{o}}\mathbf{p}_1 = \hat{\mathbf{o}}\mathbf{p}_1.\end{aligned}$$

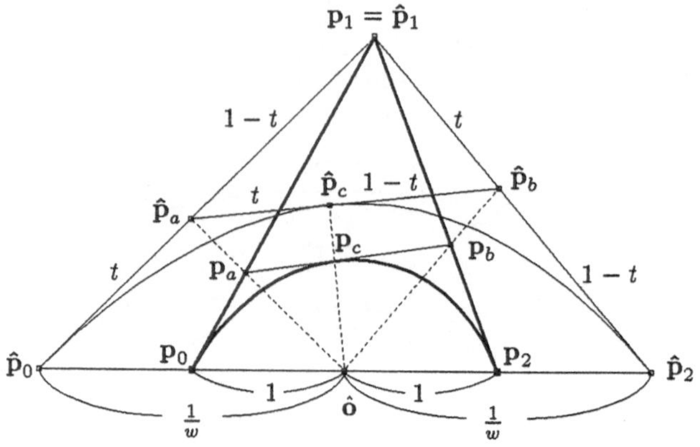

(a) Division of positive weight curve

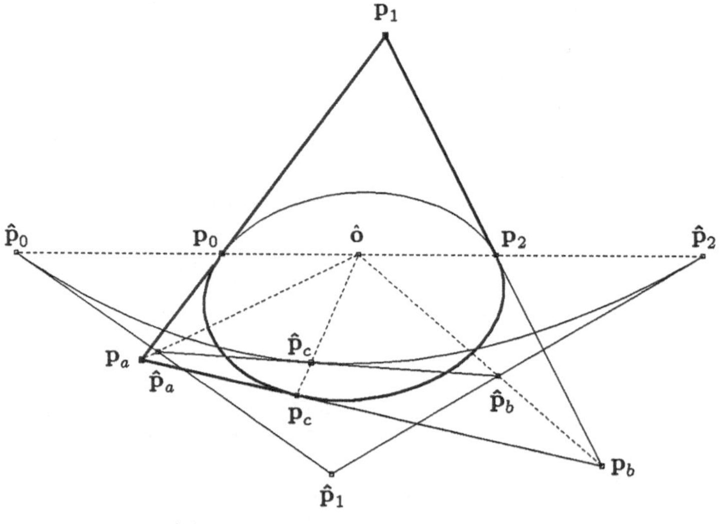

(b) Division of negative weight curve

Figure 12.20 Division of a rational polygon with a positive or a negative weight

The $(1/w)$-times magnification used above is just for convenience.

By the rational mapping from a canonical Bézier curve $\hat{\mathbf{r}}_0(t;2)$ produced from the above polygon to a rational one $\mathbf{r}(t;2)$, we obtain the rational Bézier curve starting from $\mathbf{p}_0$ and ending at $\mathbf{p}_2$ in the convex hull $(\mathbf{p}_0\ \mathbf{p}_1\ \mathbf{p}_2)$. This is $\mathbf{r}(t,w)$. Division of the B polygon $(\hat{\mathbf{p}}_0\ \hat{\mathbf{p}}_1\ \hat{\mathbf{p}}_2)$ at t produces two polygons $(\hat{\mathbf{p}}_0\ \hat{\mathbf{p}}_a\ \hat{\mathbf{p}}_c)$ and $(\hat{\mathbf{p}}_c\ \hat{\mathbf{p}}_b\ \hat{\mathbf{p}}_2)$. Their inverse mapping produces the divided rational B polygons $(\mathbf{p}_0\ \mathbf{p}_a\ \mathbf{p}_c)$ and $(\mathbf{p}_c\ \mathbf{p}_b\ \mathbf{p}_2)$ of the rational B polygon $(\mathbf{p}_0\ \mathbf{p}_1\ \mathbf{p}_2)$.

A similar dividing procedure is applied for the rational B polygon with a negative weight. Now to make $\mathbf{r}(t, -w)$, we construct the $(1/w)$-times magnified canonical Bézier polygon with the weight of its second control point changed from w to $-w$. Refer to Fig. 12.20(b). Then we have

$$\begin{aligned} \hat{\mathbf{o}}\hat{\mathbf{p}}_0 &= (1/w)\hat{\mathbf{o}}\mathbf{p}_0, \quad \hat{\mathbf{o}}\hat{\mathbf{p}}_2 = (1/w)\hat{\mathbf{o}}\mathbf{p}_2, \\ \hat{\mathbf{o}}\hat{\mathbf{p}}_1' &= (1/w)(-w)\hat{\mathbf{o}}\mathbf{p}_1 = -\hat{\mathbf{o}}\mathbf{p}_1. \end{aligned}$$

The points $\hat{\mathbf{p}}_a$ and $\hat{\mathbf{p}}_b$ are the dividing points of $\hat{\mathbf{p}}_0\hat{\mathbf{p}}_1'$ and $\hat{\mathbf{p}}_1'\hat{\mathbf{p}}_2$ in the ratio $t : 1-t$. And $\hat{\mathbf{p}}_c$ is the dividing point of $\hat{\mathbf{p}}_a\hat{\mathbf{p}}_b$ in the same ratio.
The intersecting point of $\hat{\mathbf{o}}\hat{\mathbf{p}}_a$ and the extension of $\mathbf{p}_1\mathbf{p}_0$ is $\mathbf{p}_a$.
The intersecting point of $\hat{\mathbf{o}}\hat{\mathbf{p}}_b$ and the extension of $\mathbf{p}_1\mathbf{p}_2$ is $\mathbf{p}_b$.
The intersecting point of $\mathbf{p}_a\mathbf{p}_b$ and the extension of $\hat{\mathbf{o}}\hat{\mathbf{p}}_c$ is $\mathbf{p}_c$.

The polygons $\mathbf{p}_0\mathbf{p}_a\mathbf{p}_c$ and $\mathbf{p}_c\mathbf{p}_b\mathbf{p}_2$ are the divided polygon of $\mathbf{p}_0\mathbf{p}_1\mathbf{p}_2$ with the weight $-w$ in the ratio $t : 1-t$. They are the divided rational B polygons $(\mathbf{p}_0^a\ \mathbf{p}_1^a\ \mathbf{p}_2^a)$ and $(\mathbf{p}_0^b\ \mathbf{p}_1^b\ \mathbf{p}_2^b)$.

12.7.4 Distance Between a Conic and a Point Near It

When a given point whose coordinate values are (ξ, η) is near a conic curve $f(x, y) = 0$, we can evaluate its nearest distance from the curve.

Let us define differences $\delta x, \delta y$ and a criterion F to be minimized:

$$\begin{aligned} \delta x &= x - \xi, \qquad \delta y = y - \eta, && (12.61) \\ F &= \delta x^2 + \delta y^2 + 2\lambda f(x, y), && (12.62) \end{aligned}$$

where λ is an auxiliary variable. We assume δx and δy are small values.

From $\partial F/\partial(\delta x) = 0$, $\partial F/\partial(\delta y) = 0$ and $\partial F/\partial\lambda = 0$, we obtain

$$\begin{aligned} \delta x + \lambda f_x &= 0, && (12.63) \\ \delta y + \lambda f_y &= 0, && (12.64) \\ f(x, y) &= f(\xi + \delta x, \eta + \delta y) \\ &= f(\xi, \eta) + f_x(\xi, \eta)\delta x + f_y(\xi, \eta)\delta y = 0. && (12.65) \end{aligned}$$

Eliminating λ, we have

$$\begin{aligned} f_y\delta x &- f_x\delta y = 0, && (12.66) \\ f_x\delta x &+ f_y\delta y = -f, && (12.67) \end{aligned}$$

from which we get

$$\delta x = \frac{-ff_x}{f_x^2 + f_y^2}, \qquad \delta y = \frac{-ff_y}{f_x^2 + f_y^2} \qquad (12.68)$$

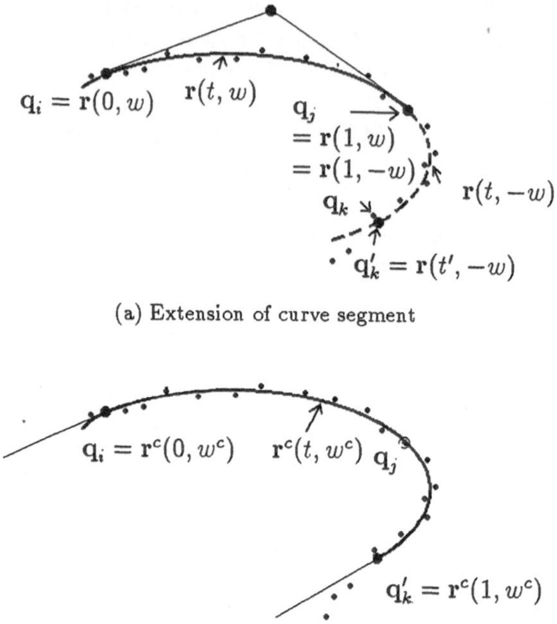

(a) Extension of curve segment

(b) New segment obtained by curve division

Figure 12.21 Extension of a conic segment for minimum number of fitting segments

and from eq. (12.61), the nearest point x, y on the curve is determined. The distance d is given by

$$d = \sqrt{(\delta x)^2 + (\delta y)^2} = \frac{f(\xi, \eta)}{\sqrt{f_x(\xi, \eta)^2 + f_y(\xi, \eta)^2}}. \tag{12.69}$$

This is used as a measure to select points within a given tolerance.

12.7.5 Curve Fitting by Conics

When a series of points $\mathbf{q}_i$ of a shape outline is given and we want to make a curve expression to approximate the shape for storage and processing, then to reduce the storage space, the curve is approximated by the least number of connected conics keeping the shape of the curve within a given tolerance. We use the range extension technique of a conic given in Sect. 12.7.3 and the tolerance evaluation technique in the previous section.

The outline of the curve fitting procedure is (refer to Fig. 12.21):.

1. From the given five consecutive points, determine the tangent direction at the mid-point and set the first control point at the mid-point. Alternatively, the points may be given by the expression for the previous segment.
2. Approximate determination of location of the third control point and the tangent direction there.
3. Determine the second control point from the intersection of the tangents at the first and the third control points. Then estimate the weight of the curve segment $\mathbf{r}$ using eq. (12.53).
4. Extension of the coverage of the conic:
   (a) Determine the last point in the given sequence which is within the tolerance, using the eq. (12.69).
   (b) Revise the old control points and the weight for the new rational Bézier $\mathbf{r}(t, w)$ by the method explained in Sect. 12.7.3.
5. Determine the tangent direction from $d\mathbf{r}/dt$ at the end point of the extended segment. This is also the direction of the tangent of the starting point of the next segment. Repeat from the beginning.

When a point sequence

$$(\cdots \mathbf{q}_{i-2}\, \mathbf{q}_{i-1}\, \mathbf{q}_i\, \mathbf{q}_{i+1}\, \mathbf{q}_{i+2} \cdots)$$

is given, a tangent direction $\mathbf{t}_i$ at $\mathbf{q}_i$ is approximated by linear combination of two vectors $(\mathbf{q}_{i+1} - \mathbf{q}_i)$ and $(\mathbf{q}_i - \mathbf{q}_{i-1})$:

$$\mathbf{t}_i = (1 - c_i)(\mathbf{q}_i - \mathbf{q}_{i-1}) + c_i(\mathbf{q}_{i+1} - \mathbf{q}_i),$$

where $(1 - c_i) : c_i$ is the ratio of contribution of the adjacent chord vectors to the tangent vector at their junction $\mathbf{q}_i$. The constant c_i is assumed to be given by

$$c_i = \frac{area(\mathbf{q}_i\mathbf{q}_{i+1}\mathbf{q}_{i+2})}{area(\mathbf{q}_i\mathbf{q}_{i-1}\mathbf{q}_{i-2})} = \frac{|(\mathbf{q}_{i+2} - \mathbf{q}_{i+1}) \times (\mathbf{q}_{i+1} - \mathbf{q}_i)|}{|(\mathbf{q}_i - \mathbf{q}_{i-1}) \times (\mathbf{q}_{i-1} - \mathbf{q}_{i-2})|}$$

The second control point $\mathbf{p}_1$ is determined by the intersection of $\mathbf{t}_0$ and $\mathbf{t}_2$. The weight of the conic is determined from eq. (12.53). We take in the new points to extend the conic and calculate the distances d between the conic $f(x, y) = 0$ and the given pointsusing eq. (12.69) until the distance exceeds the tolerance. Then the new control points and the weight are calculated for the new rational Bézier $\mathbf{r}(t, w)$ which covers $\mathbf{q}_0, \cdots, \mathbf{q}_j$ using the method in Sect. 12.7.3. To increase the accuracy of the fitting, we recalculate weight values w_i for each points using eq. (12.52) and eq. (12.53) and then we take their weighted mean as the new weight. The weighted mean is calculated by

$$\begin{aligned} w &= \frac{\sum e_i w_i}{\sum e_i} \\ \text{where } e_i &= \partial f(x_i, y_i)/\partial w|_{w=w_i}. \end{aligned}$$

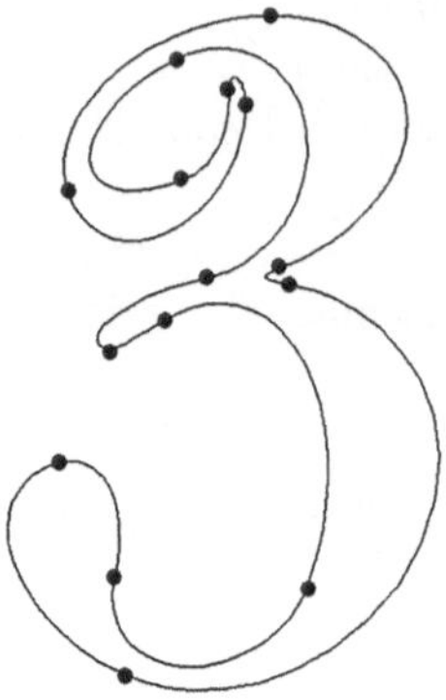

Figure 12.22 An example of a vector font (boundaries of curve segments are shown with dots)

The value e_i is the sensitivity of pseudo-distance change from the conic $f(x, y) = 0$, which is given from eq. (12.42) with the weight w_i. This value becomes small near the end controlpoints where error of w becomes large.

Figure 12.22 shows an application of this method to the construction of a vector font. The initial number of contour points produced from the data by the scanner is 695 and the number of segments determined by the method is 15.

13 Non-regular Connections of Four-Sided Patches and Roundings of Corners

13.1 Introduction

Usually in design of free-form surfaces, four-sided patches are connected to synthesize required shapes. For this purpose, S nets can be used advantageously because of their simplicity and the ease of shape control with the automatic adjustment of defined continuity along the boundaries of the patches.

However, shapes with special features cannot always adequately be expressed with S nets. For instance, box-like shapes, or fillets and roundings along the intersecting boundaries are difficult. To cope with them, we introduce another technique of smooth connection and synthesis of non-four-sided patches. They are produced by non-regular connection of B patches, blending of patches, boundary correction patches and triangular patches. These techniques are treated in this chapter and the following two chapters.

In the next three sections, we describe methods of the non-regular connection of Bézier patches to cover non-four-sided regions. Though only continuity of the tangential planes on the boundaries between adjacent patches is guaranteed, the non-regular connection of patches is useful for expressing rounding corners of various types which appear in products or for aesthetic representation in design. To easily determine the locations of control points of patches, we introduce polyhedra as the basic forms, from which the desired shape of patches can be constructed.

First, we explain notations used in this chapter as well as a basic technique and its simple application to make a closed surface with the minimum number of patches of degree two. We keep the degree of patches as low as possible to make calculation easy. After treating three types of non-regular regions, we explain the rolling ball method to blend a connecting region with an envelop surface of a rolling ball.

13.2 General $C^{(1)}$ Connection of B Patches

Let two Bézier patches $\mathbf{s}(u,v)$ and $\mathbf{s}'(u,v)$ connect along a common boundary which is expressed by $\mathbf{s}(0,v)$ or $\mathbf{s}'(0,v)$ and the parameter u of the both patches increase in the opposite directions along their $v = const$ curves. The connection of two B control nets of degree two is schematically shown in Fig. 13.1. We use the same symbols for the same geometric quantities of the patches, discriminating one from the other by attaching a prime mark.

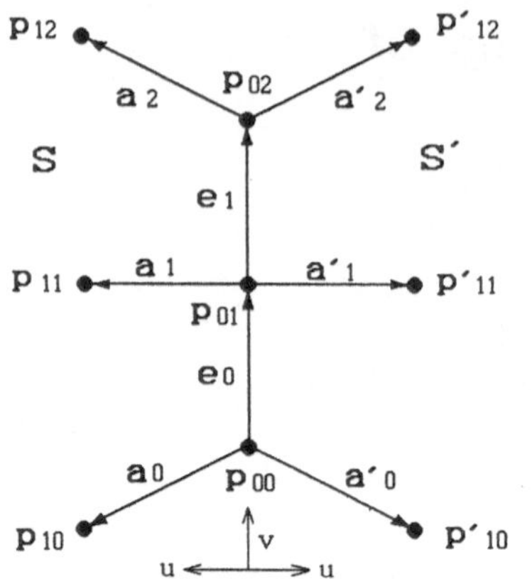

Figure 13.1 Notation of control points and edges for connection of B patches

The common B control points of the boundary of both patches are $\mathbf{p}_{00}$, $\mathbf{p}_{01}$, $\mathbf{p}_{02}$ or $\mathbf{p}'_{00}$, $\mathbf{p}'_{01}$, $\mathbf{p}'_{02}$ and their corresponding points coincide. The expressions of both patches are given by

$$\mathbf{s}(u,v) = (1-u+\mathsf{E}u)^2(1-v+\mathsf{F}v)^2\mathbf{p}_{00}, \tag{13.1}$$
$$\mathbf{s}'(u,v) = (1-u+\mathsf{E}u)^2(1-v+\mathsf{F}v)^2\mathbf{p}'_{00}, \quad 0 \le u,v \le 1. \tag{13.2}$$

On each boundary of a patch, we define the two tangent vectors: the *cross-boundary tangent* and the *along-boundary tangent.* The cross-boundary tangents are given by $\mathbf{s}_u(0,v)$ and $\mathbf{s}'_u(0,v)$ on the common boundary:

$$\mathbf{s}_u(0,v) = 2(1-v+\mathsf{F}v)^2\mathbf{a}_0, \tag{13.3}$$
$$\mathbf{s}'_u(0,v) = 2(1-v+\mathsf{F}v)^2\mathbf{a}'_0, \tag{13.4}$$

where $\mathbf{a}_j = \mathbf{p}_{1j} - \mathbf{p}_{0j}$ and $\mathbf{a'}_j = \mathbf{p'}_{1j} - \mathbf{p'}_{0j}$ for $j = 0,1,2$, which are the control vectors of the cross-boundary tangents. The along-boundary tangents on the common boundary $u = 0$ are given by

$$\mathbf{s}_v(0,v) = \mathbf{s'}_v(0,v) = 2(1-v+\mathsf{F}v)\mathbf{e}_0, \tag{13.5}$$
$$\text{where } \mathbf{e}_j = \mathbf{p}_{0(j+1)} - \mathbf{p}_{0j}, \quad j = 0,1. \tag{13.6}$$

The conditions of smooth connection of the tangential planes of both patches at any points on their boundary curve are:

(i) the cross-boundary tangent vectors of both patches are collinear, or
(ii) the two cross-boundary tangent vectors and the along-boundary tangent vector are coplanar.

For these conditions, the following equations (13.7) and (13.8) must hold for any values of v,

$$\lambda\mathbf{s}_u(0,v) + \lambda'\mathbf{s'}_u(0,v) = \{\alpha v + (1-v)\beta\}\mathbf{s}_v(0,v), \tag{13.7}$$
$$\lambda + \lambda' = 1, \tag{13.8}$$

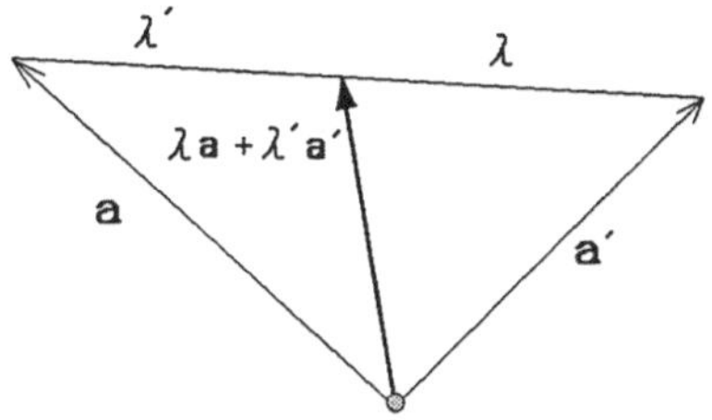

Figure 13.2 Geometric relation of vectors : $\lambda\mathbf{a} + \lambda'\mathbf{a}'$

where α, β, λ and λ' are constants to be determined. From condition (i), α and β must be 0, and from (ii) a scalar triple product $[\mathbf{s}_u, \mathbf{s}'_u, \mathbf{s}_v]$ is required to be 0. The latter condition is easily proved by substituting the left-hand side of eq. (13.7) into $\mathbf{s}_v$ in the triple product, in which two elements become the same one. Substituting equations (13.3) and (13.4) for $\mathbf{s}_u(0, v)$ and $\mathbf{s}'_u(0, v)$ into eq. (13.7), and equating the coefficients of the terms of $(1-v)^2, v(1-v)$ and v^2 to zero, we obtain three equations:

$$\lambda\mathbf{a}_0 + \lambda'\mathbf{a}'_0 = \alpha\mathbf{e}_0, \tag{13.9}$$

$$\lambda\mathbf{a}_1 + \lambda'\mathbf{a}'_1 = (\alpha\mathbf{e}_1 + \beta\mathbf{e}_0)/2, \tag{13.10}$$

$$\lambda\mathbf{a}_2 + \lambda'\mathbf{a}'_2 = \beta\mathbf{e}_1. \tag{13.11}$$

Referring to Fig. 13.2, the tails of $\mathbf{a}_i$ and $\mathbf{a}'_i$ are made coincident for $i = 0, 1, 2$. Because of the relation (13.8) between λ and λ', the left-hand sides of equations (13.9)–(13.11) are vectors, whose heads are at the dividing point of the line joining the heads of $\mathbf{a}_i$ and $\mathbf{a}'_i$, $i = 0, 1, 2$, in the ratio $\lambda' : \lambda$. Giving appropriate values for the constants α, β and λ, we can determine the control vectors $\mathbf{e}_0$, $\mathbf{e}_1$, $\mathbf{a}_i$ and $\mathbf{a}'_i$ and the tangent plane along the boundary of the patches.

13.3 Example of Closed Surface of Minimum Number of Patches

As a simple example of the patch connection which cannot be expressed by an S net, we show six identical four-sided patches making a closed surface with $C^{(1)}$ connection. This is not to make a sphere, which is made simply by two rational patches of degree two as shown in Sect. 12.6.3. Setting $\alpha = -1/2$, $\beta = 1/2$ and $\lambda = \lambda' = 1/2$ in equations (13.9)–(13.11), we have

$$\mathbf{a}_0 + \mathbf{a}'_0 = -\mathbf{e}_0 \tag{13.12}$$

$$\mathbf{a}_1 + \mathbf{a}'_1 = (-\mathbf{e}_1 + \mathbf{e}_0)/2 \tag{13.13}$$

$$\mathbf{a}_2 + \mathbf{a}'_2 = \mathbf{e}_1 \tag{13.14}$$

and we make the absolute values of vectors $\mathbf{e}_0, \mathbf{e}_1, \mathbf{a}_0$ and $\mathbf{a}_2$ all equal.

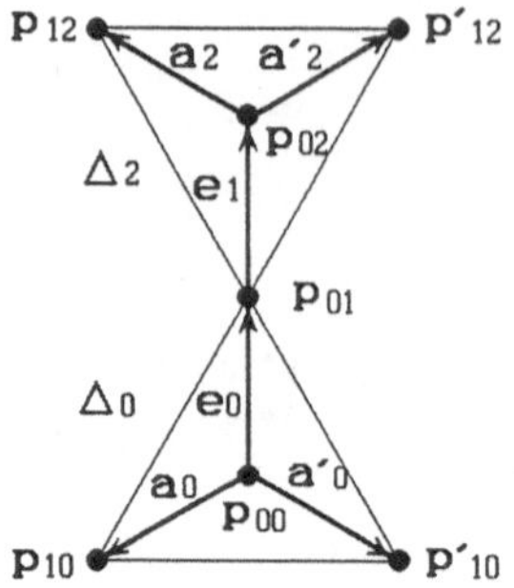

Figure 13.3 Locations of control points

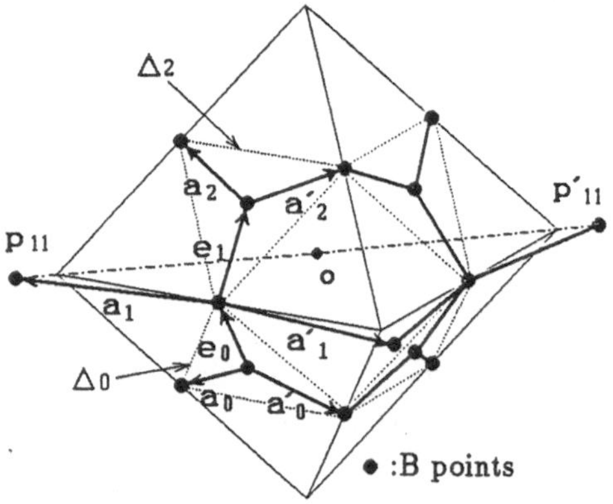

Figure 13.4 A regular octahedron and locations of control points

Referring to Fig. 13.3, since vectors $\mathbf{a}_0, \mathbf{a}_0'$ and $\mathbf{e}_0$ must be coplanar and their absolute values are equal, they make an angle of 120 degrees with each other. The control points $\mathbf{p}_{10}$, $\mathbf{p}_{10}'$ and $\mathbf{p}_{01}$ make an equilateral triangle Δ_0 with $\mathbf{p}_{00}$ at its centroid. Three vectors $\mathbf{a}_2, \mathbf{a}_2'$ and $-\mathbf{e}_1$ also are coplanar, hence $\mathbf{p}_{12}$, $\mathbf{p}_{12}'$ and $\mathbf{p}_{01}$ make an equilateral triangle Δ_2 with $\mathbf{p}_{02}$ at its centroid. Since six patches of the same shape enclose a space, there are eight corner points where three patches meet. The eight corner points of the patches become the centroids of the eight triangles Δ_i. Suppose a given regular octahedron with eight equilateral triangular faces. See Fig. 13.4. We set each of the corner points $\mathbf{p}_{00}$, $\mathbf{p}_{02}$, $\mathbf{p}_{20}$, $\mathbf{p}_{22}$, etc., at each centroid of the faces of the octahedron, and we put the vertices of Δ_i at the three mid-points of the edges of the face.

On each face of the octahedron, a small triangle like Δ_0 exists and its vertices are common with those of the similar small triangles on the adjacent faces. There are eight such triangles which control the connection of B patches. If we can determine six points $\mathbf{p}_{11}$ to satisfy eq. (13.13), we can obtain all the B points of the connected six patches in $C^{(1)}$. Let the length of an edge of the octahedron

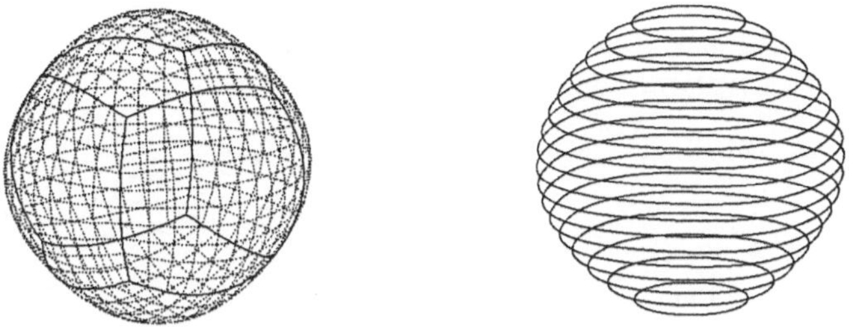

(a) (b)

Figure 13.5 Shape of connected patches: (a) parameter lines, (b) contour curves

be $2c$. Refer to Fig. 13.4. Extending outwards a line joining the center $\mathbf{o}$ of the octahedron and one of its vertices by length $c/(3\sqrt{2})$, we set a point $\mathbf{p}_{11}$ at its end. Since $\mathbf{a}_1 = \mathbf{p}_{11} - \mathbf{p}_{01}$ and $\mathbf{a}'_1 = \mathbf{p}'_{11} - \mathbf{p}_{01}$, the left-hand side of eq. (13.13) is a vector from the center $\mathbf{o}$ to $\mathbf{p}_{01}$ with the absolute value $c/3$. The right-hand side of eq. (13.13) becomes also the same vector, because the absolute values of $\mathbf{e}_1$ and $\mathbf{e}_0$ are $(c/\sqrt{3})$.

Thus all the B points of the patches of the second degree are determined. Figure 13.5(a) shows the constituent patches with constant parameter curves. The contour curves of the synthesized surface in Fig. 13.5(b) show smooth connection of the patches though their parametric curves have corners on the boundaries.

13.4 Three or Five-Sided Patch in Regular Patch Nets

Next we will explain methods to determine B points when there are elements of three or five-sided meshes in a regular four-sided S net. We treat cases appearing frequently in the rounding of corner parts.

13.4.1 Rounding of a Convex Region

We explain a rounding procedure of a region which is roughly approximated by the shape around a convex vertex with three convex edges. Refer to Fig. 13.6. By cutting each of the convex edges by a plane parallel to its edge, and also cutting off the central pyramid generated by the previous cutting procedures, we can make a primary body as shown in Fig. 13.6. In the center, there is a triangular region shown in grey, which is to be treated specially for the rounding of the corner. The surface outside of this region is easily rounded by using the S net. The rounded surface of the central triangular region is synthesized by the three connected B patches B_1, B_2 and B_3. We assume all B patches are of second degree and the dividing ratios of each S net in the u and v directions are both $1:1$ for simplicity. Figure 13.7 is its schematic diagram indicating the

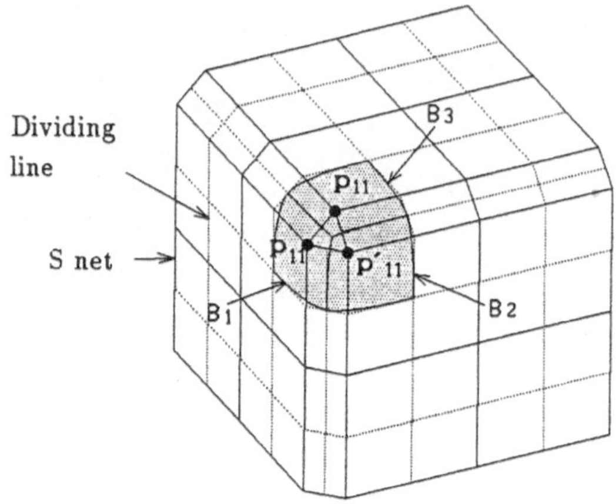

Figure 13.6 A triangular region at a convex corner

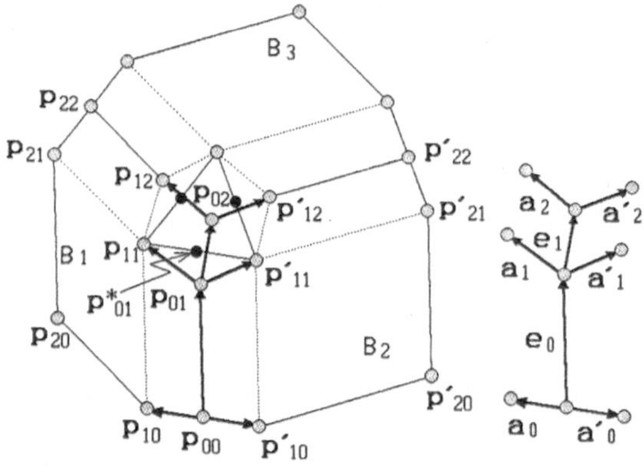

Figure 13.7 Locations of B points of three patches

locations of central control points $\mathbf{p}_{11}$ and the constituent B nets B_1, B_2 and B_3. We are to determine the locations of the control points of each B patch from the positions of the vertices of the cut block.

The B control points of B_1, B_2 and B_3 are determined by the following procedures:

1. The control points $\mathbf{p}_{11}$ and $\mathbf{p}'_{11}$ of B_1 and B_2 are two vertices of the central triangle made by the cutting process, and the third vertex corresponds to $\mathbf{p}_{11}$ of B_3.
2. Refer to Fig. 13.7. Let a mid-point of the edge of the triangle be $\mathbf{p}^*_{01}$. A lower corner control point of $\mathbf{p}_{00}$, which is the same as $\mathbf{p}'_{00}$, is fixed on the symmetrical line as shown in the figure. A dividing point between $\mathbf{p}^*_{01}$ and

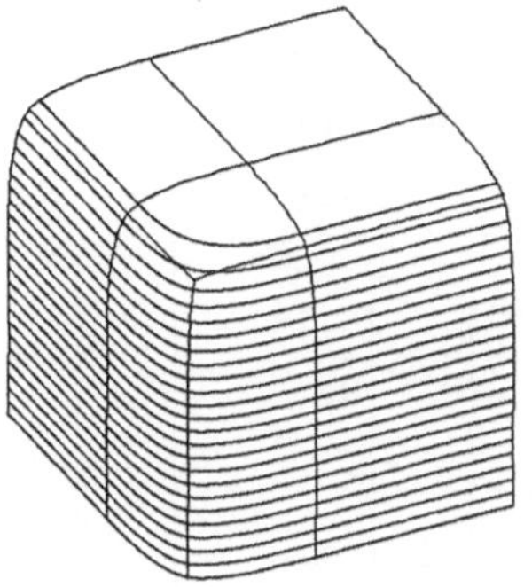

Figure 13.8 Contour curves of connected patches

$\mathbf{p}_{00}$ in the ratio (1 : 4) becomes $\mathbf{p}_{01}$. Similar points to this $\mathbf{p}_{01}$ exist between B_2 and B_3 and between B_3 and B_1. They are the control points $\mathbf{p}'_{12}$ and $\mathbf{p}_{12}$.

3. The centroid of these three points $\mathbf{p}_{01}, \mathbf{p}'_{12}$ and $\mathbf{p}_{12}$ becomes $\mathbf{p}_{02}$, which is the meeting point of the three boundaries of B patches B_1, B_2 and B_3.
4. The control points $\mathbf{p}_{10}, \mathbf{p}_{20}$ and $\mathbf{p}'_{10}, \mathbf{p}'_{20}$ of B_1 and B_2 are determined from the surrounding S net. Those of B_3 are similarly determined.

A proof of the $C^{(1)}$ connection of these patches is: on the three boundaries of B_1, B_2 and B_3, equations (13.9)–(13.11) with values of the constants

$$\lambda = \lambda' = 1/2, \ \alpha = 0, \ \beta = 1/2,$$

become

$$\begin{aligned} \mathbf{a}_0 + \mathbf{a}'_0 &= 0, & (13.15)\\ (1/2)\mathbf{a}_1 + (1/2)\mathbf{a}'_1 &= (1/4)\mathbf{e}_0, & (13.16)\\ \mathbf{a}_2 + \mathbf{a}'_2 &= \mathbf{e}_1. & (13.17) \end{aligned}$$

For instance, we consider the connection of patches B_1 and B_2. Since $\mathbf{p}_{02}$ is the centroid of three points $\mathbf{p}_{01}, \mathbf{p}_{12}$ and $\mathbf{p}'_{12}$ of the patches B_1 and B_2, eq. (13.17) holds. The left-hand side of eq. (13.16) is equal to $\mathbf{p}_{01}\mathbf{p}^*_{01}$ and the right-hand side is a quarter of $\mathbf{p}_{00}\mathbf{p}_{01}$, hence they are equal by the definition of the location of $\mathbf{p}_{01}$. Equation (13.15) holds naturally. An example is shown in Fig. 13.8. Smoothness check by the contour curves gives good result.

The above is the case when the dividing line ratios $\lambda : \lambda'$ of the surrounding S net are taken as 1 : 1, but even when the dividing ratios of the net are $(\lambda_1 : \lambda'_1)$, $(\lambda_3 : \lambda'_3)$, $(\lambda_5 : \lambda'_5)$ as indicated in Fig. 13.9, we can determine B points of three B patches connected in $C^{(1)}$, if the following condition holds among the dividing ratios:

$$(\lambda_1/\lambda'_1)(\lambda_3/\lambda'_3)(\lambda_5/\lambda'_5) = 1. \qquad (13.18)$$

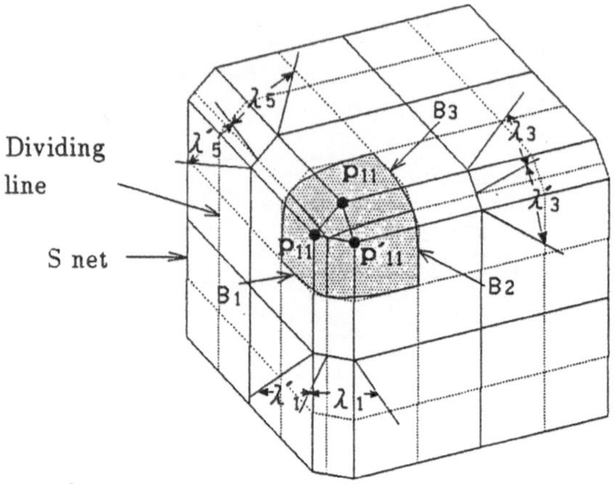

Figure 13.9 A triangular region at a convex corner for non-equal λ (cf. Fig. 13.6)

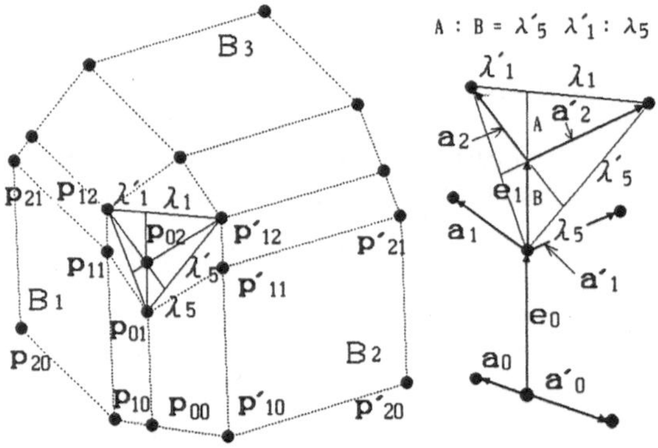

Figure 13.10 Location of B point of three patches for non-equal λ (cf. Fig. 13.7)

Figures 13.9 and 13.10 correspond to Figures 13.6 and 13.7. The proof is given in Sect. 13.7.1.

13.4.2 Rounding of a Convex-Concave Mixed Region 1

We explain a method of rounding around a concave vertex with mixed concave and convex edges. See Fig. 13.11. For simplicity, we assume the two edges are convex and the one concave, and the shape is symmetrical with respect to the bisecting plane of the concave angle. We want to cover this region by five B patches of degree two connected in $C^{(1)}$. This process is rather complicated. If we are allowed to use patches of rational type, we can divided the region into two instead of five(refer to Chap. 14).

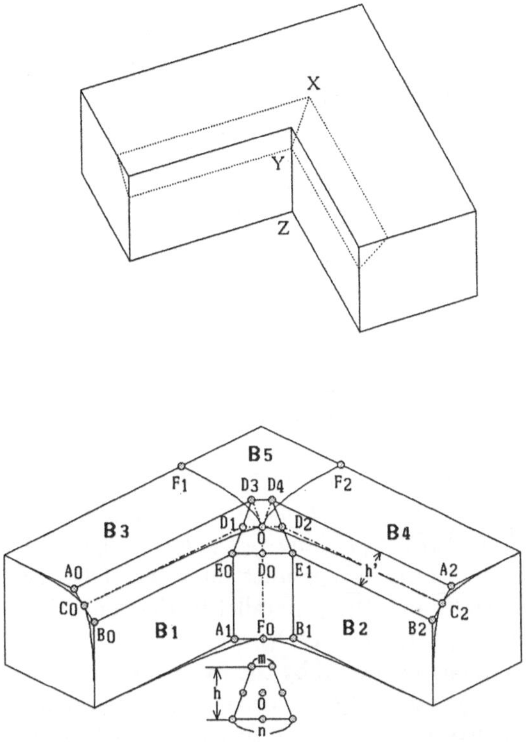

(a)

(b)

Figure 13.11 A concave corner and five patches: (a) basic block, (b) modified block

Before making five B patches, we modify the basic block shown in Fig. 13.11(a) to adapt to suitable B polygons as follows.

1. Make two new faces by cutting off the convex edges to decrease the convexity. The new faces are parallel to their original convex edges. The new planes intersect each other making a new concave edge $\overline{XY}$ which connects with the original concave edge $\overline{YZ}$. Fill the concave space along the connected concave edges with a triangular prismatic column and a truncated pyramid to decrease their concavity. See Fig. 13.11(b).
2. Four new faces are produced with the above procedures: in the central part, a trapezoidal face (E_0E_1, D_4D_3) whose height is h; in the left and right-hand sides two bands of faces (A_0D_3, B_0E_0) and (A_2D_4, B_2E_1) parallel to the respective convex edges, and let their widths be h'. We call them the convex edge-bands; in the lower vertical part there is a face (E_0A_1, E_1B_1) parallel to the original concave edge, which is the concave edge-band. When the slopes of the convex edge-bands and that of the trapezoidal face are equal, their widths are equal, $h = h'$. These edge-bands connect with the

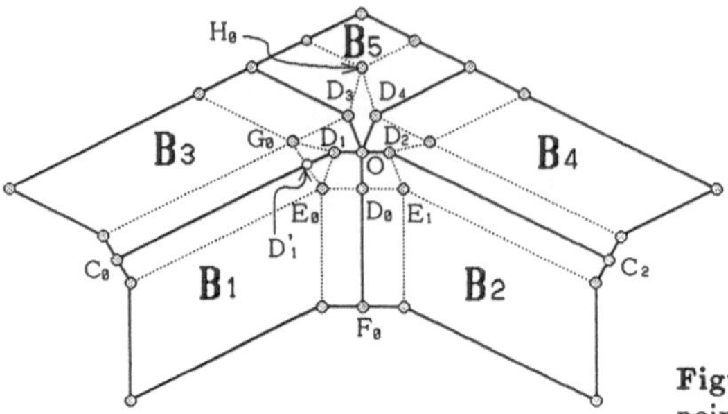

Figure 13.12 Locations of all control points

trapezoid along $\overline{E_0E_1}$, $\overline{E_0D_3}$ and $\overline{E_1D_4}$. Let the length of the upper edge $\overline{D_3D_4}$ of the trapezoid be m and that of the lower edge be n, and let the mid-points of $\overline{E_0E_1}$, $\overline{E_0D_3}$ and $\overline{E_1D_4}$ be D_0, D_1 and D_2, and the mid-point of $\overline{D_1D_2}$ be O. The ratio $m : n$ can be arbitrary, but if we take $1 : 3$, the values of various geometric relations become simple.

3. We draw mid-lines $\overline{F_0D_0}, \overline{C_0D_1}$ and $\overline{C_2D_2}$ in the edge-faces and extend them into the trapezoidal face. These lines intersect at the center point O. We draw also lines $\overline{D_3F_1}$ and $\overline{D_4F_2}$, parallel to the convex edges, that is, $\overline{C_2D_2}$ and $\overline{C_0D_1}$. Connect $\overline{OD_3}$, and $\overline{OD_4}$.
4. There are five regions separated by the five boundaries: (OD_0F_0), (OD_1C_0), (OD_3F_1), (OD_4F_2) and (OD_2C_2). We name these regions B_1, B_2, B_3, B_4 and B_5. From these five regions we are to make B nets which define smoothly connected B surface patches.

Figure 13.12 shows the completed disposition of their control points. The equations (13.9)–(13.11) are applied for each boundary of a pair of the adjacent B patches. Since the two bottom vectors $\mathbf{a}_0$ and $\mathbf{a}'_0$ are collinear (refer to Fig. 1), $\alpha=0$ in eq. (13.9). The equations which stipulate the connections for each pair of B_i are:

$$\lambda\mathbf{a}_0 + \lambda'\mathbf{a}'_0 = 0 \tag{13.19}$$
$$\lambda\mathbf{a}_1 + \lambda'\mathbf{a}'_1 = \beta\mathbf{e}_0/2 \tag{13.20}$$
$$\lambda\mathbf{a}_2 + \lambda'\mathbf{a}'_2 = \beta\mathbf{e}_1. \tag{13.21}$$

Though we can take arbitrary values for h, h' and m and n, hereafter we assume $h = h'$ and $3m = n$, for simplicity of explanation. Then the slopes of the convex edge bands and the central trapezoidal part are equal and ratio of the upper edge and the lower edge of the trapezoid is $1 : 3$. This makes determining the locations of B points simple. General cases are described in Sect. 13.7.2. We treat three pairs of connection: (B_1 and B_2), (B_2 and B_3) and (B_3 and B_5). The other connections (B_2 and B_4) and (B_4 and B_5) are the same as (B_2 and B_3) and (B_3 and B_5). We explain the method of locating the control points of each pair.

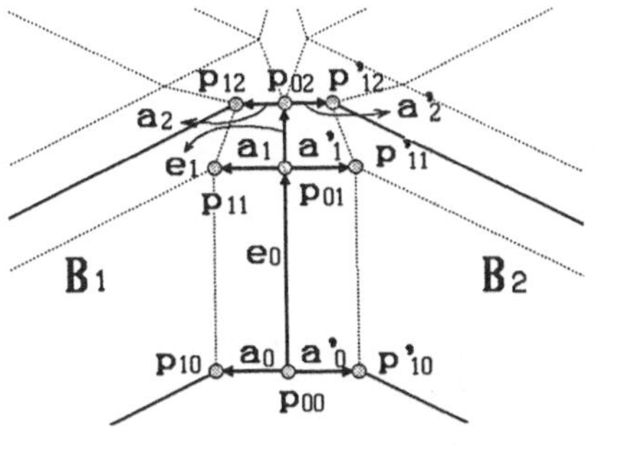

Figure 13.13 Connection of B_1 and B_2

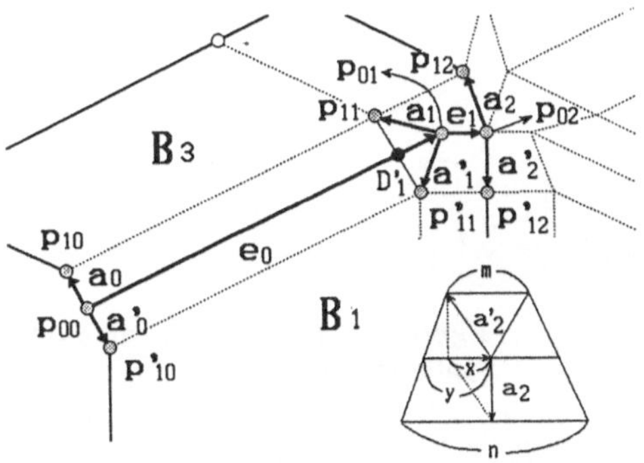

Figure 13.14 Connection of B_1 and B_3

- We treat first the pair (B_1 and B_2). Refer to Figures 13.13 and 13.12. The boundary is (OD_0F_0). The points E_0 and E_1 become $\mathbf{p}_{11}$ and $\mathbf{p}'_{11}$. The point D_0 is $\mathbf{p}_{01}$ and F_0 is $\mathbf{p}_{00}$ which is determined from outside conditions. Then D_1, D_2 become the control points $\mathbf{p}_{12}, \mathbf{p}'_{12}$, mid-point of which is O or $\mathbf{p}_{02}$. For this case, $\lambda = \lambda' = 0.5$, $\beta = 0$ in equations (13.19)–(13.21).
- We treat the pair (B_1 and B_3). Refer to Figures 13.14 and 13.12. The prime mark of a vector indicates the vector belong to B_1. Their common boundary is (OD_1C_0). The control vectors are $\mathbf{e}_1 = \overline{D_1O}$, $\mathbf{e}_0 = \overline{C_0D_1}$, $\mathbf{a}_2 = \overline{OD_3}$, $\mathbf{a}'_2 = \overline{OD_0}$. From the configuration of the central trapezoid, the constants λ, λ' and β in eq. (13.21) are determined:

$$\lambda = \lambda' = 1/2, \quad \beta = -1/4. \tag{13.22}$$

Accordingly, from eq. (13.20) whose left-hand side is equal to $\overline{D_1D'_1}$, the length of $\mathbf{e}_0$ is determined as $8\overline{D_1D'_1}$. Since $\mathbf{p}'_{11}$ is at E_0, the location of $\mathbf{p}_{11}$ is its symmetrical point G_0 with respect to $\overline{C_0D_1}$. Thus all the B points are fixed.
- We consider the pair B_3 and B_5. Refer to Figures 13.15 and 13.12. The prime mark belongs to the patch B_3. The boundary is (OD_2F_1). The control point $\mathbf{p}'_{11}$ of B_3 is the same location G_0 as that of $\mathbf{p}_{11}$ of B_3 of the pair (B_1, B_3).

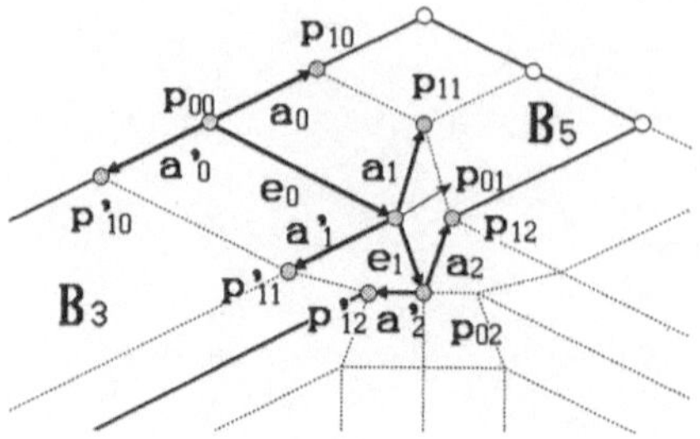

Figure 13.15 Connection of B_3 and B_5

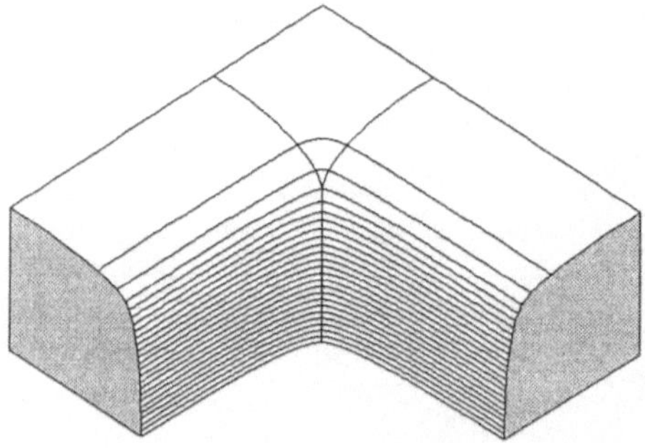

Figure 13.16 Contour curves of the synthesized region

The control edges are $\mathbf{e}_0 = \overline{D_3F_1}$, $\mathbf{e}_1 = \overline{D_3O}$ and $\mathbf{a}_2' = \overline{OD_1}$, $\mathbf{a}_2 = \overline{OD_4}$. We have $\mathbf{p}_{02} = O$ and $\mathbf{p}_{01} = D_3$, $\mathbf{p}_{11}' = G_0$ and $\mathbf{p}_{11} = H_0$, where H_0 is located on the line whose distance from $\mathbf{a}_2$ is the same as that of $\mathbf{p}_{11}'$ and $\mathbf{e}_2$ and in the symmetrical bisecting plane of the initial concave angle. From these relations and (13.21) we can determine the following values:

$$\lambda = \lambda' = 1/2, \quad \beta = -1/2 \tag{13.23}$$

In this way all the B points of the five B nets are determined as shown in Fig. 13.12 and the $C^{(1)}$ connections of the five patches $B_1 \cdot B_5$ are attained. Figure 13.16 shows the contour curves of the synthesized surface.

13.5 Rounding of a Convex-Concave Mixed Region 2

Next we treat a case of rounding around a vertex at which six edges meet. Its block model is shown in Fig. 13.17. Six orthogonal faces intersect to make three convex edges and three concave edges interlaced. We define its three convex edges as the coordinate axes with its origin $\mathbf{o}$. Extensions of the convex edges beyond the origin coincide with the concave edges. We treat the case of orthogonal edges, but the method can be generalized for non-orthogonal cases.

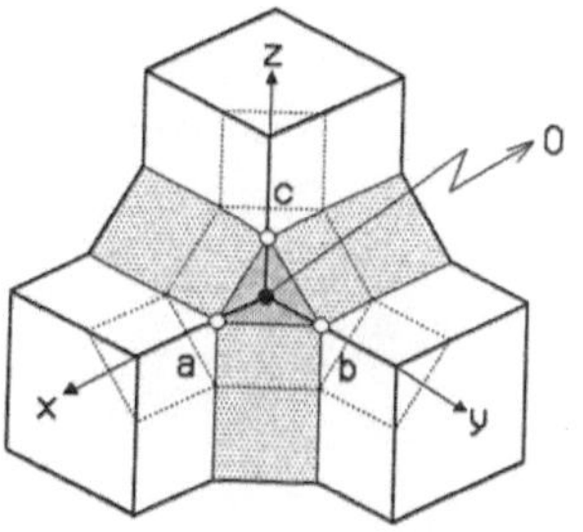

Figure 13.17 Block model of a mixed corner

A mixed region of this type cannot be synthesized by patches of degree two only, because of the insufficient number of control points of the constituent patches, but for patches of degree three there is a surplus of control points. Therefore, double control points in a patch are to be considered.

A basic block modification is performed as follows.

1. At first we apply the procedure of concave space filling and convex edge cutting to the original shape. Referring to Fig. 13.17, we fill the three concave edge spaces with the corresponding prismatic columns shown in grey. The cross section of the column is a rectangle isosceles triangle. Two side faces of the column are coplanar with the coordinates planes, and the three top edges of each of the columns make an equilateral triangle with vertices on each of the coordinate axes.
2. We cut these fillers by a plane parallel to the equilateral triangle stated above. The cutting plane reaches the side edges of three prismatic columns as shown in Fig. 13.18. Let the intersecting points be ($\mathbf{q}_6$ and $\mathbf{q}_1$), ($\mathbf{q}_2$ and $\mathbf{q}_3$), ($\mathbf{q}_4$ and $\mathbf{q}_5$).
3. Then we cut the convex edge of the original body by a plane parallel to the vertical edge and passing through the line $\mathbf{q}_6\mathbf{q}_5$, and remove the cut part of original convex edge. We call the produced face ($\mathbf{q}_6\mathbf{q}_5\mathbf{q}_{43}\mathbf{q}_{51}$) the convex edge-face and also we call a prismatic filler face ($\mathbf{q}_5\mathbf{q}_4\mathbf{q}_{32}\mathbf{q}_{41}$) the concave edge-face. Repeating the procedure for the other two convex edges, we obtain the shape shown in Fig. 13.18. In the central region, there is a hexagon with vertices $\mathbf{q}_6, \mathbf{q}_1, \mathbf{q}_2, \mathbf{q}_3, \mathbf{q}_4$ and $\mathbf{q}_5$.

Let the center of the hexagon be $\mathbf{q}_0$ and the mid-points of its edges be ($\mathbf{q}_{m6}, \mathbf{q}_{m1}, \mathbf{q}_{m2}, \mathbf{q}_{m3}, \mathbf{q}_{m4}$ and $\mathbf{q}_{m5}$); and the center lines in the edge-faces, each passing through $\mathbf{q}_{mi}$ and the center $\mathbf{q}_0$, be L_i. We are now to make six B nets connected to cover the central region, which generate B patches connected smoothly. Let the corresponding B nets be B_1, B_2, B_3, B_4, B_5 and B_6. We define their outer boundaries by the intersecting lines with the six cutting planes, each of which is parallel to one of the coordinate planes. We name the main node points on the

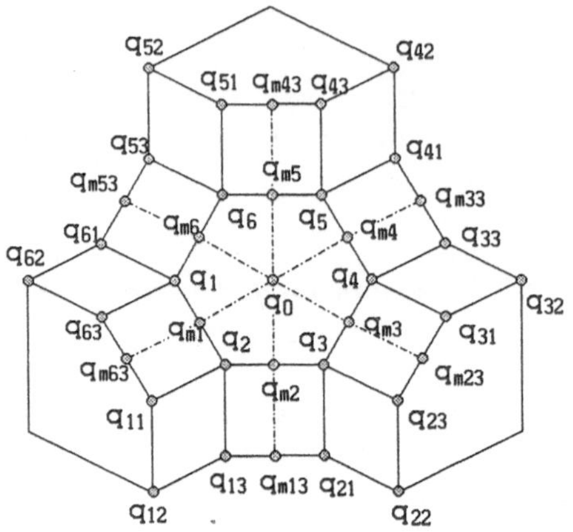

Figure 13.18 Modified block and names of specific points

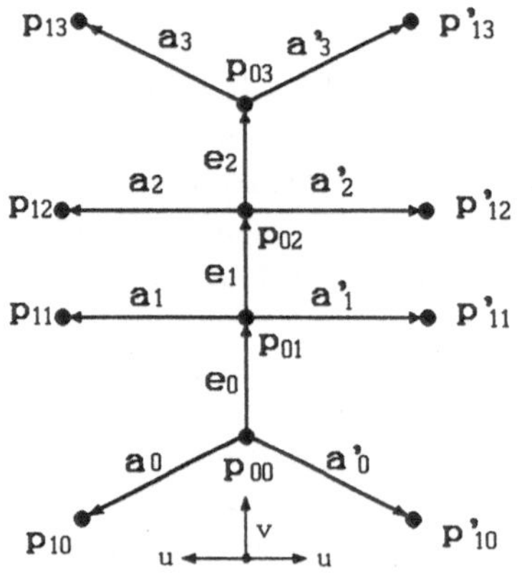

Figure 13.19 Relations between control points of connecting patches of degree three

outer boundaries. They are

$$(\mathbf{q}_{63}\ \mathbf{q}_{m63}\ \mathbf{q}_{11}\ \mathbf{q}_{12}\mathbf{q}_{13}\ \mathbf{q}_{m13}\mathbf{q}_{21}\ \mathbf{q}_{22}\ \mathbf{q}_{23}\ \mathbf{q}_{m23}\ \mathbf{q}_{31}\cdots)$$

as shown in Fig. 13.18. In this case, it is impossible to make a smooth region by connecting B patches of second degree, because no possible disposition of B points satisfies equations (13.9)–(13.11) for a shape of this type. Accordingly, we assume B patches of degree three.

We take a pair of adjacent patches to discuss their connection. Refer to Fig. 13.19. The patch expressions are given by

$$\mathbf{s}(u,v) = (1-u+\mathsf{E}u)^3(1-v+\mathsf{F}v)^3\mathbf{p}_{00} \tag{13.24}$$

$$\mathbf{s}'(u,v) = (1-u+\mathsf{E}u)^3(1-v+\mathsf{F}v)^3\mathbf{p}'_{00}. \tag{13.25}$$

On their boundary $u = 0$, the tangent vectors are

$$\begin{aligned} \mathbf{s}_u(0,v) &= 3(1-v+\mathsf{F}v)^3\mathbf{a}_0 && (13.26)\\ \mathbf{s}'_u(0,v) &= 3(1-v+\mathsf{F}v)^3\mathbf{a}'_0, && (13.27)\\ \mathbf{s}_v(0,v) &= 3(1-v+\mathsf{F}v)^2\mathbf{e}_0. && (13.28)\end{aligned}$$

For the $C^{(1)}$ connection, the following relation must hold, just as in the case of the second degree patches,

$$\lambda\mathbf{s}_u(0,v) + \lambda'\mathbf{s}'_u(0,v) = \{\alpha(1-v)+\beta v\}\mathbf{s}_v(0,v). \qquad (13.29)$$

The coefficients of $(1-v)^3, v(1-v)^2, v^2(1-v)$ and v^3 should be zero, accordingly we have

$$\begin{aligned} \lambda\mathbf{a}_0 + \lambda'\mathbf{a}'_0 &= \alpha\mathbf{e}_0 && (13.30)\\ \lambda\mathbf{a}_1 + \lambda'\mathbf{a}'_1 &= (1/3)(2\alpha\mathbf{e}_1+\beta\mathbf{e}_0) && (13.31)\\ \lambda\mathbf{a}_2 + \lambda'\mathbf{a}'_2 &= (1/3)(2\beta\mathbf{e}_1+\alpha\mathbf{e}_2) && (13.32)\\ \lambda\mathbf{a}_3 + \lambda'\mathbf{a}'_3 &= \beta\mathbf{e}_2. && (13.33)\end{aligned}$$

We set $\lambda = \lambda' = 1/2$, $\alpha = 0$, since B_0 and B_1 are symmetrical with respect to their boundary, and $\mathbf{a}_0$ and $\mathbf{a}'_0$ are collinear. The absolute values of $\mathbf{a}_3$, $\mathbf{a}'_3$ and $\mathbf{e}_2$ must be equal, and the angles between $\mathbf{a}_3$ and $\mathbf{e}_2$ and between $\mathbf{e}_2$ and $\mathbf{a}'_3$ are 120 degrees. Accordingly, $\beta = -1/2$ satisfies eq. (13.33). Then eq. (13.31) and eq. (13.32) become

$$\begin{aligned} (1/2)(\mathbf{a}_1+\mathbf{a}'_1) &= -(1/6)\mathbf{e}_0 && (13.34)\\ (1/2)(\mathbf{a}_2+\mathbf{a}'_2) &= -(1/3)\mathbf{e}_1. && (13.35)\end{aligned}$$

Referring to Fig. 13.20, at first we determine the control points of B_1 net as follows:

$$\begin{aligned} &\mathbf{p}_{00} = \mathbf{q}_{m13},\ \mathbf{p}_{01} = \mathbf{q}_{m1},\ \mathbf{p}_{02} = \mathbf{q}_{m1},\ \mathbf{p}_{03} = \mathbf{q}_0\\ &\mathbf{p}_{10} = \mathbf{q}_{13},\ \mathbf{p}_{11} = (1/6)(\mathbf{q}_{13}+5\mathbf{q}_1),\ \mathbf{p}_{12} = \mathbf{q}_1,\ \mathbf{p}_{13} = \mathbf{q}_{m6},\\ &\mathbf{p}_{20} = (1/2)(\mathbf{q}_{13}+\mathbf{q}_{12})^*,\ \mathbf{p}_{21} = (1/2)(\mathbf{q}_{12}+\mathbf{q}_1)^*, \qquad (13.36)\\ &\qquad \mathbf{p}_{22} = (1/6)(\mathbf{q}_{11}+5\mathbf{q}_1),\ \mathbf{p}_{23} = \mathbf{q}_{m6},\\ &\mathbf{p}_{30} = \mathbf{q}^*_{12},\ \mathbf{p}_{31} = (1/2)(\mathbf{q}_{11}+\mathbf{q}_{12})^*,\ \mathbf{p}_{32} = \mathbf{q}_{11},\ \mathbf{p}_{33} = \mathbf{q}_{m63}\end{aligned}$$

Control points with a superscript * mark can be taken arbitrarily. Their indicated values in eq. (13.36) are examples. The control points of B_2 net are

$$\begin{aligned} &\mathbf{p}'_{00} = \mathbf{q}_{m13},\ \mathbf{p}'_{01} = \mathbf{q}_{m1},\ \mathbf{p}'_{02} = \mathbf{q}_{m1},\ \mathbf{p}'_{03} = \mathbf{q}_0\\ &\mathbf{p}'_{10} = \mathbf{q}_{21},\ \mathbf{p}'_{11} = (1/6)(\mathbf{q}_{21}+5\mathbf{q}_2),\ \mathbf{p}'_{12} = \mathbf{q}_2,\ \mathbf{p}'_{13} = \mathbf{q}_{m2}\\ &\mathbf{p}'_{20} = (1/2)(\mathbf{q}_{23}+\mathbf{q}_{22})^*,\ \mathbf{p}'_{21} = (1/2)(\mathbf{q}_{22}+\mathbf{q}_2)^*, \qquad (13.37)\\ &\qquad \mathbf{p}'_{22} = (1/6)(\mathbf{q}_{23}+5\mathbf{q}_2),\ \mathbf{p}'_{23} = \mathbf{q}_{m2},\\ &\mathbf{p}'_{30} = \mathbf{q}^*_{22},\ \mathbf{p}'_{31} = (1/2)(\mathbf{q}_{23}+\mathbf{q}_{22})^*,\ \mathbf{p}'_{32} = \mathbf{q}_{23},\ \mathbf{p}'_{33} = \mathbf{q}_{m23},\end{aligned}$$

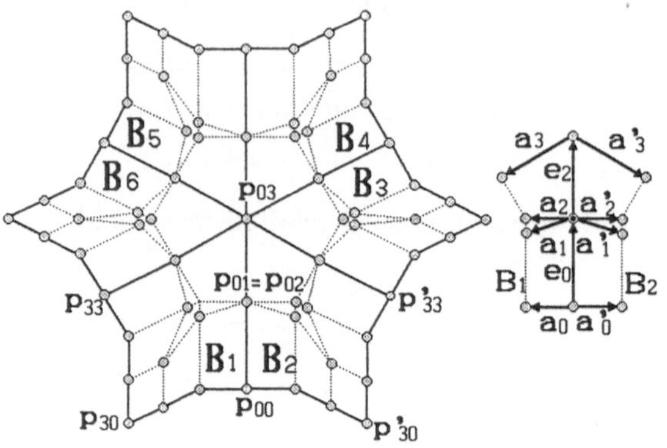

Figure 13.20 B patches in hexagonal region and their control points

The symbol * indicates arbitrary locations. From the above equations we get the relations

$$\mathbf{a}_2 = \mathbf{p}_{12} - \mathbf{p}_{02} = \mathbf{q}_1 - \mathbf{q}_{m1}, \quad \mathbf{a}_2' = \mathbf{p}_{12}' - \mathbf{p}_{02}' = \mathbf{q}_2 - \mathbf{q}_{m1} \tag{13.38}$$

Substituting the above equations in the left-hand side of eq. (13.35), we get zero vector. The right-hand side is $\mathbf{e}_1 = \mathbf{p}_{02} - \mathbf{p}_{01} = \mathbf{q}_{m1} - \mathbf{q}_{m1} = 0$. Therefore, eq. (13.35) holds. Also we have

$$\mathbf{a}_1 = \mathbf{p}_{11} - \mathbf{p}_{01} = (1/6)(\mathbf{q}_{13} + 5\mathbf{q}_1) - \mathbf{q}_{m1}, \tag{13.39}$$

$$\mathbf{a}_1' = \mathbf{p}_{11}' - \mathbf{p}_{01}' = (1/6)(\mathbf{q}_{21} + 5\mathbf{q}_2) - \mathbf{q}_{m1}. \tag{13.40}$$

Substituting the above equations in the left-hand side of eq. (13.34), we get

$$(1/3)(\mathbf{q}_{m13} + 5\mathbf{q}_{m1}) - 2\mathbf{q}_{m1} = -1/3(\mathbf{q}_{m1} - \mathbf{q}_{m13}).$$

The right-hand side is $-1/3(\mathbf{q}_{m1} - \mathbf{q}_{m13})$, so eq. (13.34) holds. Accordingly, with the control points(13.36) and (13.37) the patch B_1 and the patch B_2 connect in $C^{(1)}$. With a similar disposition of the control points, any two adjacent B patches of this arrangement connect in $C^{(1)}$. The control points $\mathbf{p}_{30}, \mathbf{p}_{20}, \mathbf{p}_{31}, \mathbf{p}_{32}$ and $\mathbf{p}_{30}, \mathbf{p}_{20}', \mathbf{p}_{31}', \mathbf{p}_{32}'$ of the patches are not concerned with this connection, hence they can be taken to adjust connections with the outside regions. Figure 13.21 shows an example of contour curves of this rounding. The curves cut by the parallel planes show smooth connection of the patches.

13.6 Rounding with a Rolling Ball

In practice, edge corners and vertices of connected patches are rounded by envelopes of a moving sphere which contacts with the bounding surfaces to eliminate their intersecting edges and vertices. We can determine the locus of the

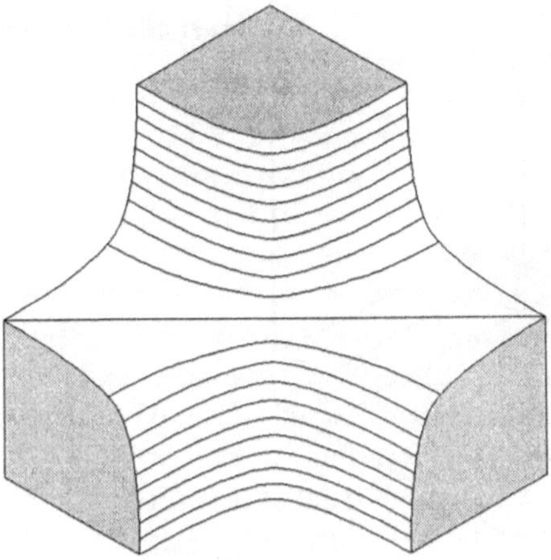

Figure 13.21 Contour curves in the mixed region

center of the ball and also the loci of contact points on the surfaces by solving the differential equation for the intersection of the offset surfaces of the original ones. The solving process is explained in Sect. 16.6.2.

The rounding procedure by a rolling ball around a vertex made by three surfaces follows the rules:

1. Around a convex vertex, usually three intersection curves made by each two of three offset surfaces meet at one point, from where the center of the ball starts along each one of the intersection curves. If the loci of the contact points of the ball and the original surfaces are continuous, an envelop of the moving ball becomes the smooth blending surface of the intersecting original surfaces. Figure 13.22 shows the locus of the center of the ball and loci of the contact points.
2. Around a convex-concave vertex, which is generated by one convex (concave) edge and two concave (convex) edges, the ball moves first along the single concave (convex) edge and then the ball moves along the consecutive convex (concave) edges touching only two surfaces, a part of which is the envelop surface already produced.

A point on the locus of the center $\mathbf{o}(s)$ of the ball and the two contacting points $\mathbf{p}_0(s)$ and $\mathbf{p}_2(s)$ with the surfaces to be rounded are functions of the arc length s of the locus of the center. Let the intersecting point of the tangents at $\mathbf{p}_0(s)$ and $\mathbf{p}_2(s)$ in the plane $\mathbf{op}_0\mathbf{p}_2$ be $\mathbf{p}_1$. Then vertices of the triangle $\mathbf{p}_0\mathbf{p}_1\mathbf{p}_2$ are considered three control points of a rational Bézier polygon of degree two, whose weights are 1, (1/2), 1. Since these three control points are functions of s, an envelop surface is determined by the moving ball.

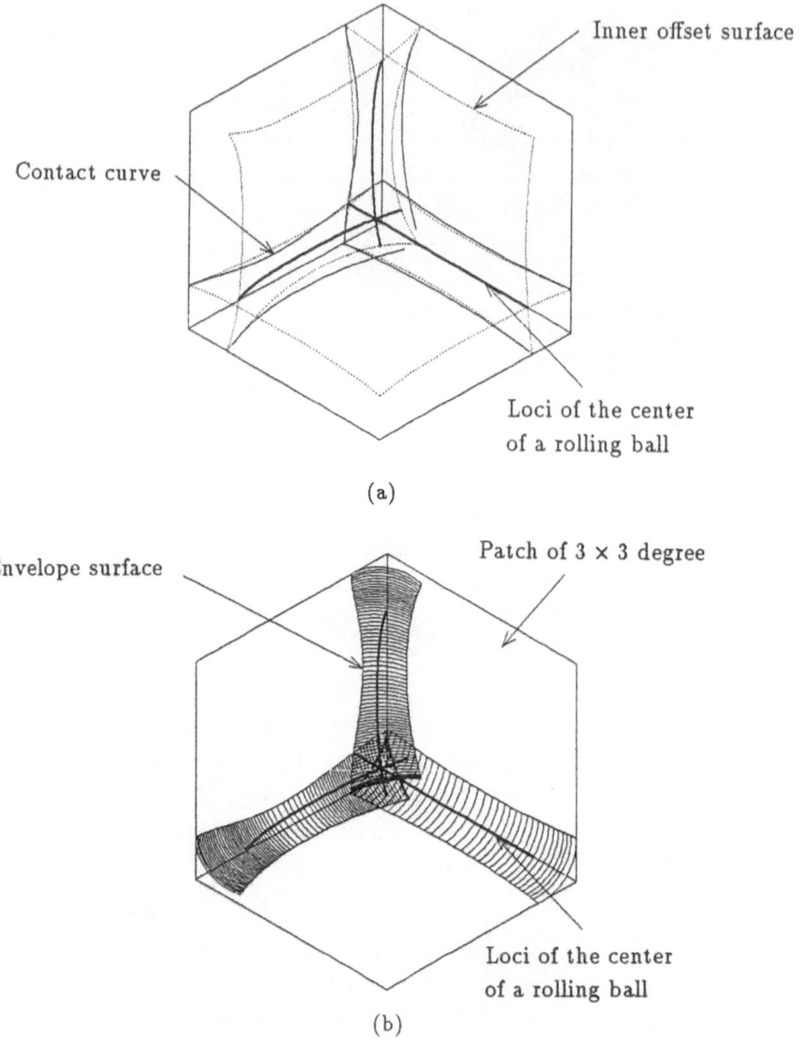

Figure 13.22 (a) Intersection curves of three inner offset surface, (b) Locus of the center of a rolling ball and its envelope surface

13.7 Appendix

13.7.1 Connection in a Triangular Region: General Case

Referring to Figures 13.9 and 13.10, the B control points of B_1, B_2 and B_3 are determined as follows:

1. The control points $\mathbf{p}_{11}$ and $\mathbf{p}'_{11}$ of B_1 and B_2 are two vertices of the central triangle made by the cutting process, and the third vertex corresponds to the control point $\mathbf{p}_{11}$ of B_3. These points are the same as those of the case of the equal λ_i's .
2. As in Fig. 13.7 the control point $\mathbf{p}^*_{01}$ is the dividing point of $\mathbf{p}_{11}$ and $\mathbf{p}'_{11}$, but its dividing ratios is $(\lambda'_1 : \lambda_1)$ instead of $(1 : 1)$, and $\mathbf{p}_{01}$ is the dividing point of $\mathbf{p}^*_{01}$ and $\mathbf{p}_{00}$ in the ratio $(2\lambda'_5 : \lambda_1\lambda_5)$ instead of $(1 : 4)$.
   A control point $\mathbf{p}_{00}$, which is the same point as $\mathbf{p}'_{00}$, is the corner point of B_1 and B_2 as shown in the figure. The control points $\mathbf{p}_{12}$ and $\mathbf{p}'_{12}$ are similar to $\mathbf{p}_{01}$. They are determined in a similar procedure, but the corresponding dividing ratios are $(\lambda'_5 : \lambda_5)$, $(2\lambda'_3 : \lambda_5\lambda_3)$ for $\mathbf{p}_{12}$, and $(\lambda'_3 : \lambda_3)$, $(2\lambda'_1 : \lambda_3\lambda_1)$ for $\mathbf{p}'_{12}$. On each side of the central triangle $(\mathbf{p}_{01}\mathbf{p}'_{12}\mathbf{p}_{12})$, there is a dividing point of the side in the ratio, each given by $(\lambda'_1 : \lambda_1)$ or $(\lambda'_3 : \lambda_3)$ or $(\lambda'_5 : \lambda_5)$ as shown in the figure, then $\mathbf{p}_{02}$ is the coincident point of three lines from each vertex to the dividing point on its opposite side, because of condition (13.18). The coincidence is guaranteed by Ceva's theorem, refer to Sect. 3.4.
3. The control points $\mathbf{p}_{10}, \mathbf{p}_{20}$ and $\mathbf{p}'_{10}, \mathbf{p}'_{20}$ of B_1 and B_2 are determined from the surrounding S net. Those of B_3 are similarly determined.

Three $\mathbf{p}_{11}$ points of each of B patches are the three vertices of the triangle in the S nets. The control points $\mathbf{p}_{10}, \mathbf{p}_{20}$ and $\mathbf{p}_{21}$ of the central three B patches are determined as in usual S net. The connection of these patches with the surrounding B patches with $C^{(1)}$ is guaranteed by the S net properties.

The positions of B points thus determined satisfy the conditions of $C^{(1)}$ connection of the three central B patches. We give its proof on the boundary of the patches B_1 and B_2.

In the triangle made by three points $\mathbf{p}_{01}$, the three vectors from the center $\mathbf{p}_{02}$ to its vertices can be looked upon as the control vectors $\mathbf{a}_2$, $\mathbf{e}_1$ and $\mathbf{a}'_2$. The following relation holds between them by definition (by applying Menelaus' theorem):

$$\lambda_1\mathbf{a}_2 + \lambda'_1\mathbf{a}'_2 = \frac{\lambda_1\lambda_5}{2\lambda'_5}\mathbf{e}_1.$$

As for $\mathbf{a}_0$ and $\mathbf{a}_1$, the following relations exist by the definition,

$$\lambda_1\mathbf{a}_0 + \lambda'_1\mathbf{a}'_0 = 0, \tag{13.41}$$

$$\lambda_1\mathbf{a}_1 + \lambda'_1\mathbf{a}'_1 = \frac{\lambda_1\lambda_5}{2\lambda'_5}\mathbf{e}_0/2. \tag{13.42}$$

These relations are the same as equations (13.11), (13.9), (13.10), which stipulate the $C^{(1)}$ connection at the boundary edge of the two patches B_1 and B_2, if we set

$$\alpha = 0, \quad \beta = \frac{\lambda_1\lambda_5}{2\lambda'_5}.$$

The $C^{(1)}$ connections between B_2 and B_3 and between B_3 and B_1 are similarly proved.

13.7.2 Connection of a Pentagonal Region: General Case

When the ratio of lengths of the upper edge and lower edge of the trapezoid is not 1 : 3 and the height h of the trapezoid and the width h' of the convex edge-band are not equal, the values given by eq. (13.22) become

$$\lambda = \lambda' = 0.5, \quad \beta = -\frac{m}{m+n},$$

then the following expressions must be used for the values:

$$\begin{aligned} |\mathbf{e}_0| &= \frac{2\overline{D_1'D_1}}{\beta}, \\ \overline{D_1'D_1} &= \sqrt{2(n-m)^2/4 + h^2 - h'^2/2}, \end{aligned}$$

and values given by eq. (13.23) are changed to

$$\begin{aligned} \lambda &= \frac{m+n}{5m+n}, \quad \lambda' = \frac{4m}{5m+n}, \\ \beta &= -\frac{m+n}{5m+n}. \end{aligned}$$

Let the distance between $\mathbf{p}'_{11}$ of B_3 and $\mathbf{e}_0$, which is the boundary vector between B_3 and B_5, be d. Then the distance between $\mathbf{p}_{11}$ of B_5 and $\mathbf{e}_0$ is $d \cdot (\lambda'/\lambda) = 4md/m+n$. The location of $\mathbf{p}_{11}$ is determined at a point which has this distance from $\mathbf{e}_0$ and also has similar distance from the boundary vector $\mathbf{e}_0$ between B_5 and B_4.

14 Connections of Patches by Blending

14.1 Introduction

Generally, when we have to define a surface patch, its degree of freedom is not always equal to that of given external conditions. If the latter is greater than the former, there must be some relations among the given conditions, or some methods must be provided to increase the degree of freedom of the patch to satisfy the given conditions.

For a polynomial surface $\mathbf{s}(u, v)$ of degree 3×3 of tensor product type, there are sixteen constant vectors to be determined corresponding to the terms $u^i v^j$ $\{i, j = 0, 1, 2, 3\}$, or the sixteen control points, or the conditions of one position vector, two tangent vectors and one twist vectors at each of the four corner points of a patch. If two curves and two cross-boundary tangent vectors expressed in degree three are given on the boundaries at $u = 0$ and $u = 1$, all the constants of a patch are determined, because the sum of degrees of freedom of the given boundary conditions is equal to sixteen. Hence we cannot give the conditions of the cross-boundary tangent vectors along the boundaries $v = 0$ and $v = 1$. They are automatically fixed. Accordingly, so long as a patch is expressed by a 3×3 polynomial, we cannot specify all the boundary conditions even up to the first derivative vectors.

To construct a patch with such boundary conditions, we have to devise some methods to increase the degree of freedom. We treat four methods. First, Coons' approach is explained concisely, for it is a classic one, though not practical. It uses a Boolean sum of three patches to meet given boundary conditions. Another method is to blend two patches, each of which satisfies boundary conditions on one pair of boundaries only, with weighting functions multiplied to eliminate unwanted effects. Then methods of using patches with variable corner twist vectors or variable inner Bézier points are introduced. Finally, a method of using correcting patches, which compensate the derivative vectors along the boundary curves of a patch to satisfy given conditions, is explained.

14.2 Coons' Patch

Coons' patch expression consists of three blending patches. It is not used today, though its name is famous [1]. So we explain it concisely. Refer to Fig. 14.1. On the boundaries 1 and 3 which correspond to the parameter values $u = 0$ and $u = 1$, the boundary curves $\mathbf{r}_1(v)$ and $\mathbf{r}_3(v)$, the cross-boundary tangents $\mathbf{t}_1(v)$ and $\mathbf{t}_3(v)$ are given and also on the boundaries 2 and 4 corresponding to

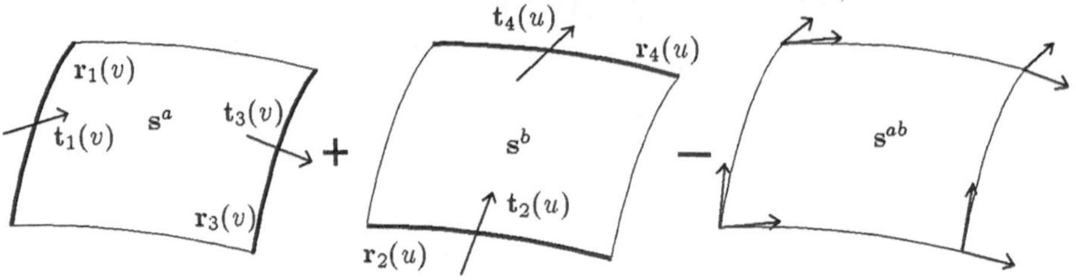

Figure 14.1 Boundary curves and cross-boundary tangents of Coons' patch

$v = 0$ and $v = 1$, the boundary curves $\mathbf{r}_2(u)$, $\mathbf{r}_4(u)$ and the cross-boundary tangents $\mathbf{t}_2(u)$ and $\mathbf{t}_4(u)$ are specified. By adding two patches which partially satisfy the given conditions, and subtracting a patch, which compensates the surplus boundary conditions, Coons constructed a patch expression satisfying all the boundary conditions.

For a fixed v on the boundaries 1 and 3, we consider points $\mathbf{r}_1(v)$ and $\mathbf{r}_3(v)$, and cross-boundary tangents $\mathbf{t}_1(v)$ and $\mathbf{t}_3(v)$, which give the boundary conditions of a curve segment $\tilde{\mathbf{r}}(u)$ in the u direction. This is because we can express $\tilde{\mathbf{r}}(u)$ by a weighted sum of the positions and the tangent vectors at its end points. Their weighting functions were already given by eq. (8.19), which are expressed in 1×4 matrix form by

$$H(t) = \{f_0(t) \quad f_1(t) \quad g(t) \quad h(t)\}, \qquad \text{for } t = u, v. \tag{14.1}$$

For the variable v moving from 0 to 1, the curve segment $\tilde{\mathbf{r}}(u)$ becomes an expression of a patch $\mathbf{s}^a(u, v)$:

$$\left.\begin{aligned} \mathbf{s}^a(u, v) &= \{\mathbf{r}_1(v) \quad \mathbf{r}_3(v) \quad \mathbf{t}_1(v) \quad \mathbf{t}_3(v)\}H(u)^T \\ &= H(u)\{\mathbf{r}_1(v) \quad \mathbf{r}_3(v) \quad \mathbf{t}_1(v) \quad \mathbf{t}_3(v)\}^T. \end{aligned}\right\} \tag{14.2}$$

Similarly, from the data on the boundaries 2 and 4, we have a surface patch $\mathbf{s}^b(u, v)$:

$$\mathbf{s}^b(u, v) = \{\mathbf{r}_2(u) \quad \mathbf{r}_4(u) \quad \mathbf{t}_2(u) \quad \mathbf{t}_4(u)\}H(v)^T. \tag{14.3}$$

On the boundaries 2 and 4, the position and cross-boundary tangent vectors of $\mathbf{s}^a$ are automatically fixed for $j = 0$ and 1 as

$$\left.\begin{aligned} \mathbf{s}^a(u, j) &= \{\mathbf{r}_1(j) \quad \mathbf{r}_3(j) \quad \mathbf{t}_1(j) \quad \mathbf{t}_3(j)\}H(u)^T, \\ \mathbf{s}^a_v(u, j) &= \{\mathbf{r}_{1v}(j) \quad \mathbf{r}_{3v}(j) \quad \mathbf{t}_{1v}(j) \quad \mathbf{t}_{3v}(j)\}H(u)^T. \end{aligned}\right\} \tag{14.4}$$

On the boundaries 1 and 3, those of $\mathbf{s}^b$ are fixed for $i = 0, 1$ as

$$\left.\begin{aligned} \mathbf{s}^b(i, v) &= \{\mathbf{r}_2(i) \quad \mathbf{r}_4(i) \quad \mathbf{t}_2(i) \quad \mathbf{t}_4(i)\}H(v)^T, \\ \mathbf{s}^b_u(i, v) &= \{\mathbf{r}_{2u}(i) \quad \mathbf{r}_{4u}(i) \quad \mathbf{t}_{2u}(i) \quad \mathbf{t}_{4u}(i)\}H(v)^T. \end{aligned}\right\} \tag{14.5}$$

Using equations (14.4) as the boundary conditions, we can define a surface patch $\mathbf{s}^{ab}(u,v)$ which is given by

$$\mathbf{s}^{ab}(u,v) = H(v)\{\mathbf{s}^a(u,0) \;\; \mathbf{s}^a(u,1) \;\; \mathbf{s}^a_v(u,0) \;\; \mathbf{s}^a_v(u,1)\}^T. \tag{14.6}$$

Substituting equations (14.4) into the above equation and letting matrix M^{ab} be

$$M^{ab} \equiv \begin{bmatrix} \mathbf{r}_1(0) & \mathbf{r}_3(0) & \mathbf{t}_1(0) & \mathbf{t}_3(0) \\ \mathbf{r}_1(1) & \mathbf{r}_3(1) & \mathbf{t}_1(1) & \mathbf{t}_3(1) \\ \mathbf{r}_{1v}(0) & \mathbf{r}_{3v}(0) & \mathbf{t}_{1v}(0) & \mathbf{t}_{3v}(0) \\ \mathbf{r}_{1v}(1) & \mathbf{r}_{3v}(1) & \mathbf{t}_{1v}(1) & \mathbf{t}_{3v}(1) \end{bmatrix}, \tag{14.7}$$

we have

$$\mathbf{s}^{ab}(u,v) = H(v)M^{ab}H(u)^T. \tag{14.8}$$

Similarly from equations (14.5), we can define a surface patch which is given by

$$\begin{aligned} \mathbf{s}^{ba}(u,v) &= H(u)\{\mathbf{s}^b(0,v) \;\; \mathbf{s}^b(1,v) \;\; \mathbf{s}^b_u(0,v) \;\; \mathbf{s}^b_u(1,v)\}^T \\ &= \{\mathbf{s}^b(0,v) \;\; \mathbf{s}^b(1,v) \;\; \mathbf{s}^b_u(0,v) \;\; \mathbf{s}^b_u(1,v)\}H(u)^T. \end{aligned} \tag{14.9}$$

Letting M^{ba} represent

$$M^{ba} \equiv \begin{bmatrix} \mathbf{r}_2(0) & \mathbf{r}_2(1) & \mathbf{r}_{2u}(0) & \mathbf{r}_{2u}(1) \\ \mathbf{r}_4(0) & \mathbf{r}_4(1) & \mathbf{r}_{4u}(0) & \mathbf{r}_{4u}(1) \\ \mathbf{t}_2(0) & \mathbf{t}_2(1) & \mathbf{t}_{2u}(0) & \mathbf{t}_{2u}(1) \\ \mathbf{t}_4(0) & \mathbf{t}_4(1) & \mathbf{t}_{4u}(0) & \mathbf{t}_{4u}(1) \end{bmatrix}, \tag{14.10}$$

we have

$$\mathbf{s}^{ba}(u,v) = H(v)M_{ba}H(u)^T. \tag{14.11}$$

The two patches $\mathbf{s}^{ab}$ and $\mathbf{s}^{ba}$ are expressed by the properties at their four corners only, which are the elements of the matrices M^{ab} and M^{ba}. Now examine their properties. The elements of the upper-left 2×2 sub-matrix of M^{ab} or M^{ba} represent the four corners of the patch, those of the lower-left and upper-right 2×2 sub-matrices represent the tangent vectors in the u and the v directions. The corresponding elements of the sub-matrices are equal. The elements of the lower-right 2×2 sub-matrix represent the lateral change of the tangent vectors at the corners which are the so-called twist vectors.

Since in a regular surface patch the order of differentiation does not influence on the result, the corresponding elements of this sub-matrix are also equal. Accordingly, $M^{ab} = M^{ba}$ and consequently $\mathbf{s}^{ab} = \mathbf{s}^{ba}$.

So, the boundary conditions of the patch $\mathbf{s}^{ab}$ or $\mathbf{s}^{ba}$ are considered equal to the sum of those of the non-defined boundaries of $\mathbf{s}^a$ and $\mathbf{s}^b$. Accordingly, in the following expression of $\mathbf{s}(u,v)$, the surplus boundary conditions are canceled:

$$\mathbf{s}(u,v) = \mathbf{s}^a(u,v) + \mathbf{s}^b(u,v) - \mathbf{s}^{ab}(u,v). \tag{14.12}$$

This satisfies the given boundary conditions on all its four boundaries. This is called Coons' patch expression by Boolean sum. Usually it is difficult to give two pairs of arbitrary cross-boundary tangent vectors in the u and the v directions along the respective boundaries.

When the boundary curves and the cross-boundary tangent vectors are expressed by the zero-th and the first derivatives at their end points together with the Hermite weighting functions of degree three, the three expressions $\mathbf{s}^a(u,v)$, $\mathbf{s}^b(u,v)$ and $\mathbf{s}^{ab}(u,v)$ become the same one. So we have

$$\mathbf{s}(u,v) = \mathbf{s}^a(u,v) = \mathbf{s}^b(u,v) = \mathbf{s}^{ab}(u,v). \tag{14.13}$$

This patch of 3×3 degree is expressed by its corner vectors:

$$\mathbf{s}(u,v) = H(v) M H(u)^T, \tag{14.14}$$

where

$$M = \begin{bmatrix} \mathbf{s}(0,0) & \mathbf{s}(1,0) & \mathbf{s}_u(0,0) & \mathbf{s}_u(1,0) \\ \mathbf{s}(0,1) & \mathbf{s}(1,1) & \mathbf{s}_u(0,1) & \mathbf{s}_u(1,1) \\ \mathbf{s}_v(0,0) & \mathbf{s}_v(1,0) & \mathbf{s}_{uv}(0,0) & \mathbf{s}_{uv}(1,0) \\ \mathbf{s}_v(0,1) & \mathbf{s}_v(1,1) & \mathbf{s}_{uv}(0,1) & \mathbf{s}_{uv}(1,1) \end{bmatrix}.$$

Usually, it is still difficult to give appropriate values of the vectors $\mathbf{s}_{uv}$ at the corners of a patch.

14.3 Independent Boundary Conditions

14.3.1 Blending by Weighted Sum

There is another method of eliminating disturbing boundary values of the blending patches by multiplying with appropriate weighting functions [4]. Let each of two patches $\mathbf{s}^a(u,v)$ and $\mathbf{s}^b(u,v)$ satisfy the given boundary conditions only on the respective pair of boundaries. Therefore, the weighting functions $w^a(u,v)$ and $w^b(u,v)$ take value 1 on one pair of the boundaries and zero on another. The synthesized patch is given by

$$\mathbf{s}(u,v) = w^a(u,v)\mathbf{s}^a(u,v) + w^b(u,v)\mathbf{s}^b(u,v), \tag{14.15}$$

$$w^a(u,v) + w^b(u,v) = 1, \qquad 0 \le w^a(u,v), w^b(u,v) \le 1. \tag{14.16}$$

The weighting functions $w^a(u,v)$ and $w^b(u,v)$ must satisfy the following conditions on the boundaries:

$$\left.\begin{array}{lllllllll} w^a(0,v) & = & w^a(1,v) & = & w^b(u,0) & = & w^b(u,1) & = & 1, \\ w^a(u,0) & = & w^a(u,1) & = & w^b(0,v) & = & w^b(1,v) & = & 0, \\ w^a_v(0,v) & = & w^a_v(1,v) & = & w^b_u(u,0) & = & w^b_u(u,1) & = & 0, \\ w^a_u(u,0) & = & w^a_u(u,1) & = & w^b_v(0,v) & = & w^b_v(1,v) & = & 0. \end{array}\right\} \tag{14.17}$$

With these weighting functions, we have for $i, j = 0, 1$,

$$\left.\begin{array}{rclrcl} \mathbf{s}(i,v) & = & \mathbf{s}^a(i,v), & \mathbf{s}(u,j) & = & \mathbf{s}^b(u,j), \\ \mathbf{s}_u(i,v) & = & \mathbf{s}_u^a(i,v), & \mathbf{s}_v(u,j) & = & \mathbf{s}_v^b(u,j). \end{array}\right\} \tag{14.18}$$

These are the values we want to have. One form of the weighting functions which satisfy the conditions (14.16) and (14.17) is given by

$$w^a(u,v) = \frac{v^2(1-v)^2}{u^2(1-u)^2 + v^2(1-v)^2}, \tag{14.19}$$

$$w^b(u,v) = \frac{u^2(1-u)^2}{u^2(1-u)^2 + v^2(1-v)^2}. \tag{14.20}$$

Disturbance of smoothness by these weighting functions is a little larger than that of other methods described in the subsequent sections.

14.3.2 Two-Valued Twist Vectors and Floating Inner Control Points

The twist vectors at the corners of a patch influence the cross-boundary tangent distribution on each of its boundaries. If the twist vectors are not independent, externally defined cross-boundary tangents are not independent. Then it is difficult to construct a net of smooth connecting Coons' patches. If the twist vectors are taken null, patch connection becomes easy, but the synthesized shape is usually unsatisfactory, because of many flat regions on the surface.

A method is devised to make the twist vectors independent on each boundary of a patch. This is to replace $\mathbf{s}_{uv}(i,j)$'s by the following expressions for $i, j = 0, 1$

$$\frac{(v-j)\mathbf{s}_{uv}(i,j) + (u-i)\mathbf{s}_{uv}(i,j)}{(u-i)+(v-j)}. \tag{14.21}$$

This expression was given by A. Gregory [2]. The expression (14.21) takes a different value according to whether $u = i$ or $v = j$. For a patch expression in Bézier form of 3×3 degree, the same effect can be attained when its inner four control points $\mathbf{p}_{1+i,1+j}$ for $i, j = 0, 1$ are replaced by

$$\frac{(v-j)\mathbf{p}_{1+i,1+j}^a + (u-i)\mathbf{p}_{1+i,1+j}^b}{(u-i)+(v-j)}. \tag{14.22}$$

When parameter u takes value $u = i$ or v takes value $v = j$, eq. (14.22) becomes $\mathbf{p}_{1+i,1+j}^a$ or $\mathbf{p}_{1+i,1+j}^b$, that is, the location of a control point is different. See Fig. 14.2. For instance, the cross-boundary tangent on the boundary $v = 0$ is given by

$$\mathbf{s}_v(u,0) = 3(1-u+\mathrm{E}u)^3\mathbf{a}_0, \tag{14.23}$$

where

$$\left.\begin{array}{rclrcl} \mathbf{a}_0 & = & \mathbf{p}_{01} - \mathbf{p}_{00} & , \quad \mathbf{a}_1 & = & \mathbf{p}_{11}^b - \mathbf{p}_{10}, \\ \mathbf{a}_2 & = & \mathbf{p}_{21}^b - \mathbf{p}_{20} & , \quad \mathbf{a}_3 & = & \mathbf{p}_{31} - \mathbf{p}_{30} \end{array}\right\}. \tag{14.24}$$

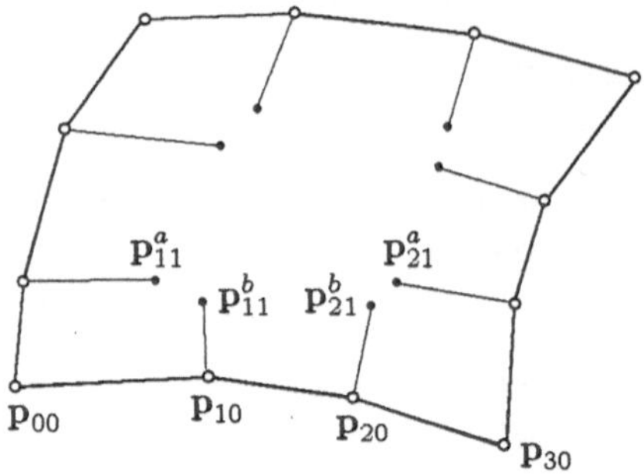

Figure 14.2 Double inner control points

The cross-boundary tangent on the boundary $u = 0$ is given by

$$\mathbf{s}_u(0, v) = 3(1 - v + \mathsf{F}v)^3 \mathbf{b}_0, \tag{14.25}$$

where

$$\left.\begin{array}{llllll} \mathbf{b}_0 & = & \mathbf{p}_{10} - \mathbf{p}_{00} & , \; \mathbf{b}_1 & = & \mathbf{p}^a_{11} - \mathbf{p}_{01}, \\ \mathbf{b}_2 & = & \mathbf{p}^a_{12} - \mathbf{p}_{02} & , \; \mathbf{b}_3 & = & \mathbf{p}_{13} - \mathbf{p}_{03}. \end{array}\right\} \tag{14.26}$$

Comparing the tangent control vectors (14.24) and (14.26), we know there is no coupling between them. So we can make the independent cross-boundary tangents on the boundaries $v = 0$ and $u = 0$. This variant of Gregory's method was introduced by H.Chiyokura [6].

14.4 Correction of Cross-Boundary Tangent Vectors

To treat the connection of a patch with its surrounding ones, we have to increase the degree of freedom of the patch to be connected. For this purpose we use a compensating patch which gives the cross-boundary derivative vectors to be added to the original ones to satisfy externally defined conditions [20][21].

The compensating patch gives null positional vectors on all the boundaries and arbitrary cross-boundary tangential and higher derivative vectors. With a composite patch of the original and compensating ones, we can satisfy all its boundary conditions. We use techniques similar to those for connection of a three-sided patch with its adjacent patches already fixed.

As for the connection in $C^{(2)}$, we treat it in the next section, since the technique applied is similar, but the expression is more complicated.

14.4.1 Connection of Four-Sided Patches

Since it is easy to define a B patch $\mathbf{s}^b(u, v)$ which satisfies only given boundary curves, we must add a correction patch $\mathbf{s}^c(u, v)$ to match the given cross-boundary tangent vectors. Here the superscript ^{b} shows a base patch and another one ^{c} indicates a compensating patch. Let the patch $\mathbf{s}^b(u,v)$ be expressed by

$$\mathbf{s}^b(u,v) = (1-u+\mathsf{E}v)^3(1-v+\mathsf{F}v)^3\mathbf{p}_{00}. \tag{14.27}$$

One form of the correction patches is given by

$$\mathbf{s}^c(u,v) = \mathbf{Q}(u,v)v(1-v)u(1-u). \tag{14.28}$$

Let the vector function $\mathbf{Q}(u,v)$ have a form:

$$\mathbf{Q}(u,v) = \frac{\mathbf{T}_1(1-u)v(1-v) + \mathbf{T}_2u(1-u)(1-v) + \mathbf{T}_3uv(1-v) + \mathbf{T}_4vu(1-u)}{\{u(1-u)+v(1-v)\}^2}, \tag{14.29}$$

where $\mathbf{T}_i$ is a cross-boundary tangent on one boundary i. Equation (14.28) vanishes on all the boundaries, but its first derivative gives independent values on each boundary of the patch:

$$\mathbf{s}^c_u(0,v) = \mathbf{T}_1,\ \mathbf{s}^c_v(u,0) = \mathbf{T}_2, \tag{14.30}$$
$$\mathbf{s}^c_u(1,v) = \mathbf{T}_3,\ \mathbf{s}^c_v(u,1) = \mathbf{T}_4. \tag{14.31}$$

Since $\mathbf{T}_1$ can be a function of v, we assume the following form:

$$\mathbf{s}^c_u(0,v) = \mathbf{T}_1 = 3(1-v+\mathsf{F}v)^3\mathbf{t}_0, \tag{14.32}$$

where $\mathbf{t}_i$, $i = 0,1,2,3$, are the control vectors of the cross-boundary tangent on the boundary $u = 0$. From equations (14.27) and (14.32), the cross-boundary tangent on $u = 0$ of the patch to be connected is given by the composite vectors:

$$\mathbf{s}_u(0,v) = \mathbf{s}^b_u(0,v) + \mathbf{s}^c_u(0,v) = 3(1-v+\mathsf{F}v)^3(\mathbf{a}_{00}+\mathbf{t}_0), \tag{14.33}$$
$$\text{where} \quad \mathbf{a}_{0j} = \mathbf{p}_{1j} - \mathbf{p}_{0j},\ j = 0,1,2,3.$$

The cross-boundary tangent can take arbitrary form so long as it does not violate the geometric constraints at the corners of the patch. For instance, at each node of boundary curve meshes of patches, a tangent vector of a compensating patch must be in the tangential planes defined there. The cross-boundary tangents on the other boundaries take forms similar to the above. If in eq. (14.32), the control tangents $\mathbf{t}_0$ and $\mathbf{t}_3$ are zero, that is, at both ends of the boundary, there is no need of tangent compensation there, and the denominator of eq. (14.29) can be reduced to $\{u(1-u)+v(1-v)\}$ instead of $\{u(1-u)+v(1-v)\}^2$.
Example 1. Modification of connection of two patches from $C^{(0)}$ to $C^{(1)}$
Figure 14.3(a) shows two connecting Bézier patches of degree 2×2, which are symmetrical with respect to the vertical plane containing their common boundary curve. Therefore, the respective cross-boundary tangent vectors make acute angles on the common boundary. The section curves of both patches cut by

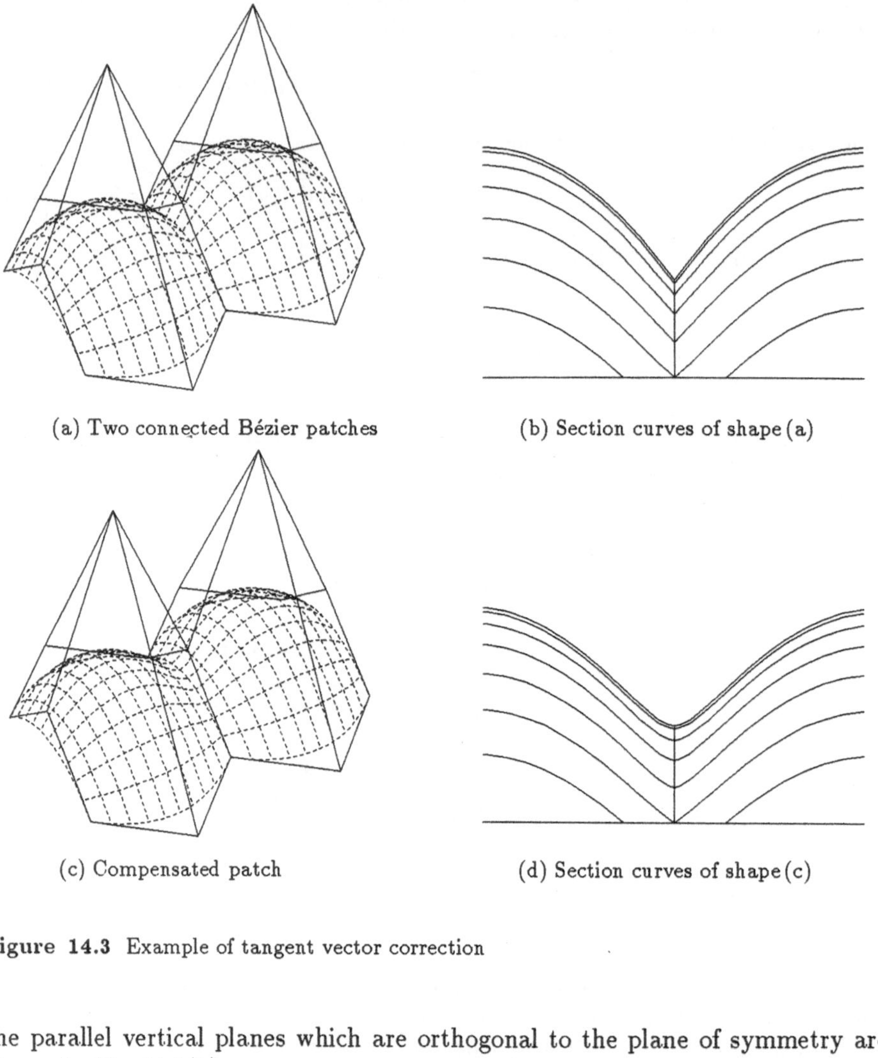

(a) Two connected Bézier patches (b) Section curves of shape (a)

(c) Compensated patch (d) Section curves of shape (c)

Figure 14.3 Example of tangent vector correction

the parallel vertical planes which are orthogonal to the plane of symmetry are shown in Fig. 14.3(b).

We add compensating patches to both patches. The form of the compensating tangent is $\mathbf{T}_1 = (1-v)^2\mathbf{t}_0 + 2v(1-v)\mathbf{t}_1 + v^2\mathbf{t}_2$. The tangent control vectors of the compensating patches are determined so as to make the composite cross-boundary tangent control vectors of the two patches the same magnitude and opposite directions:

$$\left.\begin{aligned} \mathbf{t}_i &= (\mathbf{p}_{1i} - \mathbf{p}'_{1i})/2 - \mathbf{a}_{0i}, \\ \mathbf{t}'_i &= -(\mathbf{p}_{1i} - \mathbf{p}'_{1i})/2 - \mathbf{a}'_{0i}, \quad i = 0, 1, 2. \end{aligned}\right\} \tag{14.34}$$

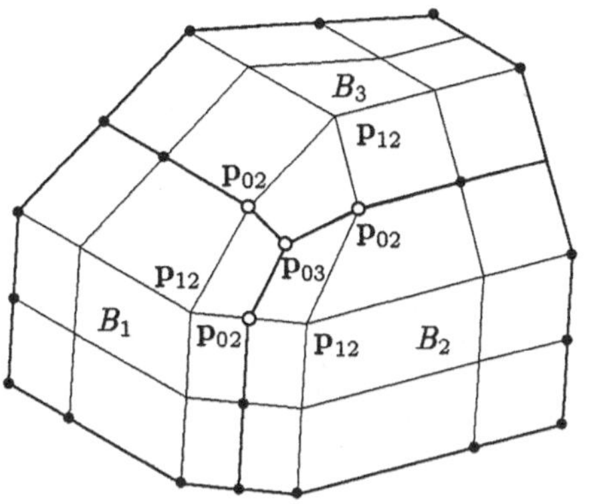

(a) Net of control points

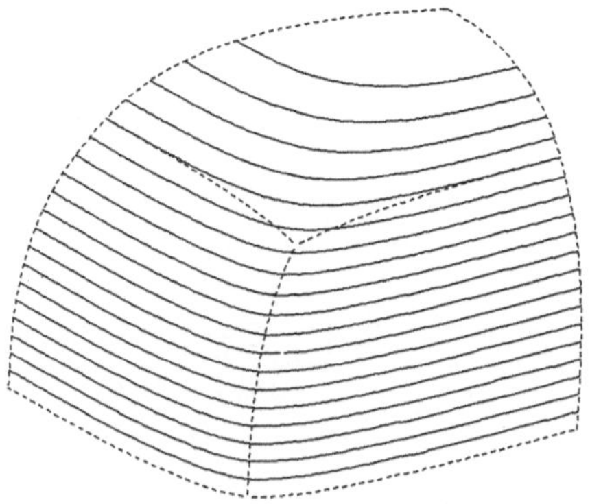

(b) Contour curves

Figure 14.4 Rounding of a convex corner

From eq. (14.33), the composite cross-boundary tangents are

$$\left.\begin{aligned} \mathbf{s}_u(0,v) &= (1-v+\mathsf{F}v)^2(\mathbf{p}_{10}-\mathbf{p}'_{10}), \\ \mathbf{s}'_u(0,v) &= (1-v+\mathsf{F}v)^2(-\mathbf{p}_{10}+\mathbf{p}'_{10}). \end{aligned}\right\} \tag{14.35}$$

Then on the common boundary, their connection become $C^{(1)}$. Figure 14.3(c) show the shape of composite patches and Fig. 14.3(d) is the section curve similar to Fig. 14.3(b). We can notice clearly the rounding effects caused by the compensating patches.

Example 2. Rounding of a convex corner

This is a similar case to Sect. 13.4.1. But use of the compensating patches makes the rounding process simple and flexible. A net of the control points of three Bézier patches B_1, B_2, B_3 of degree three is shown in Fig. 14.4(a). Let the symbols used for corresponding control points both sides of the boundary be the same as used in Chap. 13. In the central triangular region where three patches meet at $\mathbf{p}_{03}$, their common four control points, point $\mathbf{p}_{03}$ and three points $\mathbf{p}_{02}$ indicated by the small circles, are in a plane to give the $C^{(1)}$ connection.

The locations of the set of control points $(\mathbf{p}_{00},\mathbf{p}_{10},\mathbf{p}_{01},\mathbf{p}_{11})$ along the connecting boundaries of the patches can easily be arranged to make control vectors of the cross-boundary tangent such as $(\mathbf{p}_{10}-\mathbf{p}_{00})$ and $(\mathbf{p}_{11}-\mathbf{p}_{01})$ equal to the corresponding vectors in the connecting patch.

But determining the locations of the control points $\mathbf{p}_{12}$ to guarantee the $C^{(1)}$ connection is difficult. So, we use the technique of compensating patches. The form of the compensating tangent is $\mathbf{T}_1 = v^3\mathbf{t}_3 + 3v(1-v)^2\mathbf{t}_2$ which is applied to both sides of the boundaries to make the control vectors of the cross-boundary tangents equal and opposite. In Fig. 14.4(b), the contour curves of the

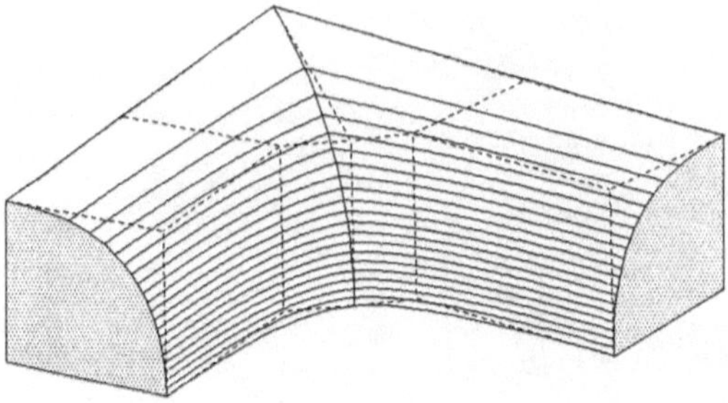

(a) Net and original shape

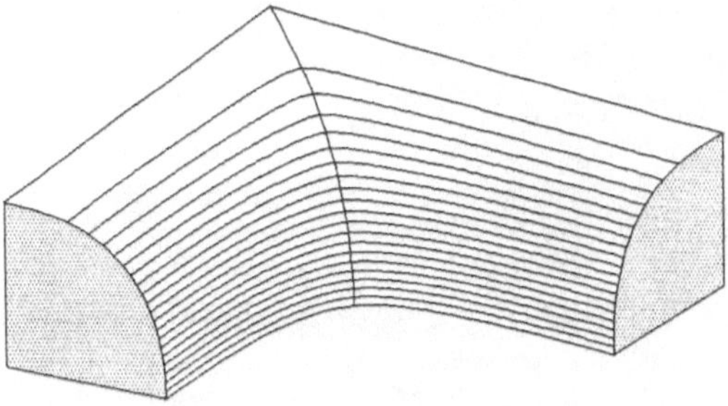

(b) Contour curves

Figure 14.5 Rounding of a convex-concave corner

compensated patch are shown for checking its smoothness.

Example 3. Rounding of a convex-concave corner
The shape of this type is usually difficult to make with simple connection of patches. One example is shown in Sect. 13.4.2. To show the ability of our method, we use two Bézier patches of degree two for the base patches and the two compensating patches whose control tangent vectors are of the form $\mathbf{T}_1 = v^2\mathbf{t}$. Figure 14.5(a) is the contour curve pattern for $C^{(0)}$ connection of the patches, while Fig. 14.5(b) is that for connection with the compensating patches to attain $C^{(1)}$. The contour curves of the former have sharp corners, but in the latter the sharp corners become rounded.

Example 4. Rounding of hexagonal region
An interlaced convex and concave hexagonal region, treated in Sect. 13.5, can be synthesized from six patches of degree two with their compensating patches. The locations of the control points of the basic patches are shown in Fig. 14.6(a) in solid lines. The tangent vectors of the compensating patch is given by the form $\mathbf{T}_1 = v^2\mathbf{t}$. The normal vectors along the their connecting boundary before and after the compensation are shown in Fig. 14.6(b) and (c). The compensation

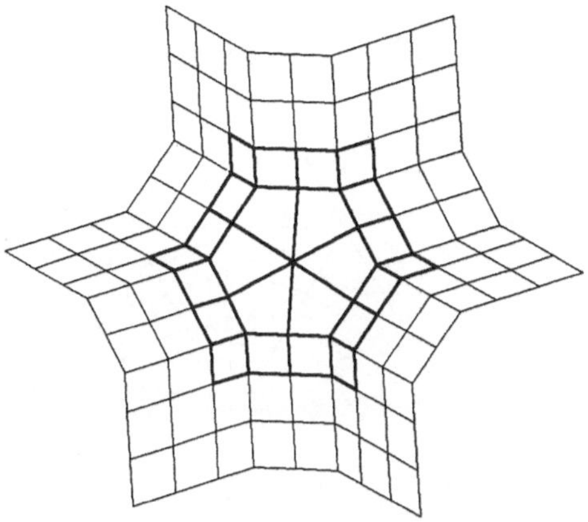

(a) Locations of nets of the base patches connected in $C^{(0)}$

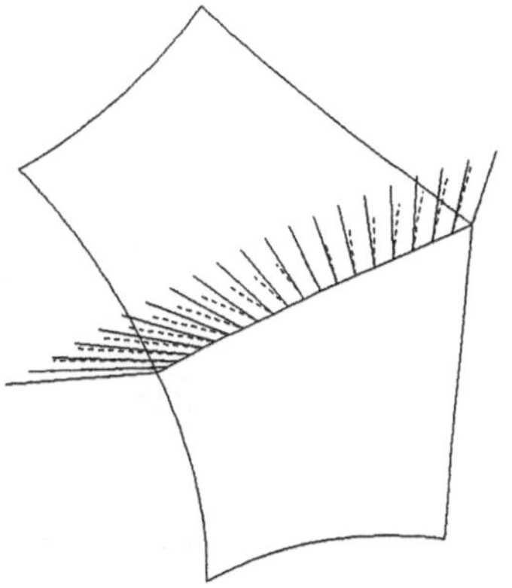

(b) Normal vectors along the boundary before compensation

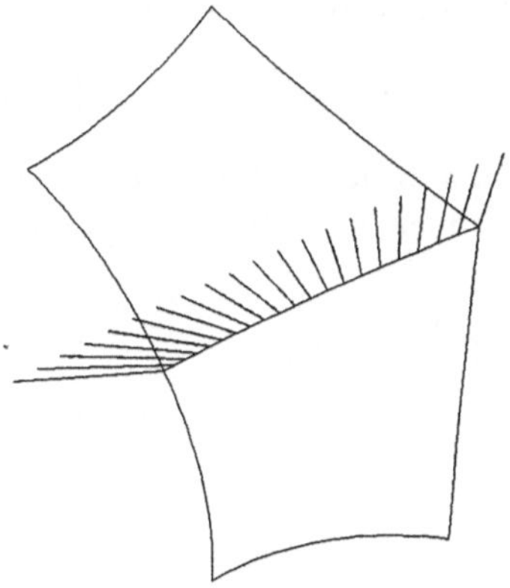

(c) Normal vectors after compensation

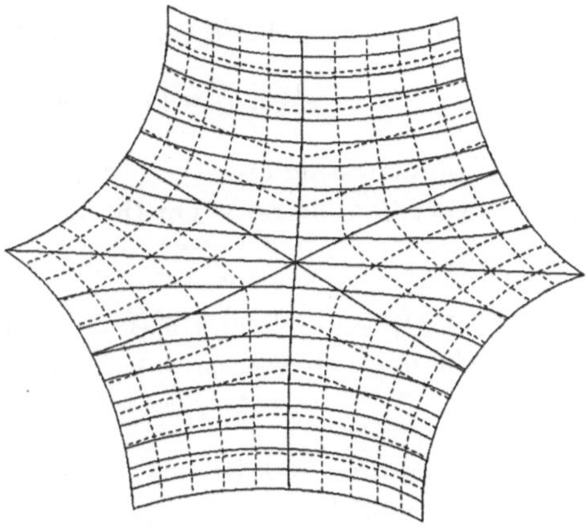

(d) Contour curve pattern of hexagonal region

Figure 14.6 Rounding of hexagonal region

makes the normal vectors equal. The contour curves of the region are shown in Fig. 14.6(d). Dotted curves are parameter constant curves.

14.4.2 Evaluation and Comparison of Methods

Owing to addition of a compensation patch $\mathbf{s}^c(u,v)$ for the $C^{(1)}$ connection, the shape of the original patch is modified together with the slope along its boundary curves. To investigate its effect on the shape, we draw a contour map

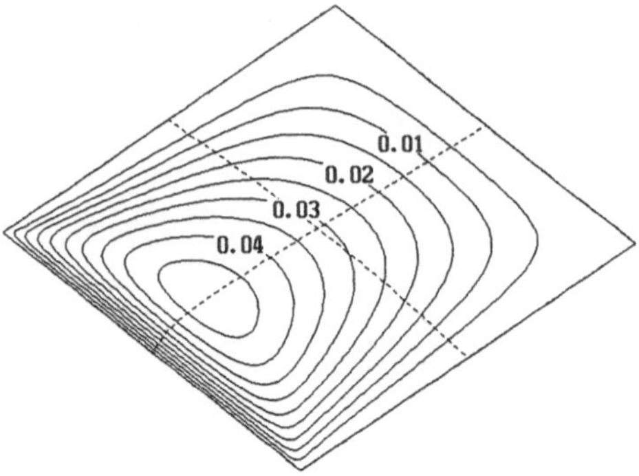

(a) Four sided patch compensation

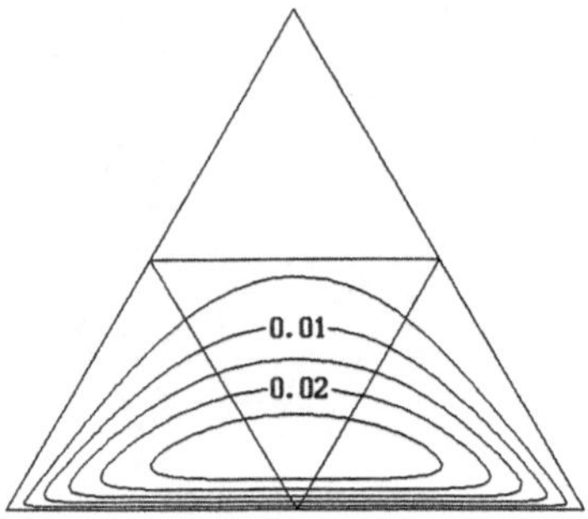

(b) Triangular patch compensation

Figure 14.7 Shape modification by compensating patch

for a compensating patch $\mathbf{s}^c(u,v)$, whose base patch is a Bézier of degree two:

$$\mathbf{s}^b(u,v) = (1-u+\mathsf{E}u)^2(1-v+\mathsf{F}v)^2\mathbf{p}_{00}. \tag{14.36}$$

Let the compensating cross-boundary vector on $u=0$ be $\mathbf{T}_1 = (1-v+\mathsf{F}v)^2\mathbf{t}_0$ and all other $\mathbf{T}_i$'s be zero. On the boundary $u=0$ the cross-boundary tangent is given by

$$\begin{aligned} \mathbf{s}_u(0,v) &= \mathbf{s}_u^b(0,v) + \mathbf{s}_u^c(0,v) = 2(1-v+\mathsf{F}v)^2(\mathbf{a}_{00}+\mathbf{t}_0/2), \\ \text{where} \quad & \mathbf{a}_{00} = \mathbf{a}_{01} = \mathbf{a}_{02} = (0.5,0,0), \\ & \mathbf{t}_0 = \mathbf{t}_1 = \mathbf{t}_2 = (0,0,1). \end{aligned} \tag{14.37}$$

The height of the patch is

$$\begin{aligned} z &= \mathbf{k}\cdot\mathbf{s}(u,v) = \mathbf{k}\cdot\mathbf{s}^b(u,v) + \mathbf{k}\cdot\mathbf{s}^c(u,v) \\ &= \frac{(1-v)^2 + 2v(1-v) + v^2 \times u(1-u)v(1-v)^2}{\{u(1-u)+v(1-v)\}^2}. \end{aligned} \tag{14.38}$$

The contour curves of the above equation are shown in Fig. 14.7(a). Though the slope on the boundary $u=0$ is modified from zero degrees to 45 degrees, its effect on shape of the patch is at most 5% of the boundary length and its variation is smooth. In Fig. 14.7(b), a similar diagram for a triangular patch for slope correction on its boundary $w=0$ is shown.

For the example of Fig. 14.7(a), Chiyokura's method cannot be applied, because it cannot modify the tangent direction at the ends of a boundary curve. So, we give the same condition of the cross-boundary tangent on one boundary of a connecting patch to compare the effects of both methods. Figures 14.8(a) and (b) are similar contour diagrams of the shapes generated by the same correction

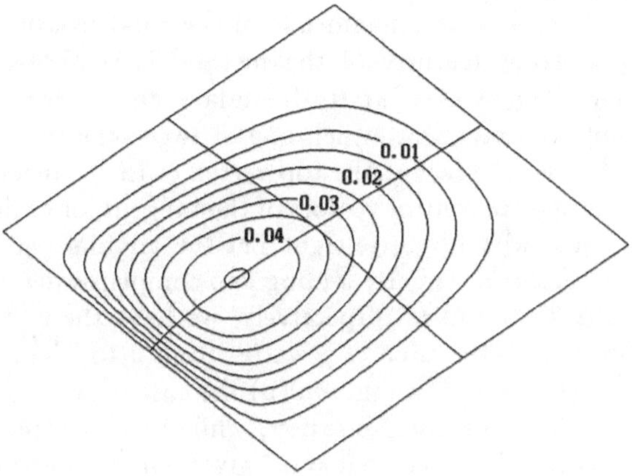

(a) Method of Sect. 14.4.1

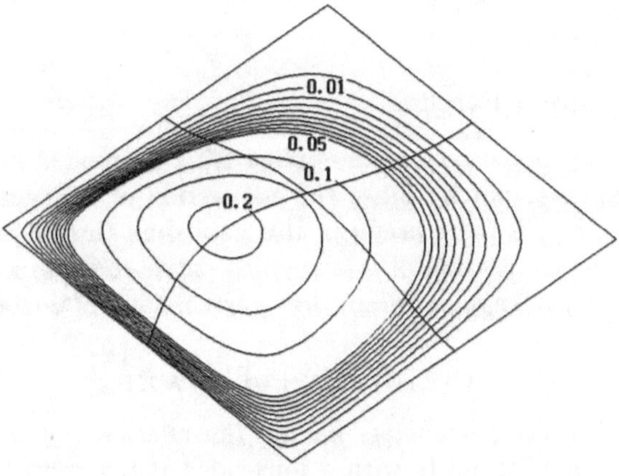

(b) Chiyokura's method

Figure 14.8 Comparison of the methods for the same cross-boundary tangent

of the cross-boundary tangent on one boundary of a Bézier patch of 3×3 degree. The shape expression by the method of this section is

$$z = \frac{9u(1-u)^2 v(1-v)^3}{\{u(1-u)+v(1-v)\}^2}. \tag{14.39}$$

That by Chiyokura's method is

$$z = \frac{9u(1-u)^2 v^2(1-v)^2(2u+1)}{(u+v)(u+1-v)}. \tag{14.40}$$

As for the influence on the shape by the correction, the disturbance of this method can be controlled by the magnitude of the compensating vectors.

One of the important features of the method is the correction ability of the cross-boundary tangent even at its boundary ends. For such a case, we examine an example to determine which one of two connecting patches should be compensated. In Fig. 14.9(a), at the top vertex of the connecting patches, we want to have two composite control vectors of the tangent for each patches, which are collinear and equal with opposite sign. Let the original two control tangent vectors be $\mathbf{a}$, $\mathbf{a}'$ and $\Delta\mathbf{a} = \mathbf{a} - \mathbf{a}'$. By adding two compensating vectors $\mathbf{t} = \mu\Delta\mathbf{a}$ and $\mathbf{t}' = (1-\mu)\Delta\mathbf{a}$ to $\mathbf{a}$ and $\mathbf{a}'$ respectively, we have the equal and opposite control tangent vectors. The value of μ varies from 0 to 1. In Fig. 14.5(b) the shape for $\mu = 0.5$ is shown and in Fig. 14.9(b) the case of $\mu = 1.0$. In the former, both patches share the same compensation, while in the latter only one patch shares the compensation. In this case $\mu = 0.5$ gives naturally good result, because the base patches are symmetrical. For general cases, more study is needed for sharing compensation and the magnitudes of the compensating vectors to attain desirable shapes.

14.4.3 Three-Sided Patches

The compensating patch technique applied for a four-sided patch is also applicable for a triangular patch. Since the connection of triangular patches is less flexible, this method is more useful in this case than that of four-sided patches. We treat here the connection of a triangular patch and a four-sided patch. For the expression of a triangular patch, the operator form is introduced in Chap. 15 such as

$$\mathbf{s}^b(u,v,w) = (u + v\mathsf{E} + \mathsf{F}w)^n \mathbf{p}_{00}, \tag{14.41}$$

where u, v and w are parameters having the relation $u + v + w = 1$. A triangular patch which connects with a four-sided patch consists of a base patch $\mathbf{s}^b(u,v,w)$ which satisfy at least the given boundary curves and a compensation patch $\mathbf{s}^c(u,v,w)$ which gives null position vectors and non-zero independent tangent vectors on its three boundaries. By adjusting independent tangent vectors of the compensation patch, we can connect it with four-sided patches smoothly. As for the detail of the triangular patch and its derivatives on the boundaries, refer to Chap. 15, we use only the results for the connection. For the base patch we use eq. (14.41). For the compensation patch, we assume the following form:

$$\mathbf{s}^c(u,v,w) = \mathbf{Q}(u,v,w)uvw, \tag{14.42}$$

The vector function $\mathbf{Q}$ takes the following forms, for a case of zero tangent values at all its vertices,

$$\mathbf{Q}(u,v,w) = \frac{\mathbf{T}_1 uv + \mathbf{T}_2 vw + \mathbf{T}_3 wu}{uv + vw + wu}, \tag{14.43}$$

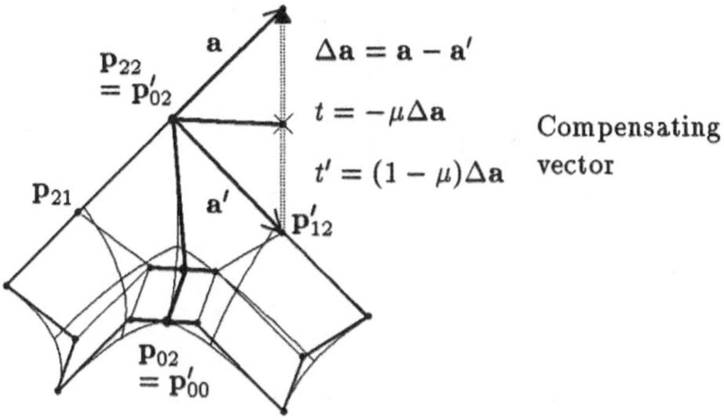

(a) Compensating and compensated vectors

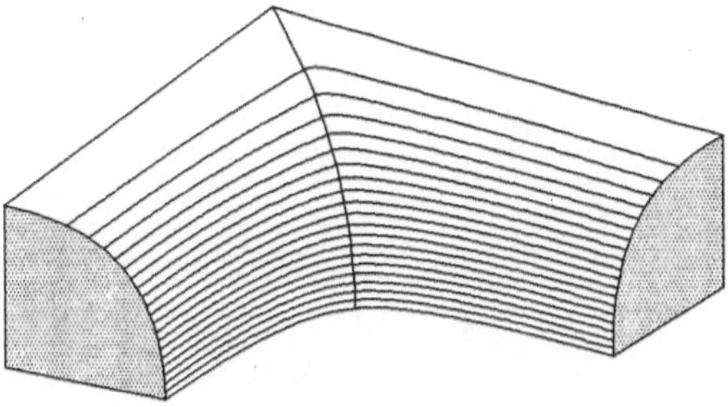

(b) Case of $\mu = 1.0$

Figure 14.9 Effects of sharing of compensation

and for a case of non-zero tangent vectors at the vertices,

$$\mathbf{Q} = \frac{\mathbf{T}_1 uv + \mathbf{T}_2 vw + \mathbf{T}_3 wu}{(uv + vw + wu)^2}. \tag{14.44}$$

On the boundary $w = 0$, derivatives of triangular patches with respect to u are:

$$\begin{aligned} \mathbf{s}_u^c(u, v, 0) &= \mathbf{s}_v^c(u, v, 0) = -\mathbf{T}_1 uv \quad \text{for eq. (14.43)}, & (14.45)\\ \mathbf{s}_u^c(u, v, 0) &= \mathbf{s}_v^c(u, v, 0) = -\mathbf{T}_1 \quad \text{for eq. (14.44)}. & (14.46) \end{aligned}$$

By using the independent tangent vectors $\mathbf{T}_i$, $i = 1, 2, 3$, we can adjust the boundary conditions for connection.

Example 5. A triangular patch surrounded by a four-sided patch array

Using a compensation patch as given by eq. (14.43), we embed a triangular patch in a net of spline surfaces. Figure 14.10(a) is a control net in the center of which is a triangular region. Figure 14.10(b) shows connected patches and Fig. 14.10(c)

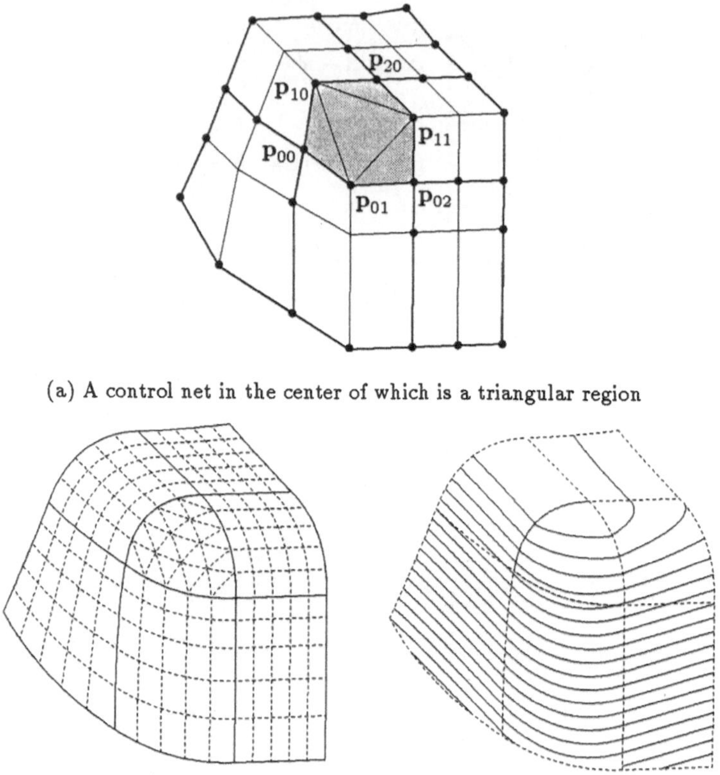

(a) A control net in the center of which is a triangular region

(b) Connected patches

(c) Contour pattern

Figure 14.10 A triangular patch surrounded by a four-sided patch array

its contour curves. The boundaries of $C^{(0)}$ connection are corrected to give $C^{(1)}$ connection.

Let a four-sided patch which is expressed by

$$\mathbf{s}'(u,v) = (1-u+\mathsf{E}u)^2(1-v+\mathsf{F}v)^2\mathbf{p}'_{00} \tag{14.47}$$

connect with a three-sided patch which is given by

$$\mathbf{s}(u,v,w) = (u+\mathsf{E}v+\mathsf{F}w)^2\mathbf{p}_{00} \tag{14.48}$$

on their common boundary curve $\mathbf{s}'(1,v) = \mathbf{s}(1-v,v,0)$. Locations of their control points on both sides of the boundary are schematically shown in Fig. 14.11, The cross-boundary tangent of the four-sided patch $\mathbf{s}'(u,v)$ on the boundary $u=1$ is

$$\mathbf{s}'_u(1,v) = 2(1-v+\mathsf{F}v)^2\mathbf{b}'_0, \tag{14.49}$$

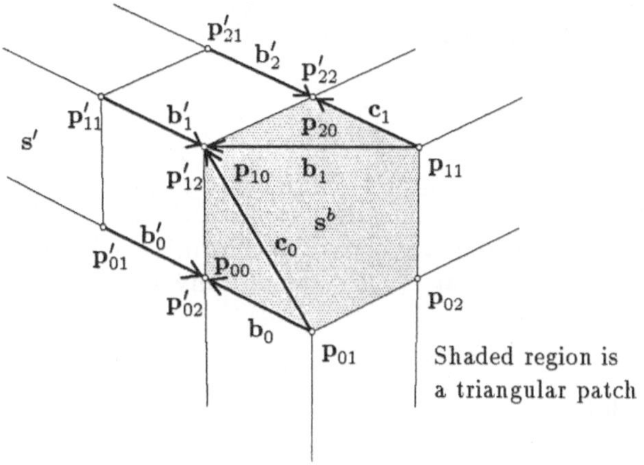

Figure 14.11 Connection of a triangular patch with a four-sided patch

where

$$\mathbf{b}'_0 = \mathbf{p}'_{20} - \mathbf{p}'_{10}, \quad \mathbf{b}'_1 = \mathbf{p}'_{21} - \mathbf{p}'_{11}, \quad \mathbf{b}'_2 = \mathbf{p}'_{22} - \mathbf{p}'_{12}.$$

The cross-boundary tangent $\mathbf{t}$ of the triangular patch $\mathbf{s}(u, v, w)$ on the boundary $w = 0$ consists of the tangent vector from the base triangular patch and that generated from the compensation patch. The former can be given by an expression $(1-v)\mathbf{s}_u^b(1-v, v, 0) + v\mathbf{s}_v^b(1-v, v, 0)$, where

$$\left.\begin{array}{rcl} \mathbf{s}_u^b(1-v, v, 0) &=& 2(1-v+\mathsf{E}v)(1-\mathsf{F})\mathbf{p}_{00} = 2(1-v+\mathsf{E}v)\mathbf{b}_0, \\ \mathbf{s}_v^b(1-v, v, 0) &=& 2(1-v+\mathsf{E}v)(\mathsf{E}-\mathsf{F})\mathbf{p}_{00} = 2(1-v+\mathsf{E}v\mathbf{c}_0, \end{array}\right\} \quad (14.50)$$

and the latter is given by $\mathbf{s}_u^c(u, v, 0) + \mathbf{s}_v^c(u, v, 0)$ from eq. (14.43), which becomes $-2v(1-v)\mathbf{T}_1$. Consequently, the cross-boundary tangent of the triangular patch is given by

$$\begin{array}{l} (1-u)\mathbf{s}_u^b(u, v, 0) + u\mathbf{s}_v^b(u, v, 0) + \mathbf{s}_u^c(u, v, 0) + \mathbf{s}_v^c(u, v, 0) = \\ \qquad 2(1-v)^2\mathbf{b}_0 + 4v(1-v)(\mathbf{b}_1 + \mathbf{c}_0/2 - \mathbf{T}_1/2) + 2v^2\mathbf{c}_1. \end{array} \quad (14.51)$$

Comparing eq. (14.51) and eq. (14.49), we can determine $\mathbf{T}_1$ to make all the coefficients of $(1-v)^2, v(1-v)$ and v^2 proportional to those control vectors of another patch to be connected in $C^{(1)}$. The result is shown in Fig. 14.10(b) and (c).

When the control tangent vectors of the triangular patch and the connecting four-sided patch are not colinear, we have to use the compensating patch given by eq. (14.44) and $\mathbf{T}_1 = 2(1-v+v\mathsf{F})^2\mathbf{t}_0$. Then, the right-hand side of eq. (14.51) is replaced by

$$2(1-v)^2(\mathbf{b}_0 - \mathbf{t}_0) + 4v(1-v)(\mathbf{b}_1 + \mathbf{c}_0/2 - \mathbf{t}_1) + 2v^2(\mathbf{c}_1 - \mathbf{t}_2). \quad (14.52)$$

This must be equal to the cross-boundary tangent of the four-sided patch, which is given by eq. (14.49), and we can determine the control tangents $\mathbf{t}_0$, $\mathbf{t}_1$ and $\mathbf{t}_2$.

14.5 Case of $C^{(2)}$ Connection

For the $C^{(2)}$ connection, a compensating patch must provide not only arbitrary second derivative vectors independently, but also arbitrary first derivative vectors on some or all of the boundaries of a compensating patch. The technique for synthesizing compensating patches is systematically given in the following.

14.5.1 Four-Sided Patches

Let a base patch be

$$\mathbf{s}^b(u,v) = (1-u+\mathsf{E}u)^3(1-v+\mathsf{F}v)^3\mathbf{p}_{00}, \tag{14.53}$$

and its compensating patch be

$$\mathbf{s}^c(u,v) = \mathbf{Q}(u,v)v(1-v)u(1-u), \tag{14.54}$$

in which a vector function $\mathbf{Q}(u,v)$ is given by

$$\mathbf{Q}(u,v) = \frac{A}{\{u(1-u)+v(1-v)\}^3} + \frac{B}{\{u(1-u)+v(1-v)\}^4}, \tag{14.55}$$

where

$$\begin{aligned} A &= (\mathbf{T}_1+\mathbf{R}_1u)\{(1-u)v(1-v)\}^2 + (\mathbf{T}_2+\mathbf{R}_2v)\{u(1-u)(1-v)\}^2 \\ &\quad +(\mathbf{T}_3+\mathbf{R}_3(1-u))\{uv(1-v)\}^2 + (\mathbf{T}_4+\mathbf{R}_4(1-v))\{vu(1-u)\}^2 \\ B &= (-3/4)[\mathbf{T}_1\{(1-u)v(1-v)\}^3 + \mathbf{T}_2\{u(1-u)(1-v)\}^3 \\ &\quad +\mathbf{T}_3\{uv(1-v)\}^3 + \mathbf{T}_4\{u(1-u)v\}^3]. \end{aligned}$$

From equations. (14.54) and (14.55), we obtain the first and the second cross-boundary derivative vectors for $u=0$:

$$\mathbf{s}^c(0,v) = 0, \quad \mathbf{s}^c_u(0,v) = (1/4)\mathbf{T}_1, \quad \mathbf{s}^c_{uu}(0,v) = 2\mathbf{R}_1. \tag{14.56}$$

where $\mathbf{T}_1$ and $\mathbf{R}_1$ are the cross-boundary first and second derivative vectors. They are independently given. We obtain similar results for the other boundaries. So we can make their connection in $C^{(2)}$ even when the original patch connection is $C^{(0)}$.

Example 6. Compensation for $C^{(2)}$ from $C^{(0)}$ connected patch

Figure 14.12(a) shows $C^{(0)}$ connected patches of degree 2×2. Figure 14.12(b) is its silhouette pattern without the compensation; its loci are discontinuous along

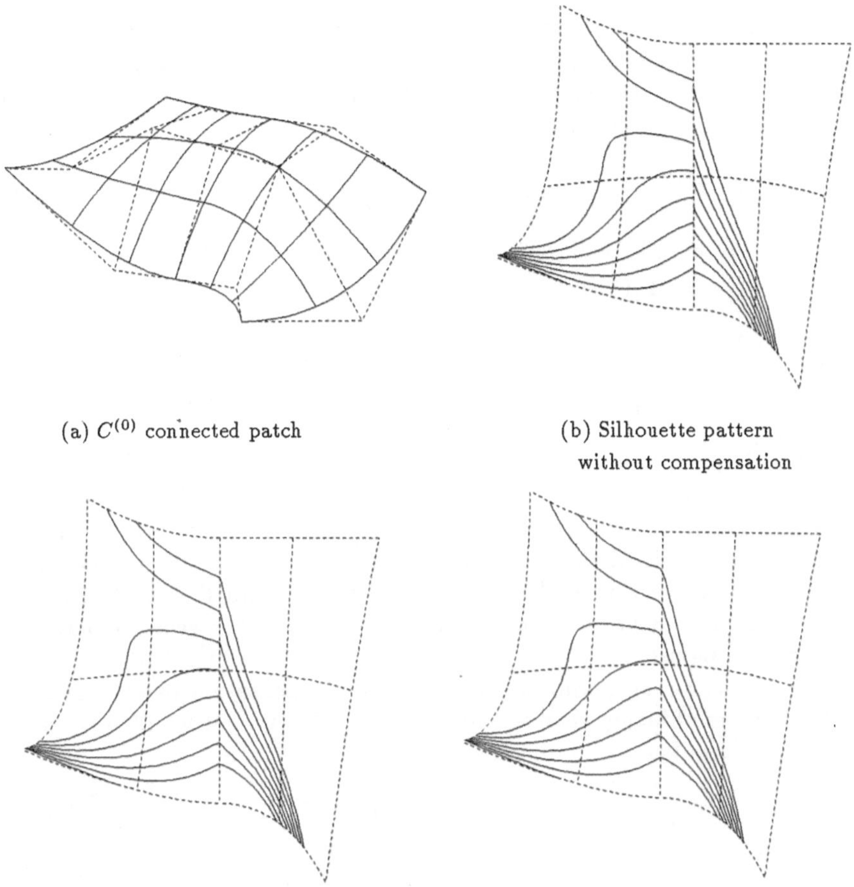

(a) $C^{(0)}$ connected patch

(b) Silhouette pattern without compensation

(c) Silhouette pattern with $C^{(1)}$ compensation

(d) Silhouette pattern with $C^{(2)}$ compensation

Figure 14.12 Compensation for $C^{(2)}$ from $C^{(0)}$ connected patch

the connected boundary. In Fig. 14.12(c) the silhouette pattern with the added compensating patches for $C^{(1)}$ are shown; the loci become continuous, but have corners on the boundary. In the same original patch with added compensation patches both for $C^{(1)}$and $C^{(2)}$, Fig. 14.12(d) shows that the discontinuity and the sharp corners of the loci on the boundary, which appeared in (b) and (c), disappear and that these compensation effects are local near the boundary. The compensating effect can be changed by the magnitudes of the compensating vectors.

14.5.2 Three-Sided Patches

To obtain independent compensating derivative vectors of the first and the second order on each boundary of a three-sided patch, we introduce the compensating patch $\mathbf{s}^c(u,v,w)$ of the same form as eq. (14.44), but with a different vector function $\mathbf{Q}(u,v,w)$ which is

$$\begin{aligned} \mathbf{Q}(u,v,w) &= \frac{A}{(uv+vw+wu)^3} - (3/4)\frac{B}{(uv+vw+wu)^4}, \qquad (14.57) \\ \text{where } A &= (\mathbf{T}_1+\mathbf{R}_1 w)(uv)^2 + (\mathbf{T}_2+\mathbf{R}_2 u)(vw)^2 + (\mathbf{T}_3+\mathbf{R}_3 v)(wu)^2, \\ B &= \mathbf{T}_1(uv)^3 + \mathbf{T}_2(vw)^3 + \mathbf{T}_3(wu)^3. \end{aligned}$$

On the boundary $w = 0$, we have

$$\mathbf{s}^c(u,v,0) = 0, \quad \mathbf{s}^c_u = \mathbf{s}^c_v = \mathbf{T}_1/4, \quad \mathbf{s}^c_{uu} = \mathbf{s}^c_{vv} = 2\mathbf{R}_1. \qquad (14.58)$$

By determining the forms of the compensating derivative vectors $\mathbf{T}_1$ and $\mathbf{R}_1$, we can make the composite derivative vectors to match the given boundary conditions for them. For the other boundaries, similar results are obtained.

When the power of the denominators of $\mathbf{Q}$ is reduced by one, the compensating derivative vectors eq. (14.58) become $\mathbf{T}_1 uv/4$ and $\mathbf{R}_1 uv$. When there is no need for compensation at the vertices of the patch, the vector function $\mathbf{Q}$ of this type can be used.

The above stated method of constructing compensation patches can be extended to cases of higher than second-degree continuity on the patch boundaries, because its form has a systematic structure.

15 Triangular Surface Patches and Their Connection

15.1 Introduction

In Chap. 13, we treated non-four-sided patches, each of which was synthesized from four-sided Bézier patches, and described their uses in special regions of shapes. In this chapter, first we explain a triangular Bézier patch defined by linear interpolations in barycentric coordinates, whose expressions in operator form are similar to that of Bézier patches and their natural extension to rational form. Then we treat methods of connection of triangular patches. Since there is no simple equivalent of a spline net as in the four-sided patch connection [10][19], we cannot control a local shape of connected triangular patches easily. But there are number of connection patterns, which are explained [12].

However, synthesizing a shape from only triangular patches of Bézier type is not easy. Accordingly, we introduce the tangent compensating patch of triangular form to increase the freedom of connection with other triangular patches. A case of a triangular patch connected with a surrounding four-sided patch net has already been treated in the previous chapter. In Sect. 15.6, we describe division and degree elevation of a triangular patch, but they are not so effective as in the case of four-sided patches.

15.2 Operator Form of a Triangular Patch

15.2.1 Triangular Bézier Patches

Just like in a Bézier surface patch, by using the shift operators E and F, a control point $\mathbf{p}_{ij}$ of a triangular patch is conveniently expressed as $\mathsf{E}^i\mathsf{F}^j\mathbf{p}_{00}$ for the purpose of simple manipulation. Refer to Fig. 15.1. A triangular plane $\mathbf{s}(u,v,w;1)$, whose three vertices are $\mathbf{p}_{00}$, $\mathbf{p}_{10}$ and $\mathbf{p}_{01}$, is expressed in the barycentric coordinates (u,v,w) by

$$\mathbf{s}(u,v,w;1) = u\mathbf{p}_{00} + v\mathbf{p}_{10} + w\mathbf{p}_{01} = (u + v\mathsf{E} + w\mathsf{F})\mathbf{p}_{00}, \tag{15.1}$$

where $u+v+w=1, \quad 0 \le (u,v,w) \le 1$.

Two of the three parameters u, v and w are independent. Values of u, v and w which determine a point in the triangle can be considered proportional to the distances from the three edges to that point. In the parameter space, the triangle is equilateral, and $u+v+w$ is proportional to its area, so by normalizing it we have eq. (15.1). If the value of one of the parameters (u,v,w) is zero, the expression represents one of its edges. That with two zero values of the

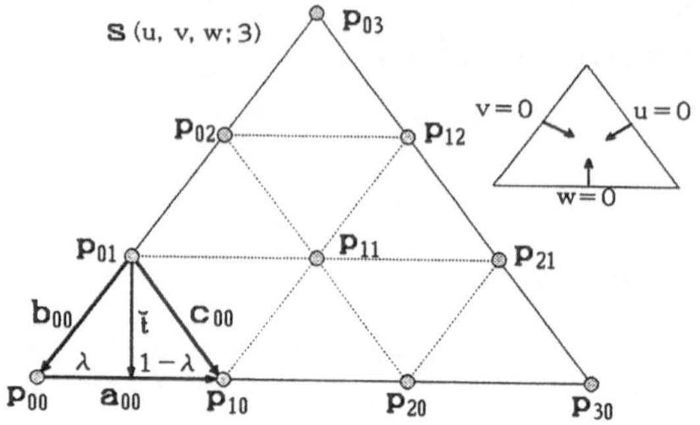

Figure 15.1 Triangular plane and barycentric coordinates

parameters corresponds to one of its vertices. For instance, $u = 0$ corresponds to the edge $\mathbf{p}_{10}\mathbf{p}_{01}$, which is

$$\mathbf{s}(0, v, w; 1) = v\mathbf{p}_{10} + w\mathbf{p}_{01} = (1 - w)\mathbf{p}_{10} + w\mathbf{p}_{01}, \tag{15.2}$$

and for $u = v = 0$, then $w = 1$,

$$\mathbf{s}(0, 0, w; 1) = w\mathbf{p}_{01} = \mathbf{p}_{01}. \tag{15.3}$$

Generalizing eq.(15.1), we define a triangular patch of degree n in the following form:

$$\begin{aligned} \mathbf{s}(u, v, w; n) &= (u + v\mathsf{E} + w\mathsf{F})^n \mathbf{p}_{00}, \\ u + v + w &= 1. \end{aligned} \tag{15.4}$$

For $w = 0$, eq.(15.4) becomes

$$\mathbf{s}(u, v, 0; n) = (u + v\mathsf{E})^n \mathbf{p}_{00} = (1 - v + v\mathsf{E})^n \mathbf{p}_{00}. \tag{15.5}$$

This ia a Bézier curve of degree n, whose control points are

$$\mathbf{p}_{00} \quad \mathbf{p}_{10} \quad \cdots \quad \mathbf{p}_{n0}.$$

Similarly, for the boundary curve of $v = 0$, its control points are

$$\mathbf{p}_{00} \quad \mathbf{p}_{01} \quad \cdots \quad \mathbf{p}_{0n}.$$

For $u = 0$, eq.(15.4) becomes

$$\mathbf{s}(0, v, w; n) = (1 - w + w\mathsf{F}\mathsf{E}^{-1})^n \mathsf{E}^n \mathbf{p}_{00} = (1 - w + w\mathsf{F}\mathsf{E}^{-1})^n \mathbf{p}_{n0}, \tag{15.6}$$

Since the operator $\mathsf{F}\mathsf{E}^{-1}$ changes an index (ij) into $((i-1)(j+1))$, its control points are

$$\mathbf{p}_{n0} \quad \mathbf{p}_{(n-1)1} \quad \cdots \quad \mathbf{p}_{0n}.$$

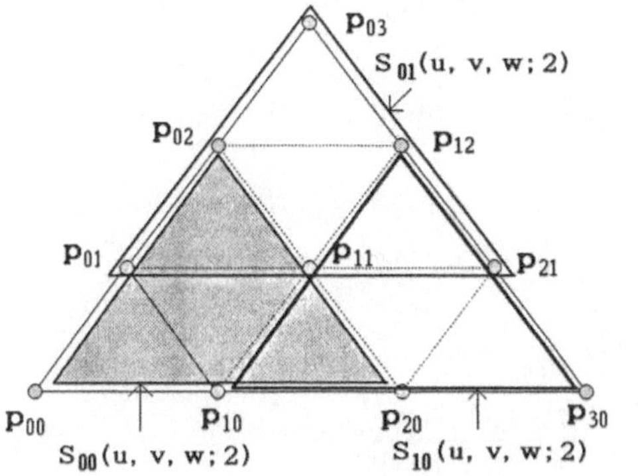

Figure 15.2 Relation of a triangular patch of degree n and triangular patches of degree $n-1$

These control points determine the shapes of the patch boundaries. The other control points $\mathbf{p}_{ij}$ with $0 < i+j < n$ determine the interior shape. The number of all the control points is $(1/2)(n+2)(n+1)$.

From eq.(15.4), we can interpret that a point of a triangular patch of degree n is on a triangular patch of degree one with its three vertices placed on the three mutually shifted triangular patches of degree $n-1$. See Fig. 15.2. Let $\mathbf{s}_{ij}(u,v,w;n)$ be a patch whose lower-left-most control point be $\mathbf{p}_{ij}$, then the shift operators E and F can operate on $\mathbf{s}_{ij}(u,v,w;n)$. Hence, eq.(15.4) can be transformed into

$$\begin{aligned}&\mathbf{s}_{00}(u,v,w;n) \qquad (15.7)\\ &= (u+v\mathsf{E}+w\mathsf{F})^n\mathbf{p}_{00} = (u+v\mathsf{E}+w\mathsf{F})\mathbf{s}_{00}(u,v,w;n-1)\\ &= u\mathbf{s}_{00}(u,v,w;n-1) + v\mathbf{s}_{10}(u,v,w;n-1) + w\mathbf{s}_{01}(u,v,w;n-1).\end{aligned}$$

15.2.2 Rational Triangular Patches

We assign a weight w_{ij} to a control point $\mathbf{p}_{ij}$ and instead of parameter u, v and w, we use $r \cdot w_{00}/w_{00}(r,s,t;1)$, $s \cdot w_{10}/w_{00}(r,s,t;1)$ and $t \cdot w_{01}/w_{00}(r,s,t;1)$, where

$$w_{00}(r,s,t;1) = r \cdot w_{00} + s \cdot w_{10} + t \cdot w_{01} = (r+s\mathsf{E}+t\mathsf{F})w_{00}. \qquad (15.8)$$

So we have a normalizing condition just like $u+v+w=1$ in the previous section,

$$r \cdot w_{00}/w_{00}(r,s,t;1) + s \cdot w_{10}/w_{00}(r,s,t;1) + t \cdot w_{01}/w_{00}(r,s,t;1) = 1. \qquad (15.9)$$

Then the rational equivalent of eq.(15.1) is

$$\mathbf{s}_{00}(r,s,t;1) = \frac{(r+s\mathsf{E}+t\mathsf{F})(w_{00}\mathbf{p}_{00})}{w_{00}(r,s,t;1)}. \qquad (15.10)$$

The interpretation of the above equation is that a point $\mathbf{s}_{00}(r,s,t;1)$ on a rational triangular patch of degree one is the rational interpolating point of three points of degree zero.

$$\mathbf{s}_{00}(r,s,t;0)=\mathbf{p}_{00}\ ,\ \mathbf{s}_{10}(r,s,t;0)=\mathbf{p}_{10}\ ,\ \mathbf{s}_{01}(r,s,t;0)=\mathbf{p}_{01}. \qquad (15.11)$$

Generalizing $w_{00}(r,s,t;1)$, which is given by eq.(15.9), we have

$$w_{00}(r,s,t;n)=(r+s\mathsf{E}+t\mathsf{F})^n w_{00}=(r+s\mathsf{E}+t\mathsf{F})w_{00}(r,s,t;n-1). \qquad (15.12)$$

Similarly, the general form of eq.(15.10) is

$$\mathbf{s}_{00}(r,s,t;n)=\frac{(r+s\mathsf{E}+t\mathsf{F})^n(w_{00}\mathbf{p}_{00})}{w_{00}(r,s,t;n)}. \qquad (15.13)$$

This is also obtained from the following recurrence formula:

$$\mathbf{s}_{00}(r,s,t;n)=\frac{(r+s\mathsf{E}+t\mathsf{F})(w_{00}(r,s,t;n-1)\mathbf{s}_{00}(r,s,t;n-1))}{w_{00}(r,s,t;n)}. \qquad (15.14)$$

The interpretation of the above equation is that a point $\mathbf{s}_{00}(r,s,t;n)$ on a rational triangular patch of degree n is a rational interpolating point of three adjacent rational triangular patches $\mathbf{s}_{00}(r,s,t;n-1)$, $\mathbf{s}_{10}(r,s,t;n-1)$, $\mathbf{s}_{01}(r,s,t;n-1)$ of degree $n-1$ with respective weights $w_{00}(r,s,t;n-1)$, $w_{10}(r,s,t;n-1)$ and $w_{01}(r,s,t;n-1)$ of degree $n-1$. Since a conic is expressed by the rational Bézier of degree two, the boundary curves of a triangular patch can be made circular arcs of the same radius and the same center. But it is possible, but not simple, to construct a part of the spherical surface by rational triangular patches of the same size[16].

15.2.3 Tangents on Patch Boundaries

On the surface of a triangular patch, there are three kinds of constant parameter curves, that is, $u=const$, $v=const$ and $w=const$. Accordingly, we can set up three basic vectors which are not independent. For simplicity, by replacing a dependent variable, for instance, w with $1-u-v$ we make a patch expression with the usual two independent variables:

$$\mathbf{s}(u,v,1-u-v)=\{u(1-\mathsf{F})+v(E-F)+F\}\mathbf{p}_{00}$$

Differentiating this expression with respect to u or v, we obtain tangent directions $\mathbf{s}_u$or $\mathbf{s}_v$ along the parameter curves $v=const$ or $u=const$ on the patch. Refer to Fig. 15.1. On one boundary $w=0$, they are

$$\mathbf{s}_u \;=\; n(1-v+v\mathsf{E})^{n-1}(1-\mathsf{F})\mathbf{p}_{00}=n(1-v+v\mathsf{E})^{n-1}\mathbf{b}_{00}, \qquad (15.15)$$
$$\mathbf{s}_v \;=\; n(1-v+v\mathsf{E})^{n-1}(\mathsf{E}-\mathsf{F})\mathbf{p}_{00}=n(1-v+v\mathsf{E})^{n-1}\mathbf{c}_{00}, \qquad (15.16)$$

where

$$\begin{aligned} \mathbf{b}_{00} &= (1-\mathsf{F})\mathbf{p}_{00} = \mathbf{p}_{00} - \mathbf{p}_{01}, \\ \mathbf{c}_{00} &= (\mathsf{E}-\mathsf{F})\mathbf{p}_{00} = \mathbf{p}_{10} - \mathbf{p}_{01}. \end{aligned} \tag{15.17}$$

The directions of these tangent vectors are not along the boundary $w = 0$. A cross-boundary tangent, which we denote $\check{\mathbf{t}}$, is given by

$$\begin{aligned} \check{\mathbf{t}} &= (1-\lambda)\mathbf{s}_u + \lambda\mathbf{s}_v \\ &= n(1-v+v\mathsf{E})^{n-1}\{(1-\lambda)\mathbf{b}_{00} + \lambda\mathbf{c}_{00}\} \\ &= n(1-v+v\mathsf{E})^{n-1}(\mathbf{b}_{00} + \lambda\mathbf{a}_{00}), \end{aligned} \tag{15.18}$$

where $\mathbf{a}_{00} = (\mathsf{E}-1)\mathbf{p}_{00} = \mathbf{c}_{00} - \mathbf{b}_{00}$, and λ is an arbitrary constant between 0 and 1.

A tangent vector along the boundary $w = 0$, which we denote $\vec{\mathbf{t}}$, is obtained from differentiation of the boundary curve $\mathbf{s}(1-v, v, 0 : n)$ by v,

$$\vec{\mathbf{t}} = n(1-v+v\mathsf{E})^{n-1}(\mathbf{p}_{10} - \mathbf{p}_{00}) = n(1-v+v\mathsf{E})^{n-1}\mathbf{a}_{00}. \tag{15.19}$$

Derivatives of higher degree are similarly obtained, for instance, on the boundary $w = 0$,

$$\begin{aligned} \mathbf{s}_{uu} &= n(n-1)(1-v+v\mathsf{E})^{n-2}(1-\mathsf{F})^2\mathbf{p}_{00} && (15.20) \\ &= n(n-1)(1-v+v\mathsf{E})^{n-2}(1-\mathsf{F})\mathbf{b}_{00} = \mathbf{b}_{00} - \mathbf{b}_{01}, \\ \mathbf{s}_{vv} &= n(n-1)(1-v+v\mathsf{E})^{n-2}(\mathsf{E}-\mathsf{F})^2\mathbf{p}_{00} && (15.21) \\ &= n(n-1)(1-v+v\mathsf{E})^{n-2}(\mathsf{E}-\mathsf{F})\mathbf{c}_{00} = \mathbf{c}_{10} - \mathbf{c}_{01}, \\ \mathbf{s}_{uv} &= n(n-1)(1-v+v\mathsf{E})^{n-2}(\mathsf{E}-\mathsf{F})(1-\mathsf{F})\mathbf{p}_{00} && (15.22) \\ &= n(n-1)(1-v+v\mathsf{E})^{n-2}(\mathsf{E}-\mathsf{F})\mathbf{b}_{00} = \mathbf{b}_{10} - \mathbf{b}_{01} = \mathbf{c}_{00} - \mathbf{c}_{01}. \end{aligned}$$

They are used for connection in $C^{(2)}$.

15.3 $C^{(1)}$ Connection of Triangular Patches

Connection of two triangular patches along a boundary in $C^{(1)}$ is not difficult, but it is difficult to cover a certain area by connected triangular patches, because there is no simple automatic control mechanism to attain globally a specified connection condition such as an S net. We have to connect a patch one by one to satisfy the $C^{(1)}$ condition. Refer to Fig. 15.3. Let $\mathbf{s}'_{00}(u, v, w; n)$ be another patch connected on the boundary $w = 0$ with $\mathbf{s}_{00}$. Its cross-boundary tangent $\check{\mathbf{t}}'$ is given similarly by

$$\check{\mathbf{t}}' = (1-\mu)\mathbf{s}'_u + \mu\mathbf{s}'_v = n(1-v+v\mathsf{E})^{n-1}(\mathbf{b}'_{00} + \mu\mathbf{a}_{00}). \tag{15.23}$$

where μ is an arbitrary constant between 0 and 1.

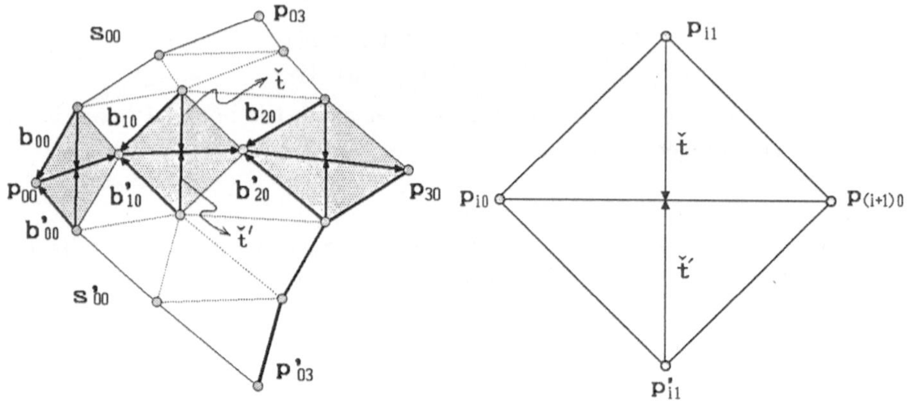

(a) Connection of patches (b) Control quadrilateral

Figure 15.3 Connection of patches and control quadrilateral

If at an arbitrary point on the common boundary, three tangent vectors $\check{\mathbf{t}}, \check{\mathbf{t}}'$ and $\vec{\mathbf{t}}$ are coplanar, then two patches $\mathbf{s}_{00}$ and $\mathbf{s}'_{00}$ are continuous with a common tangent plane. This condition of coplanarity is

$$[\check{\mathbf{t}}\ ,\ \check{\mathbf{t}}'\ ,\ \vec{\mathbf{t}}] = 0. \tag{15.24}$$

For this to be satisfied, the following conditions must hold:

$$\text{(i)} \qquad \check{\mathbf{t}} = \gamma\check{\mathbf{t}}', \tag{15.25}$$

$$\text{or (ii)} \qquad \check{\mathbf{t}} = \alpha\vec{\mathbf{t}} + \beta\check{\mathbf{t}}', \tag{15.26}$$

where α, β and γ are appropriate constants.

1. Substituting equations (15.19) and (15.23) into eq.(15.25), we obtain

$$(1 - v + v\mathsf{E})^{n-1}(\mathbf{b}_{00} + \lambda\mathbf{a}_{00}) = \gamma(1 - v + v\mathsf{E})^{n-1}(\mathbf{b}'_{00} + \mu\mathbf{a}_{00}). \tag{15.27}$$

   Developing both sides by terms $(1-v)^i v^{n-1-i}$, $i = 0 \dots n-1$, and equating their coefficients, we get

$$\mathbf{b}_{i0} + \lambda\mathbf{a}_{i0} = \gamma(\mathbf{b}'_{i0} + \mu\mathbf{a}_{i0}), \qquad i = 0 \dots (n-1). \tag{15.28}$$

   Refer to Figures 15.3 and 15.1. The simplest case to satisfy the above condition is to take $\gamma = -1$, $\lambda = \mu$. Then each quadrilateral ($\mathbf{p}_{i0}$, $\mathbf{p}_{i1}$, $\mathbf{p}_{(i+1)0}$, $\mathbf{p}'_{i1}$, $\mathbf{p}_{i0}$) is in a plane and its diagonal $\mathbf{p}_{i1}\mathbf{p}'_{i1}$ divides the other diagonal $\mathbf{p}_{i0}\mathbf{p}_{(i+1)0}$ in the same ratio for all i. We call this quadrilateral ($\mathbf{p}_{i0}$, $\mathbf{p}_{i1}$, $\mathbf{p}_{(i+1)0}$, $\mathbf{p}'_{i1}$) the *control quadrilateral.* A pair of grey triangles in Fig. 15.3 makes the control quadrilateral.

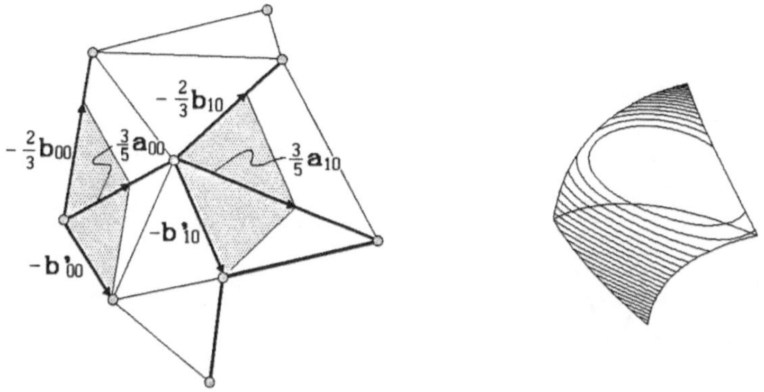

(a) Connection of patches (b) Contour curves

Figure 15.4 Connected triangular patches and contour curves

2. From equations (15.19), (15.23) and (15.26), we obtain

$$\gamma \mathbf{a}_{i0} = \alpha \mathbf{b}_{i0} + \beta \mathbf{b}'_{i0}, \qquad i = 0 \ldots (n-1), \tag{15.29}$$

where α, β and γ are constants independent of i. The connection of two patches in this pattern is shown in Fig. 15.4 together with their contour curves.
If $\mathbf{p}_{i0}$ and α, β, γ are given, $\mathbf{p}'_{i0}$ is determined from eq.(15.29). Two control points $\mathbf{p}_{i1}$ and $\mathbf{p}'_{i1}$ and an edge $\mathbf{a}_{i0}$ determine a control quadrilateral.
If α, β, γ are all equal to 1, all the control quadrilaterals become parallelepipeds.
For $n = 2$ and $\gamma = -1$, eq.(15.29) determines relations among the magnitudes of the vectors $\mathbf{b}_{00}, \mathbf{b}'_{00}, \mathbf{a}_{00}$ and $\mathbf{b}_{10}, \mathbf{b}'_{10}, \mathbf{a}_{10}$. Denoting their magnitudes by b_{00}, b'_{00}, a_{00} and b_{10}, b'_{10}, a_{10} , we have the following equations.

$$\left.\begin{array}{l} \alpha \sin\theta_0 \cdot b_{00} + \beta \sin\theta_0 \cdot b'_{00} = 0, \\ \alpha \cos\theta_0 \cdot b_{00} + \beta \cos\theta_0 \cdot b'_{00} = a_{00}, \\ \alpha \sin\theta_1 \cdot b_{10} + \beta \sin\theta_1 \cdot b'_{10} = 0, \\ \alpha \cos\theta_1 \cdot b_{10} + \beta \cos\theta_1 \cdot b'_{10} = a_{10}, \end{array}\right\} . \tag{15.30}$$

where θ_i is an angle between $(-\mathbf{b}_{i0})$ and $\mathbf{a}_{i0}$, $i = 0, 1$.
We can choose two vectors arbitrarily among $\mathbf{b}_{00}, \mathbf{b}'_{00}, \mathbf{b}_{10}$, and $\mathbf{b}'_{10}$. The remaining two vectors are determined from the above equations.

Though various shapes can be constructed from the triangular patches connected one by one so as to satisfy the above conditions, the lack of a simple mechanism of automatic adjustment of locations of the control points makes general use of triangular patches difficult.

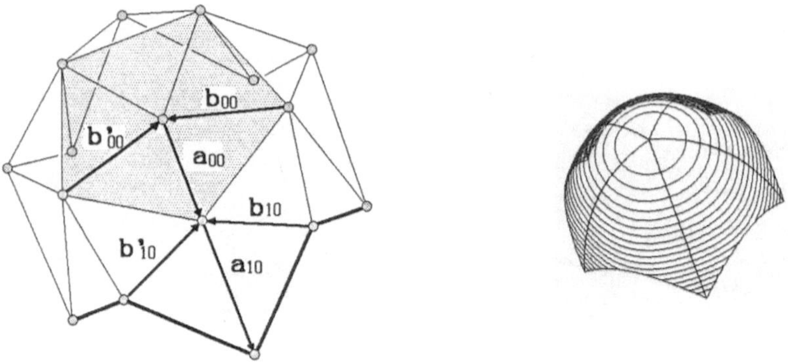

(a) Connection of patches (b) Contour curves

Figure 15.5 Special connection of triangular patches and their contour curves

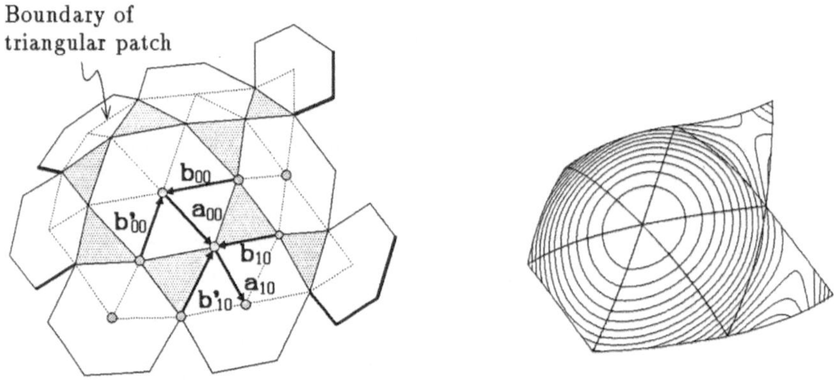

(a) Connection of patches (b) Contour curves

Figure 15.6 Surface synthesized from triangular patches

Examples. From eq.(15.30), if b_{00}, b'_{00}, a_{00} and θ_0 are given, two unknown constants α and β are obtained. With these α and β and given θ_1 and a_{10}, two unknown b_{10} and b'_{10} are determined from eq.(15.30). Using these data, we can construct the $C^{(1)}$ connected triangular patches such as shown in Figures 15.5 and 15.6. In their central regions, five or six patches are connected. In Fig. 15.5 a hatched area of pentagonal shape and in Fig. 15.6 the hexagonal areas must be planar. Further connection can be made outwards to satisfy equations (15.30) and (15.30). Figures 15.5(b) and 15.6(b) are shapes of synthesized surfaces with their contour patterns.

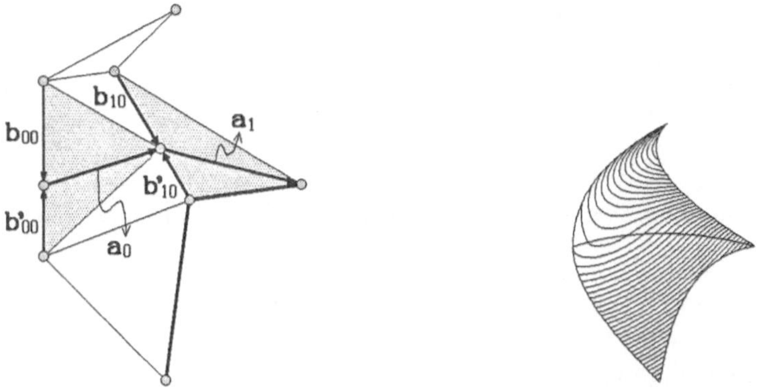

(a) Connection of patches (b) Contour curves

Figure 15.7 Control triangle and connected patches

If $\gamma = 0$, then $\mathbf{b}_{i0}$ and $\mathbf{b}'_{i0}$ must be on a straight line and the ratios of their magnitudes are constant. In this case, a control quadrilateral reduces to a triangle. See Fig. 15.7.

15.4 Arbitrary Connection of Three-Sided Patches

The connection of a three-sided patch surrounded by four-sided patches was treated in the previous Sect. 14.3 with use of the cross-boundary tangent compensating patch. Without the compensating patch, the shape synthesis by simple connection of three-sided patches is not easy, because the conditions imposed on the control quadrilaterals gathering around all the common vertices of the patches of a synthesized surfaces are difficult to satisfy. Accordingly, for easy synthesis of a shape, we need to introduce the compensating patches of the cross-boundary tangents to satisfy the conditions easily.

A triangular patch $\mathbf{s}(u,v,w;2)$ is considered consisting of a base patch $\mathbf{s}^b(u,v,w;2)$ and a compensating patch $\mathbf{s}^c(u,v,w)$:

$$\begin{aligned} \mathbf{s}(u,v,w;2) &= \mathbf{s}^b(u,v,w;2) + \mathbf{s}^c(u,v,w), \\ \text{where} \quad \mathbf{s}^b(u,v,w;2) &= (u + v\mathsf{E} + w\mathsf{F})^2\mathbf{p}_{00}, \\ \mathbf{s}^c(u,v,w) &= \mathbf{Q}uvw. \end{aligned} \tag{15.31}$$

The coefficient $\mathbf{Q}$ is a function of u, v, w and given by

$$\mathbf{Q} = \frac{\mathbf{T}_1uv + \mathbf{T}_2vw + \mathbf{T}_3wu}{(uv + vw + wu)^2}. \tag{15.32}$$

The compensating patch $\mathbf{s}^c(u,v,w)$ does not affect the boundary shape of the patch, but the derivative takes one of the relative vectors $\mathbf{T}_1$ or $\mathbf{T}_2$ or $\mathbf{T}_3$ on their boundaries $w = 0$ or $u = 0$ or $v = 0$. From eq.(15.19), the cross-boundary tangent of the base patch $\mathbf{s}^b(u, v, w; 2)$ on $w = 0$ is given by

$$\begin{aligned}\check{\mathbf{t}}^b &= (1-\lambda)\mathbf{s}_u + \lambda\mathbf{s}_v = 2(1 - v + v\mathsf{E})\{(1-\lambda)\mathbf{b}_{00} + \lambda\mathbf{c}_{00}\} \\ &= 2(1 - v + v\mathsf{E})\{\mathbf{b}_{00} + \lambda\mathbf{a}_{00}\}.\end{aligned} \tag{15.33}$$

The cross-boundary tangent of the compensating patch $\mathbf{s}^c(u, v, w)$ on $w = 0$ is given by

$$\check{\mathbf{t}}^c = (1-\lambda)\mathbf{s}_u^c + \lambda\mathbf{s}_v^c = -\mathbf{T}_1, \tag{15.34}$$

because $\mathbf{s}_u^c = -\mathbf{T}_1 = \mathbf{s}_v^c$.

Let the compensating vector $\mathbf{T}_1$ be composed of $2(1-v)\mathbf{d}_0$ and $2v\mathbf{d}_1$:

$$\mathbf{T}_1 = -2(1 - v + v\mathsf{E})\mathbf{d}_0. \tag{15.35}$$

The cross-boundary tangent $\check{\mathbf{t}}$ of the synthesized patch on the boundary $w = 0$ becomes

$$\check{\mathbf{t}} = \check{\mathbf{t}}^b + \check{\mathbf{t}}^c = 2(1 - v + v\mathsf{E})(\mathbf{b}_{00} + \mathbf{d}_0 + \lambda\mathbf{a}_{00}). \tag{15.36}$$

Since the cross-boundary tangent of the other connecting patch on the same boundary is given by

$$\check{\mathbf{t}}' = 2(1 - v + v\mathsf{E})(\mathbf{b}'_{00} + \mathbf{d}'_i + \lambda\mathbf{a}_{00}), \tag{15.37}$$

we can make two vectors (15.36) and (15.37) proportional by adjusting the compensating control vectors $\mathbf{d}_0$ and $\mathbf{d}_1$. With this technique, we can construct shapes composed of smoothly connecting triangular patches. The degree of the patches is not restricted to two. An example of a synthesized surface is shown in Fig. 15.8. Four triangular patches are connected in $C^{(0)}$ around a point. With addition of the compensating patches, their boundary regions are rounded to give the $C^{(1)}$ connection. The shape of the surface can also be modified by magnitudes of the control vectors $\mathbf{d}_0$, $\mathbf{d}_1$ and $\mathbf{d}'_0$, $\mathbf{d}'_1$.

15.5 Division of a Triangular Patch

Since a triangle can be divided into four small equilateral triangles, the corresponding triangular patch can be divided into four triangular patches. See Fig. 15.9. Therefore the number of control points of the patch increases, but there is no increase of freedom so long as the degree of the connection conditions is lowered. Since there is no such simple equivalent net as a spline net, shape control by division is difficult. But by using the connection technique described in the previous section we can replace a divided triangular patch by another triangular patch with smooth connection.

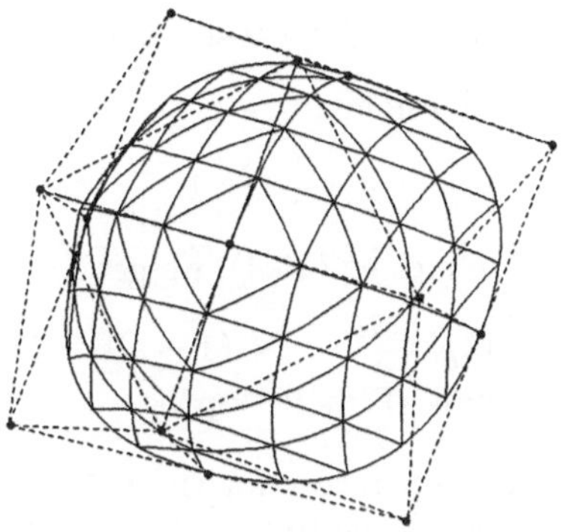

(a) Connected patches and parameter constant curves

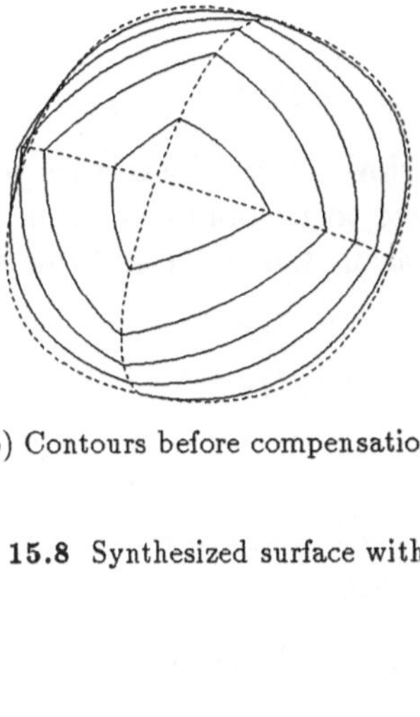

(b) Contours before compensation

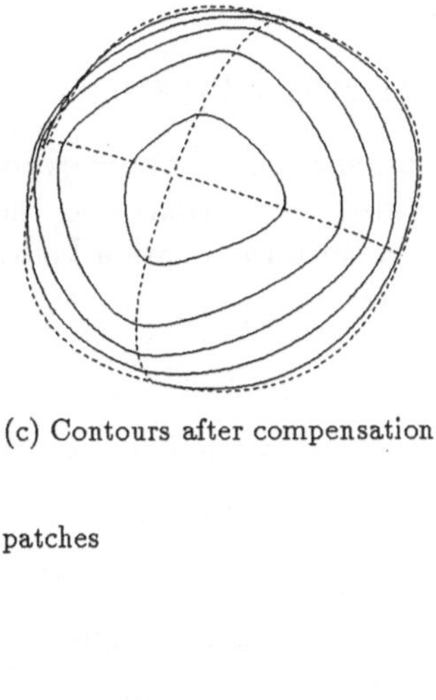

(c) Contours after compensation

Figure 15.8 Synthesized surface with compensating patches

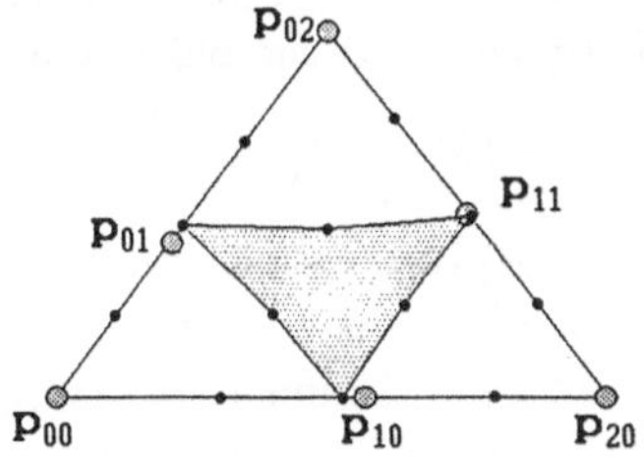

Figure 15.9 Division of a triangular patch of degree two: location of new control points

Letting r, s, t be new parameters for the parameter space of the central small triangle in Fig. 15.9, the relations between r, s, t and u, v, w are given by

$$u = (1-r)/2, \qquad v = (1-s)/2, \qquad w = (1-t)/2. \tag{15.38}$$

The following relation naturally holds:

$$r + s + t = 1. \tag{15.39}$$

Substituting these u, v and w into eq.(15.4), we get

$$\begin{aligned} \mathbf{s}(r,s,t;n) &= (1/2)^n\{1 - r + (1-s)\mathsf{E} + (1-t)\mathsf{F}\}^n \mathbf{p}_{00} \\ &= (r + s\mathsf{E}^a + t\mathsf{F}^a)^n \mathbf{p}^a_{00}, \end{aligned} \tag{15.40}$$

where E^a and F^a are the shift operators for the first and the second subscripts of the new control points $\mathbf{p}^a_{ij}$. Substituting $1 - s - t$ for r in eq.(15.40) and differentiating it i times with respect to s and j times with respect to t and setting $s = 0$, $t = 0$, we obtain the relations between the old and the new control points $\mathbf{p}_{ij}$ and $\mathbf{p}^a_{ij}$:

$$(1/2)^n(\mathsf{E}+\mathsf{F})^{n-i-j}(1-\mathsf{E})^i(1-\mathsf{F})^j\mathbf{p}_{00} = (\mathsf{E}^a - 1)^i(\mathsf{F}^a - 1)^j\mathbf{p}^a_{00}. \tag{15.41}$$

Application of the above equation is treated below. Next, the control points of the divided small triangle at the left corner are to be obtained. Referring to Fig. 15.9, since the relation between the new parameters r, s, t and the old ones u, v, w are

$$u = r/2, \qquad v = (s+1)/2, \qquad w = t/2, \tag{15.42}$$

the equation corresponding to eq.(15.40) is

$$\begin{aligned} \mathbf{s}(r,s,t;n) &= (1/2)^n\{r + (1+s)\mathsf{E} + t\mathsf{F}\}^n \mathbf{p}_{00} \\ &= (r + s\mathsf{E}^b + t\mathsf{F}^b)^n \mathbf{p}^b_{00}. \end{aligned} \tag{15.43}$$

where E^b and F^b are the shift operators for the first and the second subscripts of the new control points $\mathbf{p}^b_{ij}$.
By the same method as previously used, we obtain the equation which determines the relation between the control points of the new and the old patches :

$$(1/2)^n(\mathsf{E}+1)^{n-i-j}(\mathsf{E}-1)^i(\mathsf{F}-1)^j\mathbf{p}_{00} = (\mathsf{E}^b - 1)^i(\mathsf{F}^b - 1)^j\mathbf{p}^b_{00}. \tag{15.44}$$

From equations (15.41) and (15.44) with the given values of n, i and j, we can obtain the relation between the original control points $\mathbf{p}_{ij}$ and the new control points $\mathbf{p}^a_{ij}$ of the divided triangle patch, and the relation between $\mathbf{p}_{ij}$ and $\mathbf{p}^b_{ij}$. Since a new control points $\mathbf{p}^a_i$ can be transformed in the following way:

$$\begin{aligned} \mathbf{p}^a_{ij} &= (\mathsf{E}^a)^i(\mathsf{F}^a)^j\mathbf{p}^a_{00} = \{(\mathsf{E}^a - 1) + 1\}^i\{(\mathsf{F}^a - 1) + 1\}^j\mathbf{p}^a_{00} \qquad (15.45) \\ &= \Sigma C^i_k C^j_m (\mathsf{E}^a - 1)^k(\mathsf{F}^a - 1)^m\mathbf{p}^a_{00}, \quad k = 0\cdots i, \ m = 0\cdots j, \end{aligned}$$

and the right-hand side of the above equation can be substituted by the left-hand side of eq.(15.41), we obtain desired relation between the new and the old control points. Similarly we obtain the relation between $\mathbf{p}_i^b$ and $\mathbf{p}_i$.
Example. Division for $n = 2$.
Let the control points before the division be

$$\mathbf{p}_{00}, \mathbf{p}_{10}, \mathbf{p}_{20}, \mathbf{p}_{01}, \mathbf{p}_{11}, \mathbf{p}_{02}. \tag{15.46}$$

The locations of the new control points of the center triangle are given by

$$\begin{aligned}
\mathbf{p}_{00}^a &= (1/4)(\mathsf{E}+\mathsf{F})^2\mathbf{p}_{00} = (1/4)(\mathbf{p}_{20}+2\mathbf{p}_{11}+\mathbf{p}_{02}),\\
\mathbf{p}_{10}^a &= (1/4)(\mathsf{F}^2+\mathsf{E}+\mathsf{F}+\mathsf{EF})\mathbf{p}_{00} = (1/4)(\mathbf{p}_{02}+\mathbf{p}_{10}+\mathbf{p}_{01}+\mathbf{p}_{11}),\\
\mathbf{p}_{01}^a &= (1/4)(\mathsf{E}^2+\mathsf{E}+\mathsf{F}+\mathsf{EF})\mathbf{p}_{00} = (1/4)(\mathbf{p}_{20}+\mathbf{p}_{10}+\mathbf{p}_{01}+\mathbf{p}_{11}),\\
\mathbf{p}_{11}^a &= (1/4)(1+\mathsf{E}+\mathsf{F}+\mathsf{EF})\mathbf{p}_{00} = (1/4)(\mathbf{p}_{00}+\mathbf{p}_{10}+\mathbf{p}_{01}+\mathbf{p}_{11}),\\
\mathbf{p}_{20}^a &= (1/4)(1+\mathsf{F})^2\mathbf{p}_{00} = (1/4)(\mathbf{p}_{00}+2\mathbf{p}_{01}+\mathbf{p}_{02}),\\
\mathbf{p}_{02}^a &= (1/4)(1+\mathsf{E})^2\mathbf{p}_{00} = (1/4)(\mathbf{p}_{00}+2\mathbf{p}_{10}+\mathbf{p}_{20}).
\end{aligned} \tag{15.47}$$

The above equations show that the new control points are located at each centroid of four points of the old ones. The control point of the divided small triangle patch at the left corner are

$$\begin{aligned}
\mathbf{p}_{00}^b &= \mathbf{p}_{00},\\
\mathbf{p}_{10}^b &= (1/2)(1+\mathsf{E})\mathbf{p}_{00} = (1/2)(\mathbf{p}_{00}+\mathbf{p}_{10}),\\
\mathbf{p}_{01}^b &= (1/2)(1+\mathsf{F})\mathbf{p}_{00} = (1/2)(\mathbf{p}_{00}+\mathbf{p}_{01}),\\
\mathbf{p}_{11}^b &= (1/4)(1+\mathsf{E}+\mathsf{F}+\mathsf{EF})\mathbf{p}_{00} = (1/4)(\mathbf{p}_{00}+\mathbf{p}_{10}+\mathbf{p}_{01}+\mathbf{p}_{11}),\\
\mathbf{p}_{20}^b &= (1/4)(1+\mathsf{E})^2\mathbf{p}_{00} = (1/4)(\mathbf{p}_{00}+2\mathbf{p}_{10}+\mathbf{p}_{20}),\\
\mathbf{p}_{02}^b &= (1/4)(1+\mathsf{F})^2\mathbf{p}_{00} = (1/4)(\mathbf{p}_{00}+2\mathbf{p}_{01}+\mathbf{p}_{02}).
\end{aligned} \tag{15.48}$$

The control points of the divided patches at the right corner and the upper corner have the similar relations. Geometrical interpretation of the locations of the new control points given by equations (15.47) or (15.48) with respect to the old ones is easy.

15.6 Elevation of Degree

It is possible to raise the degree of a triangular patch without changing its shape. The method is similar to that of the four-sided patch: multiplying a factor $(u+v+w)$ with $(u+v\mathsf{E}+w\mathsf{F})^n\mathbf{p}_{00}$, which represents a triangular patch of degree n, and comparing to $(u+v\mathsf{E}'+w\mathsf{F}')^{n+1}\mathbf{p}'_{00}$, we obtain the locations of the new control points $\mathbf{p}'_{ij}$ in terms of the old control points $\mathbf{p}_{ij}$.

For $n = 2$ we obtain the following expressions:

$$\begin{aligned}
\mathbf{p}'_{00} &= \mathbf{p}_{00}, \quad \mathbf{p}'_{30} = \mathbf{p}_{20}, & \mathbf{p}'_{03} &= \mathbf{p}_{02},\\
\mathbf{p}'_{20} &= (1/3)(\mathbf{p}_{20}+2\mathbf{p}_{10}), & \mathbf{p}'_{02} &= (1/3)(\mathbf{p}_{02}+2\mathbf{p}_{01}),\\
\mathbf{p}'_{10} &= (1/3)(\mathbf{p}_{00}+2\mathbf{p}_{10}), & \mathbf{p}'_{12} &= (1/3)(\mathbf{p}_{02}+2\mathbf{p}_{11}),\\
\mathbf{p}'_{01} &= (1/3)(\mathbf{p}_{00}+2\mathbf{p}_{01}), & \mathbf{p}'_{21} &= (1/3)(\mathbf{p}_{20}+2\mathbf{p}_{11}),\\
\mathbf{p}'_{11} &= (1/3)(\mathbf{p}_{10}+\mathbf{p}_{01}+\mathbf{p}_{11}).
\end{aligned} \tag{15.49}$$

16 Surface Intersections

16.1 Introduction

Interference calculation of surfaces is an important basic procedure in CAD/CAM technology. It appears in various processes such as set operations in solid modelling, surface modelling, cutter path calculation for NC machining, and calculations in robot motion. Efficiency and robustness of its solutions are required especially for those of free-form surfaces, because the degree of intersection curves becomes so high that their analytical methods are generally hopeless and the numerical methods become unstable where the normal vectors of interfering surfaces on their intersecting curves become nearly parallel.

Recently, a great many mathematical studies on these problems have been made, but they seem still too theoretical to be used by engineers for their practical applications. A basic textbook [10] on these topics has appeared recently, but we explain our methods of intersection calculation in this section, because they are easier as well as practically proved [13].

Interference of quadrics has been treated in Chap. 5 using pencils of two quadrics and conics, but in this chapter it is included in a case of the intersection of surfaces of implicit forms. First we treat a case of the intersection between a plane and a free-form surface as a basis of the intersection of surfaces which are represented in parametric forms or implicit forms. Since the contour curves of a shape, which are its section curves by equidistant cutting planes, are very useful for understanding the shape of a curved surface without its real model, methods of producing the contour curves of free-form surfaces must be efficient and accurate. Accordingly first we treat the intersection of a curved surface and a plane, then we extend the methods to the intersection of curved surfaces.

To trace them we apply numerical methods for solving ordinary differential equations with automatic stepsize adjustment, because this provides good results without requiring attention to determine the stepsizes to the next point [14][15]. Selection of stepsizes has always been a problem for tracing curves whether the method of differential equation is adopted or not. Since near the singular points of the interfering surfaces the behavior of the curve to be traced becomes unpredictable and in other regions a large stepsize may be applied, control of the stepsizes is needed for efficient and robust calculation. If we adopt the methods of differential equation solving with quality control, we are freed from this problem. The other important problem is to find the initial starting points of the curves and all the intersection loops. Since the intersection may become a contact point or a very small loop, they may be missed if an appropriate method to find them is not given. We discuss this problem and show a practical method to solve it.

In regions where the normal vectors of interfering surfaces become nearly parallel with each other, the basic differential equation for their intersection curve becomes invalid. We explain a method of treating these regions and discuss the behavior of the intersection curves there. In this chapter, we also treat intersection curves of offset surfaces, because the problem is practically important and similar solving methods can be applied to this problem.

Some of the methods of differential equation solving for our purpose are compared and evaluated in the Appendix.

16.2 Intersection of a Curved Surface and a Plane

16.2.1 General Remarks

Intersection curves of a plane and a free-form surface are important, because they are fundamental to the interference calculation and frequently required in other problems. For example, as we cannot understand and evaluate a three-dimensional free-form shape only by its two-dimensional projections, we make drawings of the intersection curves of the surface with equi-distant parallel planes placed in various directions for visualizing its detailed shape. Since obtaining the intersection between curved surfaces frequently becomes a rather complicated unstable problem in numerical analysis, we introduce a common cutting plane for both surfaces to determine a common intersecting point of the two intersection curves of the surfaces and the plane. Or we frequently replace a curved surface by its tangential plane or a polyhedron, in which case we need the cutting procedure of the surface and faces of the polyhedron.

Let $\mathbf{n}_f$ be the unit normal vector of a plane and d its distance from the origin, then an equation of the plane is given by $\mathbf{r} \cdot \mathbf{n}_f = d$ (refer to eq. (3.11)). Let a surface patch be denoted by $\mathbf{s}(u, v)$. Since points of the plane and those of the surface patch are coincident on the intersection curve, that is, $\mathbf{s}(u, v) = \mathbf{r}$, eliminating $\mathbf{r}$ from both equations we have

$$\mathbf{s}(u, v) \cdot \mathbf{n}_f = d. \tag{16.1}$$

This is the equation of the intersection curves. Fixing a value of u or v as a constant, we can obtain the corresponding v or u of a point on the intersection by solving eq. (16.1). For patches of the first or the second degree, the equation to solve being linear or quadratic, we obtain its analytic solution easily. But for a patch of higher than the second degree, we have to turn to the numerical methods of solving high-degree algebraic equations. If only a few points are to be obtained, the finding roots of higher-degree polynomials (usually of degree three to five) is not a problem, but calculating all the points by this method is not desirable. Methods of tracing the curve from a starting point are required.

In regions where the surface normal vector is nearly parallel to that of the cutting plane, the solution becomes sensitive to numerical errors. Such cases occur frequently in practical problems, so this problem is treated in Sect. 16.6.

When the contour curves of a surface which has many peaks are to be obtained, a sophisticated method such as the extremum search curves (refer to Sect. 7.3.3) is recommended to determine the starting points for tracing the contour curves. The extremum search curves are loci of points on the surface where its basic vectors are parallel to the plane. During its tracing, the distance from the cutting plane is calculated to determine an intersecting point, where one of the intersection curves starts. Since the extremum search curves determine all the peaks, the bottom points and the saddle points as well as the paths between these points, we can find all the starting points of the loops on the same level.

For a simle surface, a standard method for tracing a contour curve is:

1. Make sure that there is a possibility of intersection between a plane F and a patch $\mathbf{s}(u, v)$ by using the convex hull property of the patch. If there is the possibility then perform the next steps.
2. Determine intersection points of the plane and the boundaries of the patch. An even number of points must be obtained including the points corresponding to double roots.
3. If the intersection points exist, then from one of the points, trace the intersection curve until it reaches another intersection point on the boundary. Determine all other intersection curve segments corresponding to the remaining intersection points on the boundaries.
4. When there is no new starting point on the boundaries, divide the patch into smaller ones by making new boundaries and test the possibility of intersection with them, and repeat the above procedure.

Since the intersection points between a plane F and a parametric curve $\mathbf{s}(u, v)$ for u or $v = const$ are obtained from solution of an algebraic equation (16.1) for u or $v = const$, those of the boundaries of the patch and the plane are easily determined using the Newton-Raphson method or as roots of the algebraic equation.

16.2.2 A Practical Method of Obtaining a Point on an Intersection Curve

When it is difficult to solve eq. (16.1), for instance in case of an offset surface, to obtain a point on the section curve we use the following iteration method to reach the true point on it. Refer to Fig. 16.1. Let the initial point be $\mathbf{p}_0 = \mathbf{s}(u_0, v_0)$ and the tangential plane there be F_t and at the point $\mathbf{p}_0$ make a normal plane F_n which is also orthogonal to the plane F. An intersection point $\mathbf{p}_1$ of the three planes F, F_t and F_n, which is given by eq. (16.11), is the next approximate point.

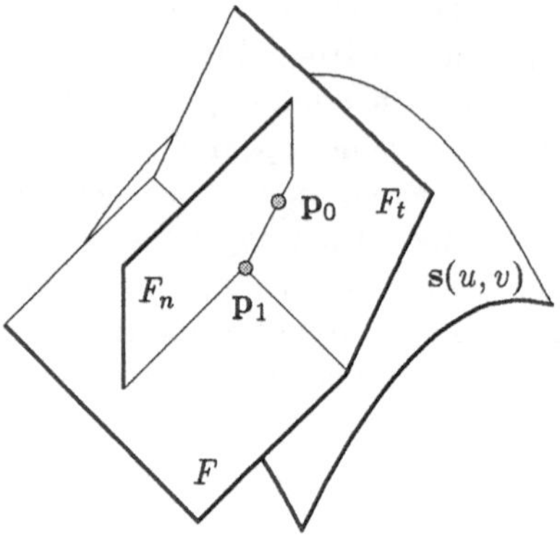

Figure 16.1 A model of obtaining a point on an intersection of a curved surface and a plane

Since this point is in the tangential plane at $\mathbf{p}_0$, a vector $\mathbf{p}_1 - \mathbf{p}_0$ can be expressed by the basic vectors of the surface at $\mathbf{p}_0$. We have

$$\mathbf{s}_u du + \mathbf{s}_v dv = \mathbf{p}_1 - \mathbf{p}_0 = \Delta\mathbf{p}, \tag{16.2}$$

where $\mathbf{s}_u$ and $\mathbf{s}_v$ are the basic vectors at $\mathbf{p}_0$. By making the inner product with a vector $\mathbf{s}_v \times \mathbf{n}_f$ and that with a vector $\mathbf{s}_u \times \mathbf{n}_f$ on both sides of the above equation, we obtain respectively

$$du = \frac{[\Delta\mathbf{p}, \mathbf{s}_v, \mathbf{n}_f]}{[\mathbf{s}_u, \mathbf{s}_v, \mathbf{n}_f]}, \quad dv = -\frac{[\Delta\mathbf{p}, \mathbf{s}_u, \mathbf{n}_f]}{[\mathbf{s}_u, \mathbf{s}_v, \mathbf{n}_f]}. \tag{16.3}$$

Then considering $\mathbf{s}(u_0 + du, v_0 + dv)$ a new $\mathbf{p}_0$, we iterate the above procedure until $\Delta\mathbf{p}$ becomes sufficiently small. Usually the convergence is very good and a point on the intersection curve is determined. When the procedure does not converge, a tangent plane at the intermediate point $\mathbf{p}$ and the plane F becomes parallel. In such a case, we can examine the shape of the patch around a point $\mathbf{q}$ which is determined on an intersection curve of $\mathbf{s}$ with the plane F slightly moved in its normal direction.

A similar but more complicated process is used for a point on the intersection of two curved surfaces.

16.2.3 Curve Tracing by Differential Equation Solving

After determining an intersection point, we start the tracing of an intersection curve from there. The conventional method to determine the next point on the intersection is to take an approximate point for $\mathbf{s}(u_0 + \Delta u, v_0 + \Delta v)$ on the tangent line of the intersection curve at the initial point. The tangent direction is given by the vector product of the normal vectors of the plane and the patch. We determine the distance between the approximate and the initial points by estimating the curvature of the intersection curve there (see Sect. 6.3) or we take a sufficiently small value for it. Since this approximate point is represented by two expressions, $\mathbf{p} = \mathbf{p}_0 + \Delta\mathbf{s}$ on the plane and $\mathbf{q} = \mathbf{s}(u + \Delta u, v + \Delta v)$ on the patch, from these two points a point on the intersection is obtained by the method described in the previous section. Repeating the same process, we obtain successive points of the intersection curve.

Instead of the conventional procedure explained above, we use well established differential equation solvers with adaptive stepsize to optimize the accuracy of the results and the calculation speed as well as to standardize the procedure for obtaining the intersection curves of all kinds of surface expressions, in which a plane is a special case. In application of numerical methods of differential equation solvers for initial-value problems, initial values have to be determined and accumulation of truncation errors has to be kept within a given tolerance. But both problems are satisfactorily solved in our case.

By taking the differential of eq. (16.1), we have

$$\phi_u du + \phi_v dv = 0, \qquad (16.4)$$

$$\text{where} \qquad \phi(u, v) = \mathbf{n}_f \cdot \mathbf{s}.$$

Considering eq. (16.4) the differential equation of a section curve, we can trace it by using one of the established numerical methods for solving initial value problems. Usually eq. (16.4) is transformed into the appropriate forms by introducing a function $f(u, v)$:

$$f(u, v) = \pm(E\phi_v^2 - 2F\phi_u\phi_v + G\phi_u^2)^{-1/2}. \qquad (16.5)$$

Then we have

$$du/ds = -f(u, v)\phi_v, \qquad dv/ds = f(u, v)\phi_u, \qquad (16.6)$$

where s is an independent variable representing the arc length of the curve and E, F, G are the fundamental magnitudes of the first order of the surface. Since we obtain values of u and v as functions of s, we can calculate the section curve $\mathbf{s}(u(s), v(s))$.

Where the normal vectors of the surface and the plane become nearly parallel, the tracing cannot be continued, because the functions ϕ_u and ϕ_v become very small. Then the behavior of the curve must be estimated from the second-order expansion of the surface at that point. This is explained in Sect. 6.7. However, if one uses a suitable numerical method of solving differential equations with adaptive stepsize control, even in a region where the difference of an angle between the normal vectors of the two surfaces becomes as small as 10^{-10} radian, the intersection curve can be traced without using the second-order terms. A practical approach for obtaining the solution is discussed in the Appendix.

Example. Figure 16.2 shows contour curves of a surface together with its extremum search curves. There are eight extremum points: three peaks, one bottom point and four saddle points. Many examples of the contour curves are

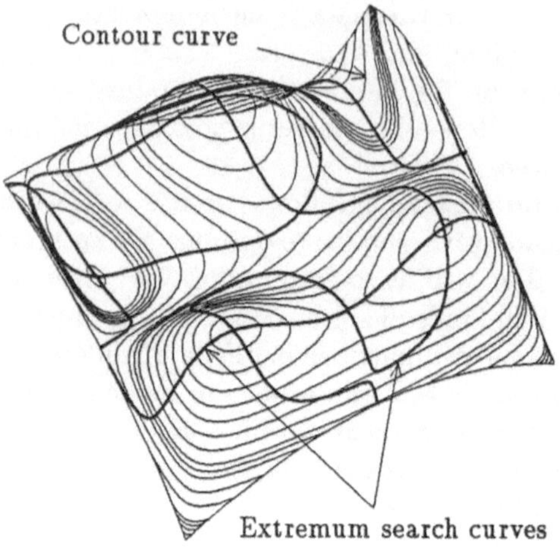

Figure 16.2 Contour curves and extremum search curves

found in this book to indicate the shapes of surfaces, or to check the smoothness of connection of patches. They are the results of this method.

16.3 Points on Intersection of Two Curved Surfaces

16.3.1 General Remarks

Determining the intersection curves of two surfaces is not so simple as in the case of a curved surface and a plane, because we have to solve three nonlinear simultaneous equations of four variables instead of solving one nonlinear equation of two variables such as eq. (16.1). A slight change of the relative positions of two intersecting surfaces may cause large a variation of the shape and the characters of their intersection curves, and singular points on the curve frequently appear in practical cases. We can determine all the singular points of the intersection curves in advance, but their calculation is not easy and we treat the problem in Sect. 16.3.3.

Intersection of two quadric surfaces, each expressed by an implicit equation such as $f(x, y, z) = 0$, has been treated theoretically in detail in Chap. 4, but the intersection of parametric surfaces equal to or higher than the second degree, which is practically important, cannot be treated with their implicit forms, because the degree of an algebraic equation of intersection curves is too high to treat with current numerical calculation techniques. For example, the degree of the equation of the intersection for parametric surfaces of degree two is 64 and that of degree three is 324. We are also required to obtain the intersection curves of surfaces which are not expressed by closed-form equations, such as offset sur-

-faces. Accordingly, we need some new ideas to solve this problem theoretically and practically.

Since differential equation solvers with adaptive stepsize can be applied to produce the intersection curves of various types of surfaces with good results, an important problem of applying them is to find all the initial starting points for all intersection loops, which in extreme cases become contact points. In obtaining the intersection curves of a plane and a free-form surface, the extremum search curves of the patch can be used successfully for determining the starting points and the loops. Curves similar to the extremum search curves can be applied for determining their initial starting points and all the loops of the intersection curve of two surface patches. This is explained in Sect. 16.3.3. But first, we explain a method which gives a point on an intersection curve from two points, one on each respective surface.

To avoid confusion with the symbols used in calculating interference of two surface patches, we use different symbols for the parameters in the expressions of the patches $\mathbf{r}(w,t)$ and $\mathbf{s}(u,v)$. Suppose we have two points $\mathbf{p}_0$ and $\mathbf{q}_0$, one of which is on each respective patch and near the probable intersection curve. Let their parameter values there be w_0, t_0 and u_0, v_0. Then we have

$$\mathbf{p}_0 = \mathbf{r}(w_0,t_0) \text{ and } \mathbf{q}_0 = \mathbf{s}(u_0,v_0). \tag{16.7}$$

Starting from the above points, we finally have to find a point on the intersection curve with parameters which satisfy the equation

$$\mathbf{s}(u,v) = \mathbf{r}(w,t). \tag{16.8}$$

The Newton-Raphson method solving nonlinear equations can be applied when the first approximate points are near the intersection curve. But this usually does not give good results unless favorable initial values are supplied. So we modify it to a more geometrically oriented method which gives a point on the curve.

16.3.2 Method with an Auxiliary Plane

A more reliable method of obtaining a point on the intersection curve of two surfaces is to introduce an auxiliary plane with which the two surfaces intersect. Moreover, for the first approximation, we replace the surfaces by their tangential planes at points near the probable intersection curve. Then we use an auxiliary plane which intersects with both tangential planes orthogonally.

Refer to Fig. 16.3. With the following definitions and procedures, we determine a point on the intersection curve of the two patches from $\mathbf{p}_0$ and $\mathbf{q}_0$.

1. Let the tangential plane and the unit normal vector at $\mathbf{p}_0$ of $\mathbf{r}(w_0,t_0)$ be F_p and $\mathbf{n}_p$, and those at $\mathbf{q}_0$ of $\mathbf{s}(u_0,v_0)$ be F_q and $\mathbf{n}_q$. Symbols d_p and d_q

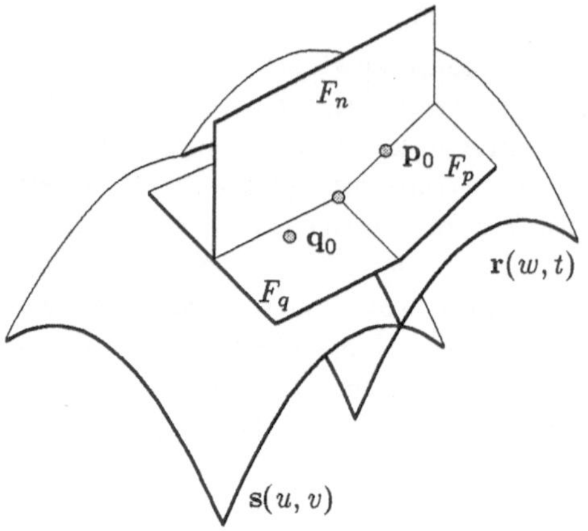

Figure 16.3 A model of obtaining a point on an intersection of two curved surfaces

denote the distances from the origin to the tangential planes F_p and F_q. Since $\mathbf{p}_0 = \mathbf{r}(w_0, t_0)$ and $\mathbf{q}_0 = \mathbf{s}(u_0, v_0)$, the distances are given by

$$d_p = \mathbf{n}_p \cdot \mathbf{r}(w_0, t_0), \qquad d_q = \mathbf{n}_q \cdot \mathbf{s}(u_0, v_0). \tag{16.9}$$

2. Make a plane F_n which passes $\mathbf{p}_0$ and is orthogonal to both F_p and F_q. Let $\mathbf{n}_n$ be the unit normal of the plane F_n, whose distance from the origin is d_n. Then we have

$$\mathbf{n}_n = \frac{\mathbf{n}_p \times \mathbf{n}_q}{|\mathbf{n}_p \times \mathbf{n}_q|}, \qquad d_n = \mathbf{n}_n \cdot \mathbf{r}(w,t). \tag{16.10}$$

Let $\mathbf{p}$ be the intersection point of the three planes F_p, F_q and F_n. It is given by

$$\mathbf{p} = \frac{d_p(\mathbf{n}_q \times \mathbf{n}_n) + d_q(\mathbf{n}_n \times \mathbf{n}_p) + d_n(\mathbf{n}_p \times \mathbf{n}_q)}{[\mathbf{n}_p, \mathbf{n}_q, \mathbf{n}_n]}. \tag{16.11}$$

(Refer to eq. (3.21) of Chap. 3.) It is an approximate point on the intersection of $\mathbf{r}(w,t)$ and $\mathbf{s}(u,v)$.

3. We are to express the difference vectors $\delta\mathbf{p}_0 = \mathbf{p} - \mathbf{p}_0$ and $\delta\mathbf{q}_0 = \mathbf{p} - \mathbf{q}_0$ by the basic vectors $\mathbf{r}_w$, $\mathbf{r}_t$ and $\mathbf{s}_u$, $\mathbf{s}_v$ at $\mathbf{p}_0$ and $\mathbf{q}_0$. We define vectors $\acute{\mathbf{r}}_w, \acute{\mathbf{r}}_t, \acute{\mathbf{s}}_u$ and $\acute{\mathbf{s}}_v$ which are orthogonal to the basic vectors of the respective surfaces and their normals:

$$\left.\begin{aligned} \acute{\mathbf{r}}_w &= \mathbf{r}_w \times \mathbf{n}_p, & \acute{\mathbf{r}}_t &= \mathbf{r}_t \times \mathbf{n}_p, \\ \acute{\mathbf{s}}_u &= \mathbf{s}_u \times \mathbf{n}_q, & \acute{\mathbf{s}}_v &= \mathbf{s}_v \times \mathbf{n}_q \end{aligned}\right\} \tag{16.12}$$

Since the following incremental relations can be assumed

$$\left.\begin{aligned} \mathbf{r}_w\delta w + \mathbf{r}_t\delta t &= \delta\mathbf{p}_0, \\ \mathbf{s}_u\delta u + \mathbf{s}_v\delta v &= \delta\mathbf{q}_0, \end{aligned}\right\} \tag{16.13}$$

we obtain the increments of the parameters by using the above-defined vectors

$$\delta w = \frac{(\acute{\mathbf{r}}_t \cdot \delta \mathbf{p}_0)}{(\acute{\mathbf{r}}_t \cdot \mathbf{r}_s)}, \quad \delta t = \frac{(\acute{\mathbf{r}}_w \cdot \delta \mathbf{p}_0)}{(\acute{\mathbf{r}}_w \cdot \mathbf{r}_t)}, \tag{16.14}$$

$$\delta u = \frac{(\acute{\mathbf{s}}_u \cdot \delta \mathbf{q}_0)}{(\acute{\mathbf{s}}_v \cdot \mathbf{s}_u)}, \quad \delta v = \frac{(\acute{\mathbf{s}}_v \cdot \delta \mathbf{q}_0)}{(\acute{\mathbf{s}}_u \cdot \mathbf{s}_v)}. \tag{16.15}$$

Then the updated initial points are given by

$$\mathbf{p}_0 = \mathbf{r}(w_0 + \delta w, t_0 + \delta t) \quad \mathbf{q}_0 = \mathbf{s}(u_0 + \delta u, v_0 + \delta v). \tag{16.16}$$

4. If $|\mathbf{p}_0 - \mathbf{q}_0|$ is not within the given tolerance, the above process from (1) is repeated until $|\mathbf{p}_0 - \mathbf{q}_0|$ becomes sufficiently small. Convergence of this method is usually very good, even when the initial points are not near. When convergent results are not obtained, information on the nearness of the two patches can be extracted.

16.3.3 Initial Starting Points and Critical Contact Points

16.3.3.1 Detection of Intersection Loops

Since the method stated above is very effective for detection of an intersection loop, if one is given several pairs of initial points, practically all the intersection loops can be detected. Figure 16.4 shows an example obtained by this method. But if one wants to make the procedure more systematic, the following procedure is recommended:

1. At first we fix a reference vector $\mathbf{n}_f$ towards which both intersecting surfaces are directed, that is, the rear sides of both surfaces can be seen from that direction. Determine the extremum search curves $\phi_u = 0$ and $\phi_v = 0$ for each of the interfering patches $\mathbf{r}$ and $\mathbf{s}$. Regions separated by these curves have monotonic gradients. Moreover, if possible, determine the points $\phi_{uu} = \phi_{vv} = 0$ on the patch $\mathbf{s}$, which we call the max-gradient points. There is one max-gradient point in each region. Otherwise we select an arbitrary point in a region in place of the max-gradient point of the region. These points are taken as $\mathbf{p}_0$.
2. For each $\mathbf{p}_0$ we fix a point $\mathbf{q}_0$ on another patch $\mathbf{r}$, which is arbitrary, but not very far from $\mathbf{p}_0$ and preferably in a region of large difference of slope from that at $\mathbf{p}_0$. We can exclude regions which surely do not interfere each other, because we can determine approximately a space in which no interference exists by the data of the extremum search curves of both surfaces.
3. From a pair $\mathbf{p}_0$ and $\mathbf{q}_0$, we can obtain a point on an intersection loop, which is traced from this point. Using all the pair points we can obtain all the intersection curves of the two patches. Usually most of points thus obtained are on the loops already determined.

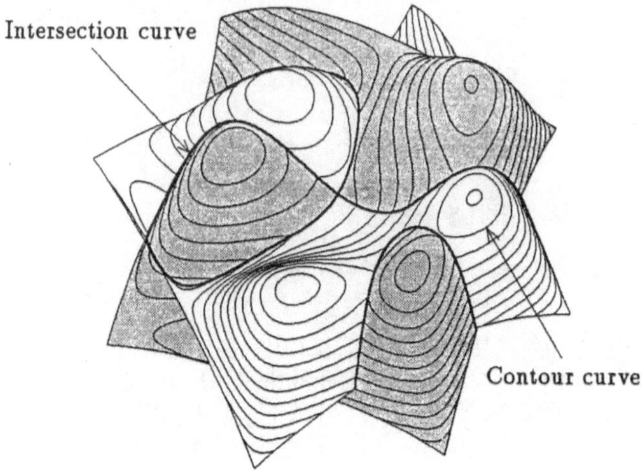

Figure 16.4 Intersection of two free-form surfaces. Contour curves are shown

4. If we reverse the role of the surfaces **s** and **r** and perform the same procedure, we can catch all the intersection loops more surely, although usually this may be omitted.

Example. Figure 16.5(a) shows intersection curve patterns of two surface patches at nine relative positions. The shapes of two patches **s** and **r** are the same as shown in Fig. 16.2, but **r** is rotated by 90 degrees anti-clockwise relative to **s** and separated by constant heights. Shapes of the intersection curves vary gradually or rather abruptly as the height changes. At certain heights, new loops are generated or disappear or one loop separates into two. These curves are obtained by the process stated above. The z value (height) of the patch **r** is changed from $h = -1.0$ to 1.0 with $\Delta h = 0.25$.

Around $h = 0$ the intersection pattern changes rather abruptly. Figure 16.5(b),(c),(d) are the intersection curves at $h = 0$, $h = 0.001$ and $h = -0.001$. Figure 16.6 is the enlarged portion at the central part. At $h = -0.775 \times 10^{-4}$ there is a branch point.

To show the effectiveness of the method, we explain a critical case. Figure 16.7 shows the intersection curves at the critical position of the two surfaces.

This position is a little lower than that of the case $h = -0.5$ of Fig. 16.5: the left small loop which is seen in the case $h = -0.5$ reduces to a point at $h = -0.65$ whose location is indicated by a small dot and the upper-right intersection curve is shown by a thick line. To test how the method of Sect. 16.3.2 can work well in this critical case, we set the point pairs of $\mathbf{p}_0$ and $\mathbf{q}_0$ and see if we can obtain a convergence point. We move the location of $\mathbf{p}_0$ to various points on **s** to determine allowable domains on **s**:

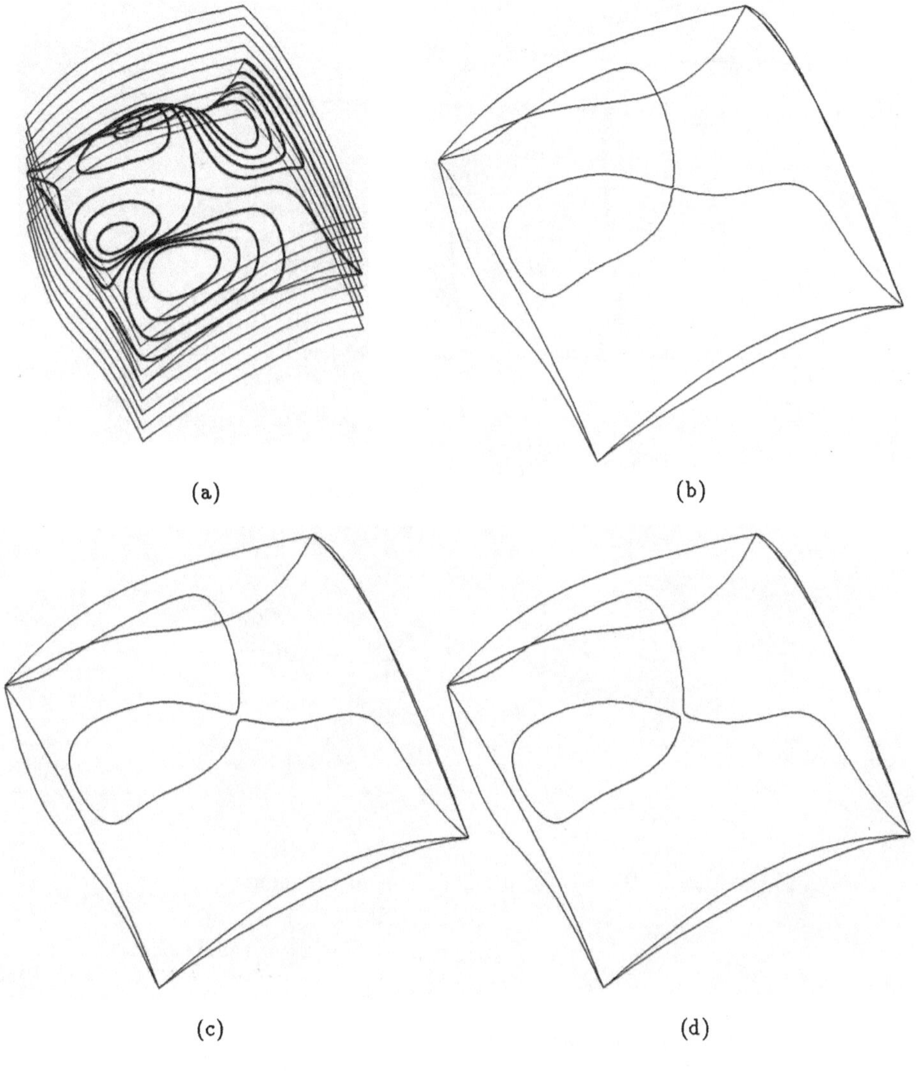

Figure 16.5 Variation of intersection curve pattern by displacement of one surface: (a) nine positions from $h = -1.0$ to $h = 1.0$ and $\delta h = 0.25$, (b) $h = 0$, (c) $h = 0.001$, (d) $h = -0.001$

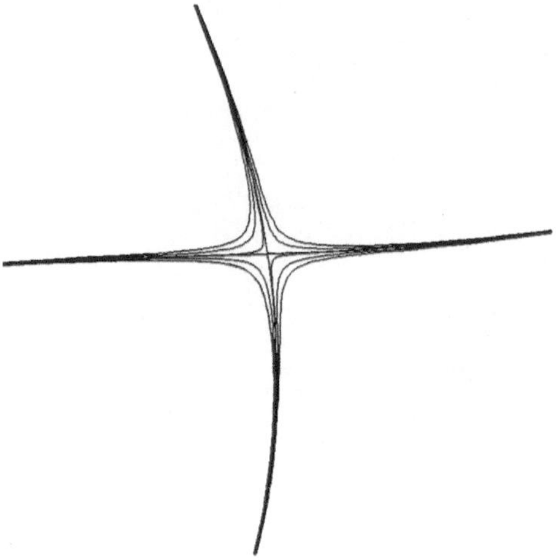

Figure 16.6 Enlarged patterns of intersection curves around $h = -0.775 \times 10^{-4}$

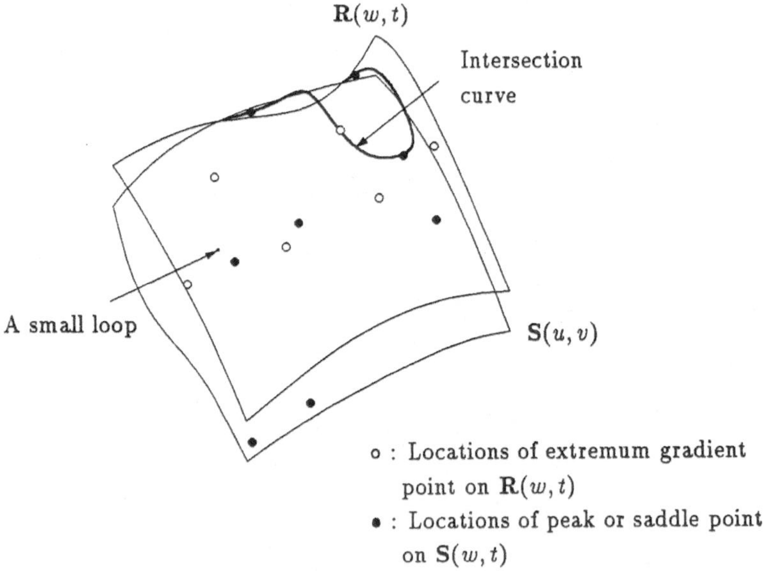

Figure 16.7 Example of very small intersection loop

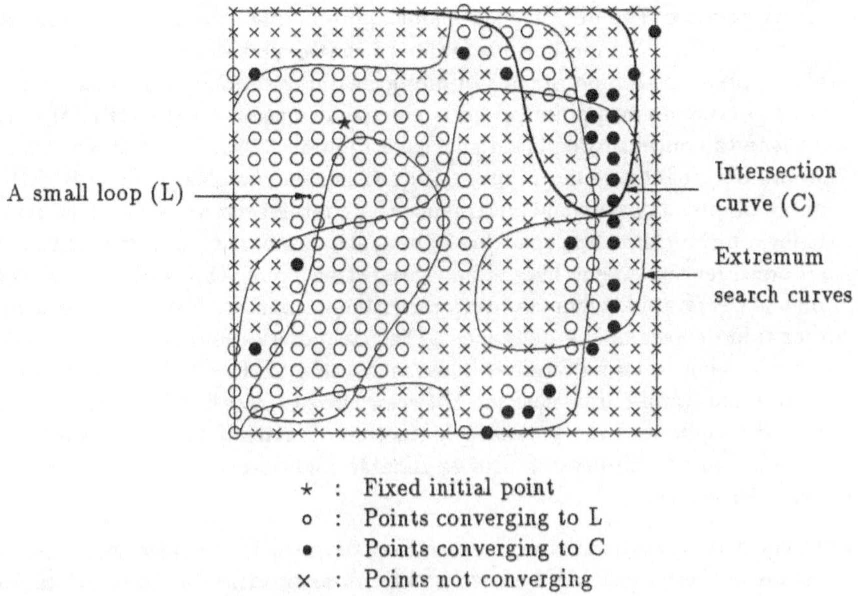

Figure 16.8 Map of starting points in one surface from Fig. 16.7.

Refer to Fig. 16.8. A max-gradient point in one region of $\mathbf{r}$ is selected as $\mathbf{q}_0$, whose projected location is indicated by a small star marker and the other point $\mathbf{p}_0$ is selected at various points on the patch $\mathbf{s}$. A blank small circle indicates locations of the point $\mathbf{p}_0$ converging to a very small intersection loop indicated by a small dot and each of black small disk mark points converges to a point on the upper-right intersection curve. A cross mark point shows a point which goes out of the boundary of the patch. In the figure the extremum search curves of the patch $\mathbf{s}$ which indicate boundaries of different slopes are also shown.

Even in such a critical case as this, the allowable area for convergence to the critical point is large. There are cases of convergence where the distance between the initial point pair is great and the distance between the final convergence point and the initial point pair is large. But the convergence is mainly determined by the distance of the pair from the converging point and also the relative slope difference of the tangent planes at the starting points.

16.3.3.2 Critical Points

So long as numerical computation is involved, we cannot find very small intersection loops or a contact point of two surfaces which are outside the precision of the computation process adopted. But if we translate one of the surfaces relative to the other in a near-normal direction or make a small offset surface of one of them, then we can detect an intersection loop, from which we can determine common normal points of the both surfaces and approximate shapes of the surfaces around these points. The method is similar to that explained in Sect. 16.5.

When applying the method of obtaining an intersection point, monitoring the distance between two points $\mathbf{p}$ and $\mathbf{q}$ gives information on the critical point without using the offset or displaced surfaces. Figure 16.9(a) shows the distance and the number of iterations. Figure 16.9(b) shows examples of routes to the convergence points. For a usual convergence to an intersection curve, a few iterations suffice, but for critically separated points, the number of iterations increases without convergence. When there is an intersecting point, the convergence to a final point is very rapid, whereas in a case of near contact, but really separating, after rapid decrese of the distance of the pair points the distance fluctuates without converging to a final points. The mean value of the distance fluctuation is an approximate minimum distance between two surfaces in this region. The above stated techniques are applied for detection of critical contact points.

The following techniques are also applicable for detection of small loops of the intersection curves.

(i) Observing the pattern of the intersection curves of one surface and a set of further surfaces, each shifted from the initial position by translation, we can estimate critical cases and also existence of the isolated loops of the intersection. See Fig. 16.5.

(ii) Since the contour curves of both surfaces are easily obtained, we can test interferences of the two sets of the contour curves, and this gives all the intersection curves of both curved surface. See Fig. 16.2.

If one wants to detect theoretically all the intersection loops including the critical points, a kind of extremum search curves can be used. For intersection curves of a patch $\mathbf{r}$ and parallel planes F with their normal $\mathbf{n}_f$, the extremum search curves, which are given by

$$\mathbf{n}_f \cdot \mathbf{s}_u = 0, \text{ and } \mathbf{n}_f \cdot \mathbf{s}_v = 0, \tag{16.17}$$

can be used to search for all the starting points for the loops of the intersection curves and the singular points at each of which one of the loops becomes a point or has a branch point. Similar extremum search curves can be considered for a set of the intersection curves of a surface $\mathbf{s}$ and offset surfaces $\mathbf{r}^+$ of $\mathbf{r}$ with various offset values e, or for a set of intersection curves of the surface $\mathbf{s}$ and the translated surfaces of $\mathbf{r}$ along a vector $\mathbf{a}$. We call here the extremum search curves for the intersection loops stated above the extended extremum search curves. When the surface $\mathbf{r}$ is a plane, the extended extremum search curves become the ordinary extremum search curves. Equations of the intersection curves of the above stated surfaces are given by

$$\mathbf{s} = \mathbf{r}^+ = \mathbf{r} + \mathbf{n}^r \cdot e, \text{ or } \mathbf{s} = \mathbf{r} + \mathbf{a} \cdot e, \tag{16.18}$$

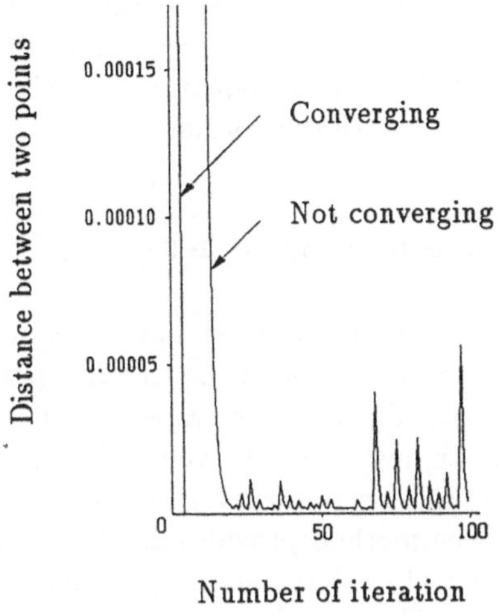

(a)

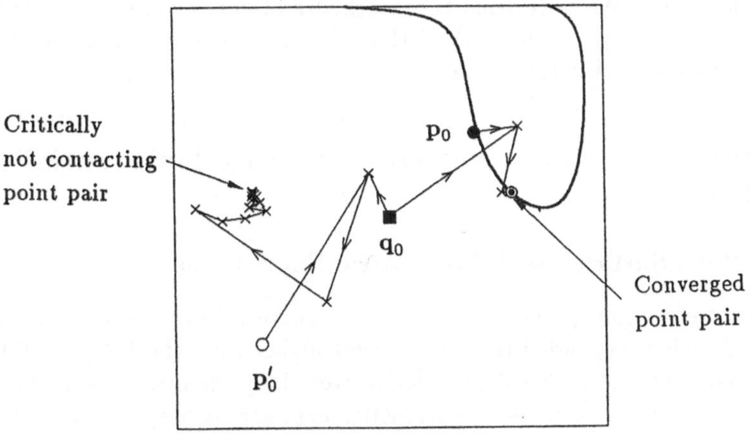

(b)

Figure 16.9 Convergence in a critical case: (a) distance and number of iterations, (b) routes to a converged point and a critically not contacting point pair where $\mathbf{n}^r$ is the normal of $\mathbf{r}(w,t)$ and $\mathbf{a}$ is the given direction.

The surface $\mathbf{s}$ is considered to be covered by the intersection curves. A certain intersection loop may converge to a point(s) or have a branch point(s) at a certain offset value or displacement value e. At these points the normal directions of the surface $\mathbf{s}$ coincide with normals of the other surfaces. Accordingly, these singular points are determined by the following two equations which are the orthogonal conditions of the basic vectors of $\mathbf{s}$ with the normal vector $\mathbf{n}^r$ of the other surface $\mathbf{r}$:

$$\mathbf{n}^r \cdot \mathbf{s}_u = 0, \qquad \mathbf{n}^r \cdot \mathbf{s}_v = 0. \tag{16.19}$$

Each of the above equations represents the extended extremum search curves, and their intersecting points give locations of the singular points.

Since the offset value e or the displacement value e is the parameter of these extremum search curves, we can determine the values of e at the singular points and the starting points of the intersection curves of $\mathbf{s}$ and $\mathbf{r}$ with the offset zero and the displacement zero. The method provides all the intersection loops, but tracing the curves given by eq. (16.19) is not easy, because the normal $\mathbf{n}^r$ is a function of the parameters w and t, which are also determined as functions of u and v through eq. (16.18). We obtain simultaneous equations with respect to the arc length for these parameters and the other parameter e for the offset or the displacement. We have tried this method and get the expected result, but compared with the previous method this one does not seem practical, because the computation load is too much.

16.4 Intersection Curves Described by Differential Equations

16.4.1 Both Surfaces with Parametric Expressions

Since the intersection curves are usually space curves, they have non-zero torsion, accordingly when expanded they have terms higher than third degree of the arc length. Derivation of higher-degree derivatives along the intersection curve from the two intersecting surfaces of parametric expressions are complicated, but if we apply the Runge-Kutta process (refer to the Appendix), we can estimate the values of the higher-degree derivatives by the weighted average of values of the first derivatives at a few sampled intervals. So we introduce the differential equations for describing the intersection curves and use reliable and efficient methods of numerical integration of the differential equations. The methods with automatic adjustment of stepsize work well for our problems. Hence, we explain these methods in this section. Surface expressions may both be of parametric form or one of them may be parametric and the other may be of implicit form, or both may be of implicit form.

On the intersection curve of two parametric expressions, the following equation must hold:

$$\mathbf{s}(u,v) = \mathbf{r}(w,t). \tag{16.20}$$

We now assume that a point on the intersection of two surface patches $\mathbf{s}(u,v)$ and $\mathbf{r}(w,t)$ has been determined. From there we start the curve tracing. The intersection curve on each surface must have the same tangent vector on their intersection, so we have from eq. (16.20),

$$\mathbf{s}_u du + \mathbf{s}_v dv = \mathbf{r}_w dw + \mathbf{r}_t dt. \tag{16.21}$$

Multiplying the normal $\mathbf{n}_1$ of $\mathbf{s}(u,v)$ and $\mathbf{n}_2$ of $\mathbf{r}(w,t)$ with eq. (16.21), we have two equations:

$$(\mathbf{n}_2 \cdot \mathbf{s}_u)(du/ds) + (\mathbf{n}_2 \cdot \mathbf{s}_v)(dv/ds) = 0, \tag{16.22}$$

$$(\mathbf{n}_1 \cdot \mathbf{s}_w)(dw/ds) + (\mathbf{n}_1 \cdot \mathbf{t}_t)(dt/ds) = 0. \tag{16.23}$$

From the square of eq. (16.21) an arc element ds of the intersection curve is given by

$$ds^2 = E_1 du^2 + 2F_1 du dv + G_1 dv^2 = E_2 dw^2 + 2F_2 dw dt + G_2 dt^2, \tag{16.24}$$

where E_i, F_i, G_i, for $i = 1, 2$ are the fundamental magnitudes of the first order of the patches $\mathbf{s}(u,v)$ and $\mathbf{r}(w,t)$.

From equations (16.22), (16.23) and (16.24), we obtain simultaneous differential equations for the intersection curve of two surfaces $\mathbf{s}(u,v)$ and $\mathbf{r}(w,t)$:

$$\frac{du}{ds} = -\phi_1(u,v,w,t) f_1, \quad \frac{dv}{ds} = \phi_1(u,v,w,t) g_1 \tag{16.25}$$

$$\frac{dw}{ds} = -\phi_2(u,v,w,t) f_2, \quad \frac{dt}{ds} = \phi_2(u,v,w,t) g_2, \tag{16.26}$$

where

$$\left.\begin{aligned} \phi_1(u,v,w,t) &= \pm\{E_1 f_1^2 - 2F_1 f_1 g_1 + G_1 g_1^2\}^{-1/2}, \\ \phi_2(u,v,w,t) &= \pm\{E_2 f_2^2 - 2F_2 f_2 g_2 + G_2 g_2^2\}^{-1/2}, \end{aligned}\right\} \tag{16.27}$$

$$\left.\begin{aligned} f_1 = (\mathbf{n}_2 \cdot \mathbf{s}_v), \quad g_1 = (\mathbf{n}_2 \cdot \mathbf{s}_u), \\ f_2 = (\mathbf{n}_1 \cdot \mathbf{r}_t), \quad g_2 = (\mathbf{n}_1 \cdot \mathbf{r}_w). \end{aligned}\right\} \tag{16.28}$$

If we adopt one of the suitable numerical methods of solving differential equations which have automatic error control and adaptive stepsize, we can trace an intersection curve of two surfaces without special attention to determining the stepsize. The steps usually ends at a boundary or returns to the starting point. If the candidate starting points still remain, the trace starts there again. During the integration process, errors can be checked against eq. (16.20), but this checking can be omitted for efficiency, because usually accumulated truncation errors of this integration can be kept within a given tolerance and the curve returns to the intial point to complete a loop.

If during marching on an intersection curve to be traced, normal vectors of the two surface becomes almost parallel, the integration program halts giving a warning of vanishing values of terms f_i and g_i. This indicates the possible existence of a singular point. Then intersection curves around a possible singular point are searched using the second-order terms of eq. (16.20). The method is described in the next section.

16.4.2 Surfaces with Implicit and Parametric Expressions

There are cases where one or both of the intersecting surfaces are expressed in implicit forms which are difficult to transform to parametric expressions. Their intersection curves can also be traced with the method of differential equation solving.

Let one of the surface be given by $f(x, y, z) = 0$ and the other by $\mathbf{s}(u, v)$. On their intersection curve we have the relations:

$$x = \mathbf{s}^x = (\mathbf{i} \cdot \mathbf{s}), \ y = \mathbf{s}^y = (\mathbf{j} \cdot \mathbf{s}), \ z = \mathbf{s}^z = (\mathbf{k} \cdot \mathbf{s}), \tag{16.29}$$

where $\mathbf{s}^x, \mathbf{s}^y$ and $\mathbf{s}^z$ are x, y, z components of $\mathbf{s}(u, v)$.

Differentiating the above relations with respect to the arc length of the intersection curve,

$$f_x \frac{dx}{ds} + f_y \frac{dy}{ds} + f_z \frac{dx}{ds} = 0, \tag{16.30}$$

$$\begin{aligned} \frac{dx}{ds} &= \mathbf{s}_u^x \frac{du}{ds} + \mathbf{s}_v^x \frac{dv}{ds}, \\ \frac{dy}{ds} &= \mathbf{s}_u^y \frac{du}{ds} + \mathbf{s}_v^y \frac{dv}{ds}, \\ \frac{dz}{ds} &= \mathbf{s}_u^z \frac{du}{ds} + \mathbf{s}_v^z \frac{dv}{ds}. \end{aligned}$$

Substituting the above three equations into eq. (16.30), we have

$$(f_x \mathbf{s}_u^x + f_y \mathbf{s}_u^y + f_z \mathbf{s}_u^z) \frac{du}{ds} + (f_x \mathbf{s}_v^x + f_y \mathbf{s}_v^y + f_z \mathbf{s}_v^z) \frac{dv}{ds} = 0. \tag{16.31}$$

Since f_x, f_y and f_z are expressed by x, y and z, which are also functions of u and v by eq. (16.29), eq. (16.31) is the differential equation of an intersection curve expressed in u and v. Hence it is solved in a similar way as equations (16.4), (16.6) and (16.5).

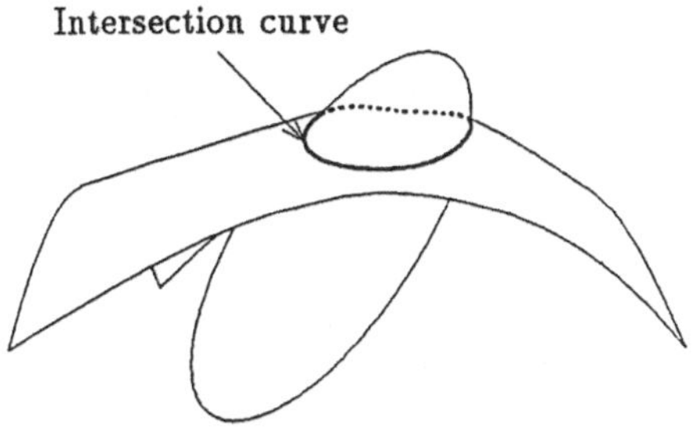

Figure 16.10 Intersection of a free-form surface and an ellipsoid (implicit expression), both of which are displayed by their silhouette curves

16.4.3 Both Surfaces with Implicit Expressions

When the other surface is also expressed by an implicit form $g(x, y, z) = 0$ as well as eq. (16.30), we have

$$g_x \frac{dx}{ds} + g_y \frac{dy}{ds} + g_z \frac{dz}{ds} = 0. \tag{16.32}$$

From equations (16.30) and (16.32), we determine a ratio $(dx/ds : dy/ds : dz/ds)$ such as

$$\begin{aligned} \frac{dx}{ds} &= \psi(x, y, z)(f_y g_z - g_y f_z), \\ \frac{dy}{ds} &= \psi(x, y, z)(f_z g_x - g_z f_x), \\ \frac{dz}{ds} &= \psi(x, y, z)(f_x g_y - g_x f_y), \end{aligned} \tag{16.33}$$

where $\psi(x, y, z)$ is the common factor, which is obtained from the normalizing condition,

$$\begin{aligned} &(\frac{dx}{ds})^2 + (\frac{dy}{ds})^2 + (\frac{dz}{ds})^2 \\ &= \psi(x, y, z)^2\{(f_y g_z - g_y f_z)^2 + (f_z g_x - g_z f_x)^2 + (f_x g_y - g_x f_y)^2\} \\ &= 1. \end{aligned} \tag{16.34}$$

Equations (16.33) together with eq. (16.34) are the differential equations of the intersection curve, which can be traced by the numerical methods.

Example 1. Figure 16.4. Intersection of two free-form surfaces of degree 4 × 4. Contour curves are also shown.

Example 2. Figure 16.10. A free-form surface and an ellipsoid with three axes of different length.

Example 3. Figure 16.11. Two surfaces of implicit forms: an ellipsoid and a hyperboloid.

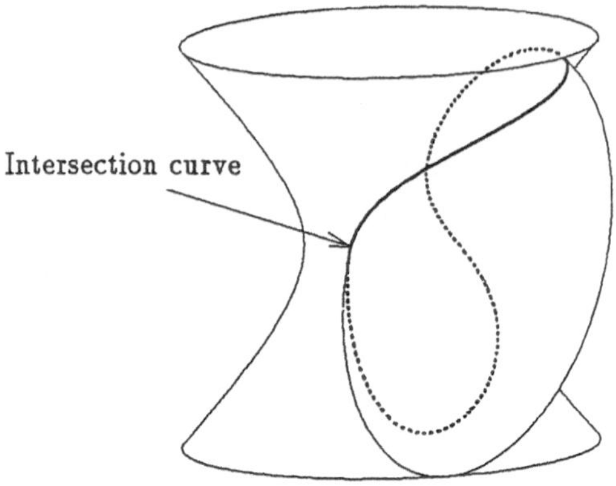

Figure 16.11 Intersection of surfaces all implicit expressions: an ellipsoid and a hyperboloid, both of which are displayed by their silhouette curves

16.5 Intersection Near a Probable Singular Point

If two surfaces share a common normal vector and a common point, their intersection curves may pass the point, or the point may be isolated. But if the two surfaces have nearly parallel normal vectors near a coincident point on the two surfaces, there are various pattern of the intersection curves. We treat this kind of problem here.

The method we use for tracing an intersection curve can be applied to a point $\mathbf{p}$ on the curve, where normal vectors $\mathbf{n}_1$ and $\mathbf{n}_2$ of the intersecting surfaces are nearly parallel, as small an angle as 10^{-6} radian or less, though at a point of equal normal direction, its differential equations (16.22) and (16.23) vanish. Accordingly, we search around the point $\mathbf{p} = \mathbf{s}(u, v) = \mathbf{r}(w, s)$ for two points $\mathbf{s}(u + du, v + dv)$ and $\mathbf{r}(w + dw, t + dt)$ whose normal vectors are coincident. Instead of the notations for normals $\mathbf{n}_1$ and $\mathbf{n}_2$, we change them to $\mathbf{n}^s$ and $\mathbf{n}^s$ for clarity. Let their common unit normal be $\mathbf{n}$ and the distance between the two surfaces along the normal be a small value δ. Then we have the relation:

$$\mathbf{s}_u du + \mathbf{s}_v dv + \mathbf{n}\delta = \mathbf{r}_w dw + \mathbf{r}_t dt, . \tag{16.35}$$

$$d\mathbf{n}^s = \mathbf{n} - \mathbf{n}^s, \; d\mathbf{n}^r = \mathbf{n} - \mathbf{n}^r, \; \Delta\mathbf{n}_0 = \mathbf{n}^r - \mathbf{n}^s. \tag{16.36}$$

So long as we can assume that each normal vector varies continuously, the locations of the desired normal direction are considered very near to the last point of the intersection curve. For a surface region with very small Gaussian curvature this assumption does not always hold (refer to eq. (6.36), then in such a region there is no critical point on intersection curves. We can proceed the integration process with smaller tolerance. Except such a case we have the equations:

$$d\mathbf{n}^s = \mathbf{n}^s_u du + \mathbf{n}^s_v dv, \; d\mathbf{n}^r = \mathbf{n}^r_w dw + \mathbf{n}^r_t dt. \tag{16.37}$$

From equations (16.36) and (16.37),

$$\Delta \mathbf{n}_0 = d\mathbf{n}^s - d\mathbf{n}^r = \mathbf{n}^s_u du + \mathbf{n}^s_v dv - \mathbf{n}^r_w dw - \mathbf{n}^r_t dt. \tag{16.38}$$

Multiplying $\mathbf{n}^r$ or $\mathbf{n}^s$ with eq. (16.35) and neglecting a small value $\delta \cdot d\mathbf{n}^s$, we obtain

$$\left.\begin{array}{lcl} \mathbf{n}^r \cdot \mathbf{s}_u du & + & \mathbf{n}^r \cdot \mathbf{s}_v dv = -\delta, \\ \mathbf{n}^s \cdot \mathbf{r}_w dw & + & \mathbf{n}^s \cdot \mathbf{r}_t dt = \delta. \end{array}\right\} \tag{16.39}$$

Multiplying $\mathbf{n}^s \times \mathbf{n}^r$ with eq. (16.35), we have

$$\begin{aligned} & [\mathbf{n}^s, \mathbf{n}^r, \mathbf{s}_u]du + [\mathbf{n}^s, \mathbf{n}^r, \mathbf{s}_v]dv \\ = \; & [\mathbf{n}^s, \mathbf{n}^r, \mathbf{r}_w]dw + [\mathbf{n}^s, \mathbf{n}^r, \mathbf{r}_t]dt - [\mathbf{n}^s, \mathbf{n}^r, \mathbf{n}]\delta. \end{aligned} \tag{16.40}$$

Since the last term of the above equation is small of second order, it can be neglected.

Multiplying $\mathbf{n}^r$ or $\mathbf{n}^s$ with eq. (16.38) and considering orthogonality conditions $\mathbf{n}^s \cdot d\mathbf{n}^s = \mathbf{n}^r \cdot d\mathbf{n}^r = 0$, we obtain

$$\left.\begin{array}{l} \mathbf{n}^r \cdot \mathbf{n}^s_u du + \mathbf{n}^r \cdot \mathbf{n}^s_v dv = \epsilon, \\ \mathbf{n}^s \cdot \mathbf{n}^r_w dw + \mathbf{n}^s \cdot \mathbf{n}^r_t dt = \epsilon \end{array}\right\}, \tag{16.41}$$

where $\epsilon = \mathbf{n}^r \cdot \Delta \mathbf{n}_0 = -\mathbf{n}^s \cdot \Delta \mathbf{n}_0$.

Solving equations (16.39) and (16.41), we have du, dv and dw, dt as functions of δ and ϵ:

$$\left.\begin{array}{lcl} du & = & \{-(\mathbf{n}^r \cdot \mathbf{n}^s_v)\delta - (\mathbf{n}^r \cdot \mathbf{s}_v)\epsilon\}/D_1, \\ dv & = & \{(\mathbf{n}^r \cdot \mathbf{n}^s_u)\delta + (\mathbf{n}^r \cdot \mathbf{s}_u)\epsilon\}/D_1, \\ dw & = & \{(\mathbf{n}^s \cdot \mathbf{n}^r_t)\delta - (\mathbf{n}^s \cdot \mathbf{r}_t)\epsilon\}/D_2, \\ dt & = & \{-(\mathbf{n}^s \cdot \mathbf{n}^r_w)\delta + (\mathbf{n}^s \cdot \mathbf{r}_w)\epsilon\}/D_2 \end{array}\right\}, \tag{16.42}$$

where

$$\left.\begin{array}{lcl} D_1 & = & \{-(\mathbf{n}^r \cdot \mathbf{n}^s_u)(\mathbf{n}^r \cdot \mathbf{s}_v) + (\mathbf{n}^r \cdot \mathbf{n}^s_v)(\mathbf{n}^r \cdot \mathbf{s}_u)\}, \\ D_2 & = & \{-(\mathbf{n}^s \cdot \mathbf{r}_t)(\mathbf{n}^s \cdot \mathbf{n}^r_w) + (\mathbf{n}^s \cdot \mathbf{r}_w)(\mathbf{n}^s \cdot \mathbf{n}^r_t)\}. \end{array}\right\} \tag{16.43}$$

Substituting the above relations into eq. (16.40), we obtain an equation for δ

$$A\delta = B\epsilon, \tag{16.44}$$

where

$$\begin{aligned} A \; = \; & D_2\{-[\mathbf{n}^s, \mathbf{n}^r, \mathbf{s}_u](\mathbf{n}^r \cdot \mathbf{n}^s_v) + [\mathbf{n}^s, \mathbf{n}^r, \mathbf{s}_v](\mathbf{n}^r \cdot \mathbf{n}^s_u)\} \\ & -D_1\{[\mathbf{n}^s, \mathbf{n}^r, \mathbf{r}_w](\mathbf{n}^s \cdot \mathbf{n}^r_t) - [\mathbf{n}^s, \mathbf{n}^r, \mathbf{r}_t](\mathbf{n}^s \cdot \mathbf{n}^r_w)\} \\ & +D_1 D_2[\mathbf{n}^s, \mathbf{n}^r, \mathbf{n}], \end{aligned} \tag{16.45}$$

$$\begin{aligned} B \; = \; & D_2\{-[\mathbf{n}^s, \mathbf{n}^r, \mathbf{r}_w](\mathbf{n}^s \cdot \mathbf{r}_t) + [\mathbf{n}^s, \mathbf{n}^r, \mathbf{r}_t](\mathbf{n}^s \cdot \mathbf{r}_w)\} \\ & +D_2\{[\mathbf{n}^s, \mathbf{n}^r, \mathbf{s}_u](\mathbf{n}^r \cdot \mathbf{s}_v) - [\mathbf{n}^s, \mathbf{n}^r, \mathbf{s}_v](\mathbf{n}^r \cdot \mathbf{s}_u)\}. \end{aligned} \tag{16.46}$$

From eq. (16.44) the distance δ between two surfaces at the common normal points, and then from eq. (16.42), du, dv, dw, dt are determined. The last term of eq. (16.45) must be neglected initially and if an iteration is needed for accuracy enhancement the term is added. Thus the common normal points $\mathbf{s}(u+du, v+dv)$ and $\mathbf{r}(w+dw, t+dt)$ of the two surfaces are obtained. In the above derivation, we assume that the distance to a common normal is so small that the Weingarten's formulas hold.

When the common normal exists, the surface shapes around these points are expressed, using their respective principal directions and the common normal as their coordinate systems (x, y, z) and (ξ, η, ζ), by

$$\left.\begin{array}{rcl} z &=& (\kappa_1 x^2 + \kappa_2 y^2)/2, \\ \zeta &=& (k_1 \xi^2 + k_2 \eta^2)/2, \\ z &=& \zeta + \delta, \end{array}\right\} \tag{16.47}$$

where κ_1, κ_2 and k_1, k_2 are their respective pairs of principal curvatures. Rotate the coordinates system (ξ, η, ζ) by an angle ϕ, which is the angle between the x and the ξ axes, around its ζ axis to eliminate ξ and η:

$$\xi = x \cos\phi + y \sin\phi, \quad \eta = -x \sin\phi + y \cos\phi. \tag{16.48}$$

Then we have an equation for x and y:

$$(\kappa_1 - K_1)x^2 + (\kappa_2 - K_2)y^2 + 2hxy = 2\delta, \tag{16.49}$$

where

$$\left.\begin{array}{rcl} K_1 &=& k_1 \cos^2\phi + k_2 \sin^2\phi, \\ K_2 &=& k_1 \cos^2(\phi + 90) + k_2 \sin^2(\phi + 90), \\ h &=& (k_2 - k_1)\cos\phi \sin\phi. \end{array}\right\} \tag{16.50}$$

The constants K_1 and K_2 are the orthogonal normal curvatures of $\mathbf{r}(w, t)$ in the principal directions of $\mathbf{s}(u, v)$ at the common normal points.

Equation (16.49), which represents the intersection curve of the two surfaces, is a quadratic curve, whose discriminant D is given by

$$\begin{array}{rcl} D &=& (\kappa_1 - K_1)(\kappa_2 - K_2) - h^2 \\ &=& \kappa_1\kappa_2 + k_1 k_2 - \kappa_1 K_2 - \kappa_2 K_1 \\ &=& K_g^s + K_g^r - 4(K_m^s K_m^r - \kappa_1 K_1 - \kappa_2 K_2), \end{array} \tag{16.51}$$

where K_g and K_m are the Gaussian curvature and the mean curvature at the point, their superscripts discriminating the respective surfaces.

We can classify the shape of the curve according to equations (16.49) and (16.51).

If $D \neq 0$, it has a center at the origin. Rotating the coordinate system (x, y, z) around the z axis by an angle

$$\theta = (\frac{1}{2})\tan^{-1}\{\frac{(k_2 - k_1)\sin 2\phi}{\kappa_1 - \kappa_2 - (k_1 - k_2)\cos 2\phi}\}, \tag{16.52}$$

we transform eq. (16.49) to

$$\lambda_1 x^2 + \lambda_2 y^2 = 2\delta, \tag{16.53}$$

where the coefficients λ_1 and λ_2 are roots of the characteristic equation for eq. (16.49) (refer to Sect. 1.9):

$$\lambda^2 - (\kappa_1 + \kappa_2 - K_1 - K_2)\lambda + D = 0,$$

Using equations (16.50) and the mean curvatures K_m^s, K_m^r, the above equation is changed to

$$\lambda^2 - 2(K_m^s - K_m^r)\lambda + D = 0. \tag{16.54}$$

The shape of the intersection curve is classified according to the root of the above equation:

- $\delta > 0$:

$$(K_m^s - K_m^r) > 0, \quad D > 0 \ : \ \text{ellipse}, \tag{16.55}$$
$$(K_m^s - K_m^r) < 0, \quad D > 0 \ : \ \text{imaginary ellipse}, \tag{16.56}$$
$$D < 0 \ : \ \text{hyperbola}. \tag{16.57}$$

  If $(\kappa_1 - K_1)(\kappa_2 - K_2) \leq 0$, the intersection curve is surely a hyperbola, as easily inferred from their geometric configuration near the origin. This case is naturally included in the above case of $D < 0$.
- $\delta = 0$:

$$D < 0 \ : \ \text{two lines}, \tag{16.58}$$
$$D > 0 \text{ and } (K_m^s - K_m^r) > 0 \ : \ \text{one contact point}, \tag{16.59}$$
$$D = 0 \ : \ \text{a parabola or two lines}. \tag{16.60}$$

A similar analysis can be applied for the intersection curve of surfaces, one or both of which is (are) expressed by implicit form(s).

Note that in calculating eq. (16.44), quantities of the denominators and the numerators are very small, and repeated calculations for approximation are needed to reduce the truncation errors.

Example. Figure 16.12 shows intersection curve around a common normal vector of two surfaces, whose distance is $\delta = 8 \times 10^{-7}$. Figure 16.12(a) is global view of the intersection curves, which are hyperbolic in the very near region of the center. To observe the shapes of the two surfaces around the points of common normal, the contour curves made by two cutting planes, whose normals are the same as the common one, are shown in Fig. 16.12(b). The distance of the two cutting planes is 0.0002, where the length of an edge of the patches is of order 1, and the figure shows clearly the triple intersecting points of the three surfaces. We can know that the principal curvatures of two surfaces are almost equal, because the shapes of the contour curves of the two surfaces are elliptical of similar size.

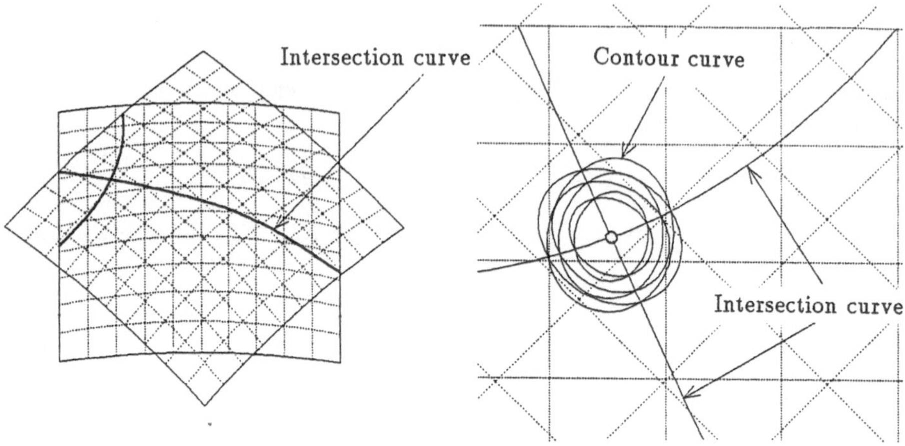

Figure 16.12 Shape of intersection curves near a critical point: (a) global view of intersection curves, (b) contour curves around a common normal point and intersection curves

16.6 Intersection of Offset Surfaces

16.6.1 Intersection with a Plane

The intersection curves of an offset surface and equidistant planes are needed when the center of a ball of a ball end-mill is controlled to move on them for machining of the original surface.

Let $\mathbf{s}^+(u,v)$ be an offset surface of a part surface $\mathbf{s}(u,v)$ to be machined by a ball-end mill of radius r,

$$\mathbf{s}^+(u,v) = \mathbf{s}(u,v) + \mathbf{n}(u,v) \cdot r, \tag{16.61}$$

where $\mathbf{n}(u,v)$ is the unit normal vector of the part surface at u and v.

If the center of the ball-end mill moves on the offset surface, it cuts the part surface. Let the unit normal vector of the plane, whose distance from the origin is d, be $\mathbf{n}_f$. Then from eq. (16.1), we have

$$\mathbf{s}^+(u,v) \cdot \mathbf{n}_f = d. \tag{16.62}$$

After the starting point of the center is determined, the intersection curve can be traced using equations (16.5) and(16.6), in which ϕ_u and ϕ_v are given by

$$\left. \begin{aligned} \phi_u &= \mathbf{s}_u^+ \cdot \mathbf{n}_f = (\mathbf{s}_u + \mathbf{n}_u \cdot r) \cdot \mathbf{n}_f, \\ \phi_v &= \mathbf{s}_v^+ \cdot \mathbf{n}_f = (\mathbf{s}_v + \mathbf{n}_v \cdot r) \cdot \mathbf{n}_f \end{aligned} \right\}, \tag{16.63}$$

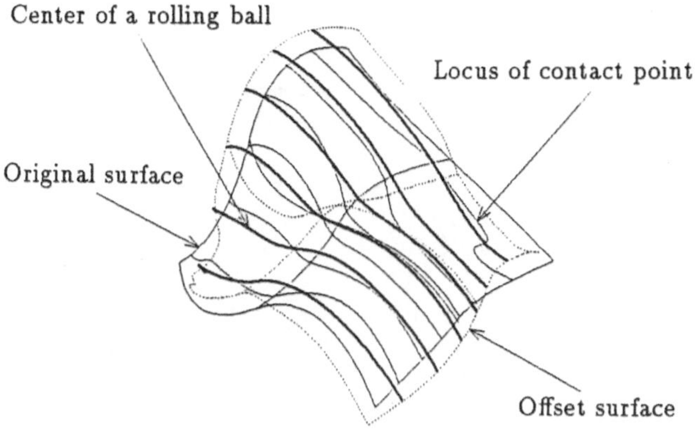

Figure 16.13 Intersection of an offset curve and parallel planes. Locus of contact point of a rolling ball is also shown

where the derivatives $\mathbf{n}_u$ and $\mathbf{n}_v$ of the normal of $\mathbf{s}(u,v)$ can be calculated by using Weingarten's formulas (refer to equations (6.28) and (7.37)–(7.39)). Figure 16.13 shows the intersection curves and the loci of the contact point of the ball. In this case the radius of the ball is too large, and though the loci of the center of the ball are smooth, the corresponding loci of the contact points on the original surface show complicated behavior around regions of large curvatures.

16.6.2 Intersection of Two Offset Surfaces

For smooth blending of two intersecting surfaces, an envelop surface of a ball rolling in contact with the two surfaces is used. A locus of the center of the ball is determined by the intersection of two offset surfaces, whose distance to the original surfaces is equal to the radius of the ball.

16.6.2.1 Two Parametric Surfaces

In this case equations (16.22) and (16.23) become

$$(\mathbf{n}_2 \cdot \mathbf{s}_u^+)(du/ds^+) + (\mathbf{n}_2 \cdot \mathbf{s}_v^+)(dv/ds^+) = 0, \tag{16.64}$$
$$(\mathbf{n}_1 \cdot \mathbf{r}_w^+)(dw/ds^+) + (\mathbf{n}_1 \cdot \mathbf{r}_t^+)(dt/ds^+) = 0, \tag{16.65}$$

where ds^+ is the line element of the intersection curve, and there is a relation

$$\left.\begin{aligned} (ds^+)^2 &= E^+{}_1 du^2 + 2F^+{}_1 du dv + G^+{}_1 dv^2, \\ (ds^+)^2 &= E^+{}_2 dw^2 + 2F^+{}_2 dw dt + G^+{}_2 dt^2. \end{aligned}\right\} \tag{16.66}$$

The symbols with a superscript $^+$ correspond to those of the original surfaces. They are given by equations (7.37)–(7.39). Equations (16.64) and (16.65) are

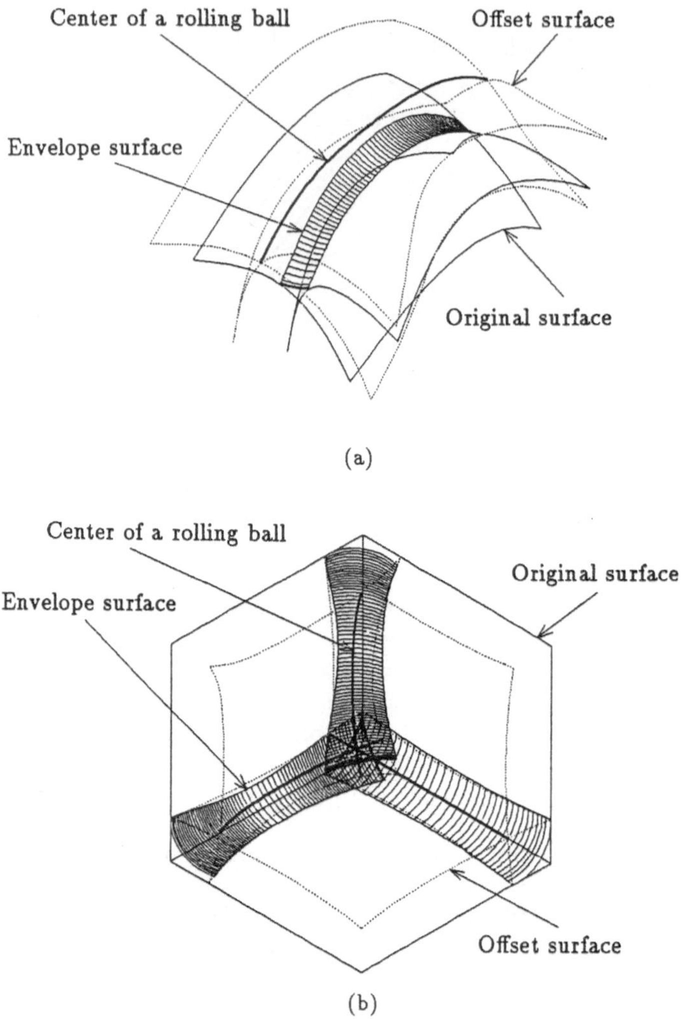

Figure 16.14 (a) Intersection curves of two offset surfaces, (b) intersecting point of three offset surfaces

the differential equations of the locus of the center of a rolling ball contacting the two surfaces; they can be traced numerically by solving the above differential equations.

Examples. Figure 16.14(a) shows the intersection curves of the two offset surfaces and (b) shows the intersecting point of the three offset surfaces.

16.6.2.2 A Parametric Surface and a Surface of Implicit Form

Let $\mathbf{p}$ be a point on a surface S_1 whose equation is $f(x, y, z) = 0$ and $\mathbf{n}_1$ a unit normal there. A point $\mathbf{q}$ offset from $\mathbf{p}$ by a distance r is given by

$$\mathbf{q} = \mathbf{p} + \mathbf{n}_1 \cdot r. \tag{16.67}$$

Let $\mathbf{s}^+(u, v)$ be an offset parametric surface. On its intersection curve with $\mathbf{q}$ an equation $\mathbf{s}^+(u, v) = \mathbf{q} = \mathbf{p} + \mathbf{n}_1 \cdot r$ holds, from which we obtain its line element vector:

$$d\mathbf{s}^+ = \mathbf{s}^+{}_u du + \mathbf{s}^+{}_v dv = d\mathbf{q} = d\mathbf{p} + d\mathbf{n}_1 \cdot r. \tag{16.68}$$

Multiplying the normal vector $\mathbf{n}_1$ of S_1 with both sides of the above equation, we have

$$(\mathbf{n}_1 \cdot \mathbf{s}_u^+)du + (\mathbf{n}_1 \cdot \mathbf{s}_v^+)dv = \mathbf{n}_1 \cdot d\mathbf{q} = 0. \tag{16.69}$$

Since $\mathbf{n}_1$ is proportional to $(f_x : f_y : f_z)$, we can determine the ratio $(du : dv)$ as a function of u, v, x, y and z.

Multiplying the normal vector $\mathbf{n}_2$ of $\mathbf{s}^+$ with both sides of eq. (16.68), we obtain

$$\mathbf{n}_2 \cdot (d\mathbf{p} + d\mathbf{n}_1 \cdot r) = 0. \tag{16.70}$$

Since $\mathbf{p}$ and $\mathbf{n}_1$ are the functions of x, y and z, eq. (16.70) is transformed to

$$\{\mathbf{n}_2 \cdot (\mathbf{i} + r\mathbf{n}_{1x})\}dx + \{\mathbf{n}_2 \cdot (\mathbf{j} + r\mathbf{n}_{1y})\}dy + \{\mathbf{n}_2 \cdot (\mathbf{k} + r\mathbf{n}_{1z})\}dz = 0, \tag{16.71}$$

where, for instance, $\mathbf{n}_{1x}$ is the derivative of $\mathbf{n}_1$ with respect to x.
Moreover we have the relation

$$df(x, y, z) = f_x dx + f_y dy + f_z dz = 0. \tag{16.72}$$

From equations (16.71) and (16.72), we obtain a ratio $(dx : dy : dz)$. As a normalizing condition, we take the line element ds^+ of the intersection curve which is given by

$$\begin{aligned} ds^{+2} &= E^+du^2 + 2F^+dudv + G^+dv^2 = \\ & (\mathbf{i} + r\mathbf{n}_{1x})^2dx^2 + (\mathbf{j} + r\mathbf{n}_{1y})^2dy^2 + (\mathbf{k} + r\mathbf{n}_{1z})^2dz^2 \\ & +2(\mathbf{i} + r\mathbf{n}_{1x})(\mathbf{j} + r\mathbf{n}_{1y})dxdy \\ & +2(\mathbf{j} + r\mathbf{n}_{1y})(\mathbf{k} + r\mathbf{n}_{1z})dydz \\ & +2(\mathbf{k} + r\mathbf{n}_{1z})(\mathbf{i} + r\mathbf{n}_{1x})dzdx. \end{aligned} \tag{16.73}$$

From the above equation and the ratios $(du : dv)$ and $(dx : dv : dz)$, we can determine $(du/ds^+), (dv/ds^+), (dx/ds^+), (dy/ds^+)$and (dz/ds^+) as functions of u, v, x, y and z. We can trace the intersection curve by solving these simultaneous differential equations.

Figure 16.15 shows an example of the intersection of offset surfaces: one is a parametric and the other is an implicit expression.

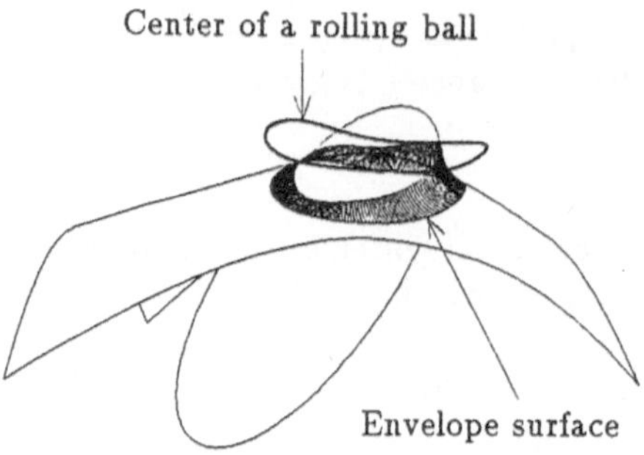

Figure 16.15 Intersection of offset surfaces, one a free-form and the other an implicit expression

16.6.2.3 Two Surfaces with Implicit Expressions

A similar method can be applied for tracing the intersection curve of two implicit surfaces S_1: $f(x,y,z) = 0$ and S_2 :$g(u,v,w) = 0$, in which symbols u, v and w are used instead of x, y and z for discriminating purpose only. A point $\mathbf{q}$ on an intersection curve of the offset surfaces is given by

$$\mathbf{q} = \mathbf{p}_1 + \mathbf{n}_1 \cdot r = \mathbf{p}_2 + \mathbf{n}_2 \cdot r, \tag{16.74}$$

where $\mathbf{p}_i$ and $\mathbf{n}_i$,$(i = 1,2)$ are a point and its unit normal vector on the original surface S_1 or S_2. Considering $\mathbf{n}_i \cdot d\mathbf{p}_i = 0$ and $\mathbf{n}_i \cdot d\mathbf{n}_i = 0$, we obtain the relations:

$$\left.\begin{array}{ll} \mathbf{n}_2 d\mathbf{p}_1 + \mathbf{n}_2 d\mathbf{n}_1 \cdot r = 0, & df = 0, \\ \mathbf{n}_1 d\mathbf{p}_2 + \mathbf{n}_1 d\mathbf{n}_2 \cdot r = 0, & dg = 0 \end{array}\right\}. \tag{16.75}$$

From the above equations, we obtain the ratios $(dx : dy : dz)$ and $(du : dv : dw)$. The line element of the intersection curve is given from eq. (16.74) by

$$\begin{aligned} ds^{+2} &= (\mathbf{i} + r\mathbf{n}_{1_x})^2 dx^2 + (\mathbf{j} + r\mathbf{n}_{1_y})^2 dy^2 + (\mathbf{k} + r\mathbf{n}_{1_z})^2 dz^2 \\ &\quad + 2(\mathbf{i} + r\mathbf{n}_{1_x})(\mathbf{j} + r\mathbf{n}_{1_y}) dx\, dy \\ &\quad + 2(\mathbf{j} + r\mathbf{n}_{1_y})(\mathbf{k} + r\mathbf{n}_{1_z}) dy dz \\ &\quad + 2(\mathbf{k} + r\mathbf{n}_{1_z})(\mathbf{i} + r\mathbf{n}_{1_x}) dz dx, \\ &= (\mathbf{i} + r\mathbf{n}_{2_x})^2 dx^2 + (\mathbf{j} + r\mathbf{n}_{2_y})^2 dy^2 + (\mathbf{k} + r\mathbf{n}_{2_z})^2 dz^2 \\ &\quad + 2(\mathbf{i} + r\mathbf{n}_{2_x})(\mathbf{j} + r\mathbf{n}_{2_y}) dx\, dy \\ &\quad + 2(\mathbf{j} + r\mathbf{n}_{2_y})(\mathbf{k} + r\mathbf{n}_{2_z}) dy dz \\ &\quad + 2(\mathbf{k} + r\mathbf{n}_{2_z})(\mathbf{i} + r\mathbf{n}_{2_x}) dz dx. \end{aligned} \tag{16.76}$$

From the above equations, six simultaneous differential equations of the first order with six dependent variables, whose left-hand sides are (dx/ds^+), (dy/ds^+),

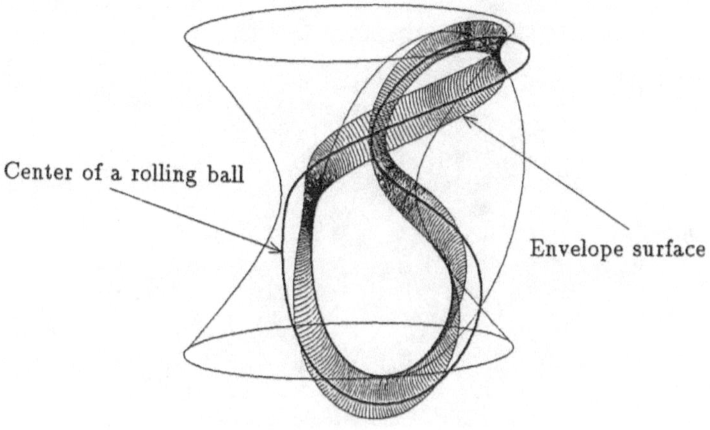

Figure 16.16 Intersection of two offset surfaces, two implicit expressions

(dz/ds^+) and (du/ds^+), (dv/ds^+), (dw/ds^+), are determined. By solving them we can obtain the intersection curve and the contacting curves on the original surfaces.

To solve these differential equations, we need initial conditions or a starting point. Since the starting point is on the intersection curve, a method similar to that given in Sect. 16.3.2 can be applied.

Figure 16.16 shows an example of the intersection of offset surfaces of the two original surfaces with implicit expressions.

17 Applications of the Theories in Industry

17.1 Introduction

Various theories and methods explained in the previous chapters were originally developed to meet the practical requirements of Japanese industries. In this chapter their application examples are described: one is style design and stamping die design and manufacture in the integrated CAD/CAM system of a large motor car company. The other is also style design, of small audio devices in a large electronic appliance company.

In these systems, computer-internal models or descriptions of the design objects were needed in place of engineering drawings. Accordingly, we first confirm the conventional roles of engineering drawings in design and manufacturing and then explain the reasons for replacing engineering drawings by the internal models. Next an example of this replacement is given from the motor car industry. Since readers of this book may not be experts on the car industry, we explain concisely the conventional processes of style design of a car body and manufacture of its stamping dies and then we show how the internal models are constructed, manipulated and used in a new system with the underlying philosophy of the conversion from the conventional system.

The other example is the design and manufacture of free-form injection-mold products for in-the-ear headphones in audio devices. In this case the shape of the products is a very important factor to attract customers together with their performance. In both cases their engineering drawings are not enough to convey information for manufacturing and to perform prior evaluation of the products. Comparing the scale and complexity of the products, the design and manufacture of motor cars include far more difficult factors than those of electric appliances, but the latter must be also excellent in design and cost, because the performance of the product is almost the same as that of similar competitive products on the market. Anyway there are common factors in design methods in both industries.

In this chapter we treat mainly the case of motor-car style design and stamping die manufacture and then touch on the case of the electric appliance. As the scale of the latter is small, it may be easy for understanding.

17.2 Engineering Drawings and Geometric Models

Engineering design is a kind of information-generating activity for manufacturing an object which performs specified tasks and satisfies given aims. For this purpose, the designers have to set clearly the specification of the requirements for the aims, and then develop it to specify every detail of the object, to determine components and materials to be used. And at the same time they are required to optimize the conflicting conditions for performance, reliability, cost and marketability within the given constraints of time, manpower and money.

Manufacturing activities include preparing materials and components, deciding machines and their operations for production of the objects and assembling the components which are specified by the design process. They include generating control information for manufacturing and performing the instructions thus produced and finally inspecting whether the manufactured objects satisfy the specifications defined in the early stages. Of course these activities are under the constraints of cost and production time.

In design and manufacturing before the advent of computers in these fields, information generation and storage and transfer were mainly performed through engineering drawings. Since their roles were and still are very important, frequently the design activity was considered equivalent to the making of engineering drawings.

Design engineers represent their design intention and designed objects mainly by engineering drawings, in which materials to be used, shapes of designed objects, and various information on meeting the specifications given to the designed products are described. Through the process of making engineering drawings, designers put their various ideas in order and select the best ones to optimize the conflicting conditions. Completed engineering drawings are media to be communicated correctly to other people and they must carry basic information for manufacturing and inspection. There are various types of engineering drawings according to their uses and they are guaranteed to have no errors and no contradiction. Through them, other people not only understand the designed objects, but also they can manufacture the real objects the drawings specify.

These engineering drawings are produced by persons who understand the objects to be manufactured and have their images in their minds and know the rules for how to express the objects. Persons who can read them not only know the rules of the drawings but have the ability to construct in their minds the images of the objects the drawings represent.

Even today the computer cannot understand engineering drawings or produce them automatically in the way that expert designers can. However, if the computer can produce, according to our instructions, information which the engineering drawings contain and can process its meanings, that information is equivalent to our explicit knowledge obtained from the engineering drawings. Without formation of this type of information, the computer is merely a machine for simple data processing in design activities.

The source of this explicit information which the engineering drawings hold is the geometric models or product models which are produced in the computer by relevant commands given by designers who can interact with the models and extract necessary information from the models. If such internal models can be constructed, the computer has the possibility to support all the design activities and control the manufacturing processes without human intervention for data transfer and conversion.

In the construction of this kind of model, to fully exert the designers' skill and talent, their way of thinking and practices must not be greatly altered so long as more convenient and helpful alternatives cannot be offered to them. Usually this is the most difficult requirement for a computer system designed to accept experts' methods directly.

For advanced CAD/CAM systems which expert designers can use willingly, the human interface with the systems is very important. The interface means not only the physical interface, but also the mental interface between the designers' mental model of the objects to be designed and the computer's internal information model of the object to be constructed. In design and manufacturing activities, a great part of this internal information model is related to the geometric model of the object to be designed and manufactured.

One of the main objectives of this book is to present theories and methods of free-form surfaces which are easily and intuitively understandable to practical engineers, some of whom are users of the systems or their developers. Another objective is to show that an excellent user interface is an essential component of a successful advanced CAD/CAM system.

17.3 Examples of Integration

As the integration of CAD and CAM is ideal, but difficult to construct, we show one real example in this chapter and explain how it was planned and how it has been developed. The example is taken from car body development processes in a car manufacturing company in Japan. For ease of understanding, first the conventional processes and then the converted ones are explained.

17.3.1 Conventional processes

Before the advent of integrated CAD/CAM, in conventional processes of car body development there were repeated transfers of information from one drawing of the previous process to further drawings or real models of current processes and from one real model produced in the previous process to further drawings or models of the current processes. These tasks required considerable manpower and degraded the accuracy of the data. Figure 17.1 shows the conventional car body engineering processes, each of which is explained in the following.

- *Style design*
  Style designers produce full-scale clay models based on their conceptual sketches. After these models are authorized by the Board of Directors, the models are measured to obtain 3D digitized data, which are used for production of full-sized external drawings. They are used as the basis of the external

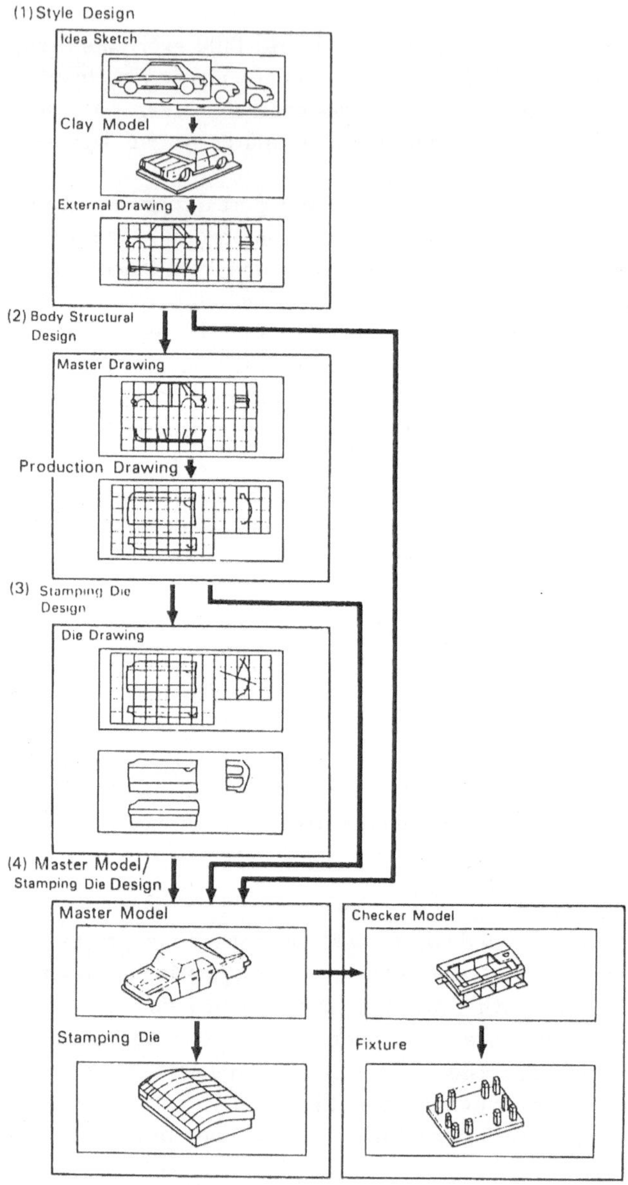

Figure 17.1 Conventional design and manufacturing processes of car body and its stamping die (courtesy of *Toyota* Motor Corp.)

shape of the cars. Drawings for interior components are produced in the same manner.

- *Body structural design*
  Based on the full-sized external drawings and the base plan of the structure, the structural design of the car is investigated. A full-sized master drawing is made, based on which product drawings of each part of the body are produced. They contain all the information needed for their manufacture.
- *Stamping die design*
  Based on the drawings produced in the previous processes, the face shape of the stamping die as well as the structure of the die are designed. The die-face design includes special shapes called blank-holder surfaces, step draws and draw beads. They perform important roles during the stamping action of the pressing machine to deform a steel sheet, which is set in the cavity of the die, into the designed shape without surface distortion or fracture of the sheet.
- *Master model and stamping die manufacture*
  The master model which becomes the standard for the following manufacturing processes is produced based on the previously defined shapes expressed by the drawings. The stamping dies are produced and checked against the master model. Their manufacture requires special skills and manpower, and correction of the die-face shape is frequently needed following a try-out with the manufactured die. This takes very much time and effort.
- *Checkers and jig manufacture*
  Copies of the master model are made for the sub-contractors of making welding fixture jigs and checkers.

In this example, since the relevant information was transferred by analog copying and analog checking, and data were interpreted by persons. Therefore, the manpower and time required in copying and interpreting were great and the management of data accuracy was a serious problem. After the introduction of computers, this data transfer could not be abolished because the computers were used for calculation and making drawings by instruction and needed separate data for each independent task. There were no common databases or internal models accessed from various tasks.

17.3.2 New Integrated Processes

Refer to Fig. 17.2 which shows the structure of an integrated CAD/CAM system developed to eliminate the above stated analog data storage and transfer through engineering drawings and the manufacture of analog real models. Simple conversion of the conventional system explained in the previous subsection to the conceptually new system was impossible, because construction of the indispensable internal model which replaces engineering drawings was quite different from making the engineering drawings and clay models, which are directly connected to the skills and intuition of designers and modelers. Since the clay models are

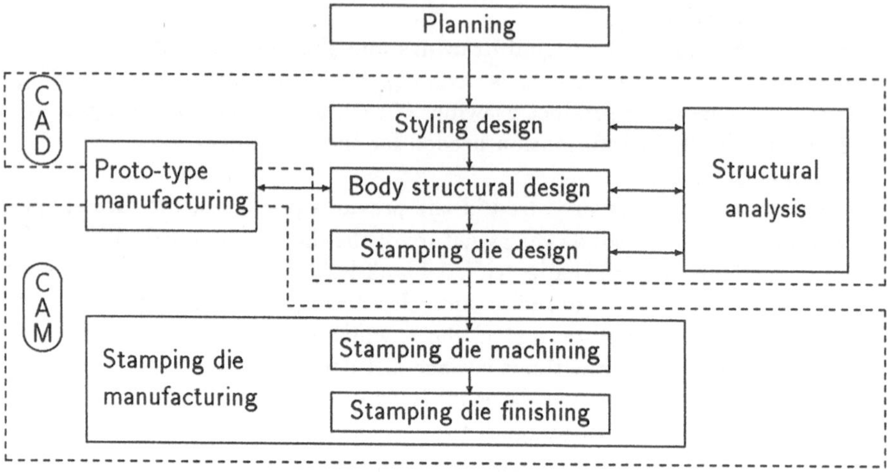

Figure 17.2 Integrated CAD/CAM system replacing a conventional system (courtesy of *Toyota* Motor Corp.)

easily understood by others, and features of the shape which the designers want are built into the clay models by collaboration of the designers and the modelers, their checking and evaluation are apparently easy. Accordingly, from the standpoint of the style designers, use of the internal model must be as easy as that of the engineering drawings and the clay models in their construction and evaluation. And all the complexity of the expression and handling of the internal models should be hidden and transparent to them.

In the new system shown in Fig.17.2 the most remarkable characteristic is the use of the internal models, which become the databases, and interactive graphic terminals, both of which replace the engineering drawings and the clay models. In style design, the designer's conceptual shapes depicted on drawing sheets are supplied to the system as input data, which are manipulated interactively by the designers to construct the internal model of the car body shape instead of making a clay model. On the other hand a clay model is produced by the machine using the internal model data and is used to evaluate the overall style design produced from the internal model. The body structure design subsystem uses the shape data for its design base and the designers create structures and product drawings with interactive display terminals using the body structure data base. Large scale FEM calculations are performed in a dedicated system to which input data are supplied from this subsystem.

In the stamping die design subsystem, the necessary information is supplied from the above stated subsystems and from its own database, and the die-face design is carried out by the designer at an interactive terminal observing simulated stamping processes. In the master model and the stamping die manufacture subsystem, the various real master models are replaced by the internal model, from which required shape information can be accessed freely to control machines producing stamping dies and checking their accuracy. Also from the internal model, various real master models and fixtures are manufactured for supply to the sub-contractor part makers where conventional manufacturing methods are still used.

The important breakthrough of the system is the successful construction and use of style design and die-face design subsystems. Of course, the body structural design subsystem is the largest and central one, but this is the powerful extension of a CAD system for making engineering drawings. So, we explain in the following sections the style design and die-face design subsystems which are more characteristic of the advanced CAD/CAM systems.

17.4 Style Design System

17.4.1 Importance of Shape Design

Different design processes for the shapes of products have been developed in various fields of engineering. When the aesthetic aspect of the products is a main factor for their value as commercial products, specialists called style designers have the leading role in the shape design process. They have special talents and characters, which are different from those of engineers who engage in the design of functions or structures of the products. For example, in motor cars, their exterior and interior design are very attractive factors for their users together with the car's performance. Accordingly, their style designers have been working as professional specialists and their tasks have not been treated directly by the computer. And most of them think of themselves as artists and that their works are among those least adaptable to computerization.

However, it has been recognized at the Toyota motor car company in Japan that this part is one of the bottlenecks to manufacturing high quality products and to attaining better productivity in the whole process of new car development. There, the style design process and evaluation methods for design quality were analyzed with the collaboration of the style designers. And with the result of this analysis an advanced interactive style design system has been developed, which the style designers have been pleased to use. The style designers feel that with this system their ability is enhanced by making various design alternatives available to them and by decreasing the time they consume in their routine work such as drawing and clay model making.

17.4.2 Two Aspects of Style Design

We give a concise check list of motor car style design. In it there are two aspects: motor cars may be seen as engineering products and also as aesthetic objects.

Items of these aspects are:

- *Aesthetic aspect*
  Newness. Level of class in feeling. Proportion and balance of style.
  Composition of surfaces.
  View of surfaces; high light, shading, feel of volume.
- *Engineering product aspect*
  Ease of use, sense of comfort.
  Structure of body, strength and rigidity, aerodynamic characteristics, new mechanisms, new materials.
  Law and regulations.
  Manufacturability (stamping, assembling and surface processing).
  Cost.

The design tasks proceed with these various factors in both aspects being considered. The style design processes consist of

(a) the process of converting from images to real shapes,
(b) the process of adapting to the requirements from the engineering product aspect.
(c) the process of checking and evaluating the created shapes.

These processes are iterative ones, such as creating idea sketches and rendering drawings, making various engineering drawings and producing clay models of one-fifth size, and returning to the previous stages as a result of the checking processes in between. Finally, these processes conclude with complete clay models of real size. So far, up to this stage in the design processes, computers have not been used, because the tasks were considered to be a realm for the sensitive talents of the specialists of style and model making.

17.4.3 Computer-Aided Style Design

To promote and rationalize the tasks performed by the style designers and also to produce the precise data for the models used in the other sections, computers have to be introduced in this design area from the very beginning of shape design. Manual drawings and clay models have to be replaced by interactively constructed models which can be displayed on CRT with appropriate conversion from the internal description. The constructed internal models must be evaluated and corrected from both aspects, as aesthetic objects and engineering products. If the checks are passed, then the models provide the fundamental data for the succeeding design tasks, and finally they can replace the solid master models used in the manufacturing stages.

The following items have been realized in the system to attain the above stated objectives.

(1) Easy input method for three-dimensional curve segments from designers' plan and elevation drawings. Conversion to 3D curve segment expressions.

(2) Evaluation of quality of constructed curves and their modification.
(3) Systematized methods of making surfaces which correspond to the manual procedures for making clay models.
(4) Evaluation and correction methods for defined surface segments.
(5) Various display techniques and object manipulation techniques.
(6) High-quality display and high-speed response time of the system.
(7) High-speed NC machining for a real-sized clay model from its internal model data and preparation of its engineering drawings.

The surface construction process is shown by the example in Fig. 17.3.

(a) shows input data of character lines of the orthogonal side views. These character lines are converted to 3D curves in the system.
(b) shows a cross-section shape which a designer want to make,
(c) shows construction of a side window using the data of the cross-section.
(d) shows the base surfaces of the cabin constructed from the above data.
(e) shows the trimmed base surfaces. With the base surfaces the object is expressed roughly. Then the boundary regions of the base surface are modified by the approach surfaces and the abrupt change of curvature is rounded by the fillet surfaces.
(f) and (g) show the construction of approach surfaces and fillet surfaces and their trimming.
(h) shows the final state of the designed surface.

As shown above, the base, approach and fillet surfaces are hierarchical classes of surfaces.

In the procedures for defining surfaces, which is the item (3), the types of the surfaces which have the specified highlight features are selected besides their classes. With this systematization, designers can express their intentions clearly and can converge quickly to their intended styles with the highlight features.

In items (2), (4) and (5), by the functions for evaluation of curves and surfaces, designers can evaluate the quality of their designed curves and surfaces objectively and with the functions for shape control of the system they can modify the shapes without violating their intended features.

In item (7), finally, the model is approved by the Board of Directors of the company. Then the information on the model is transferred and used in the various departments of design and manufacturing as the shape database. This is the outline of the computer-aided style design.

17.4.3.1 Input of Simplified Drawings

The style designers and modelers of motor cars use orthogonal projection drawings to make a solid model of a car with clay, in which the main character lines and vertical sectional curves of the car body are depicted. They make these drawings from their various ideas and rough sketches of the car they want to

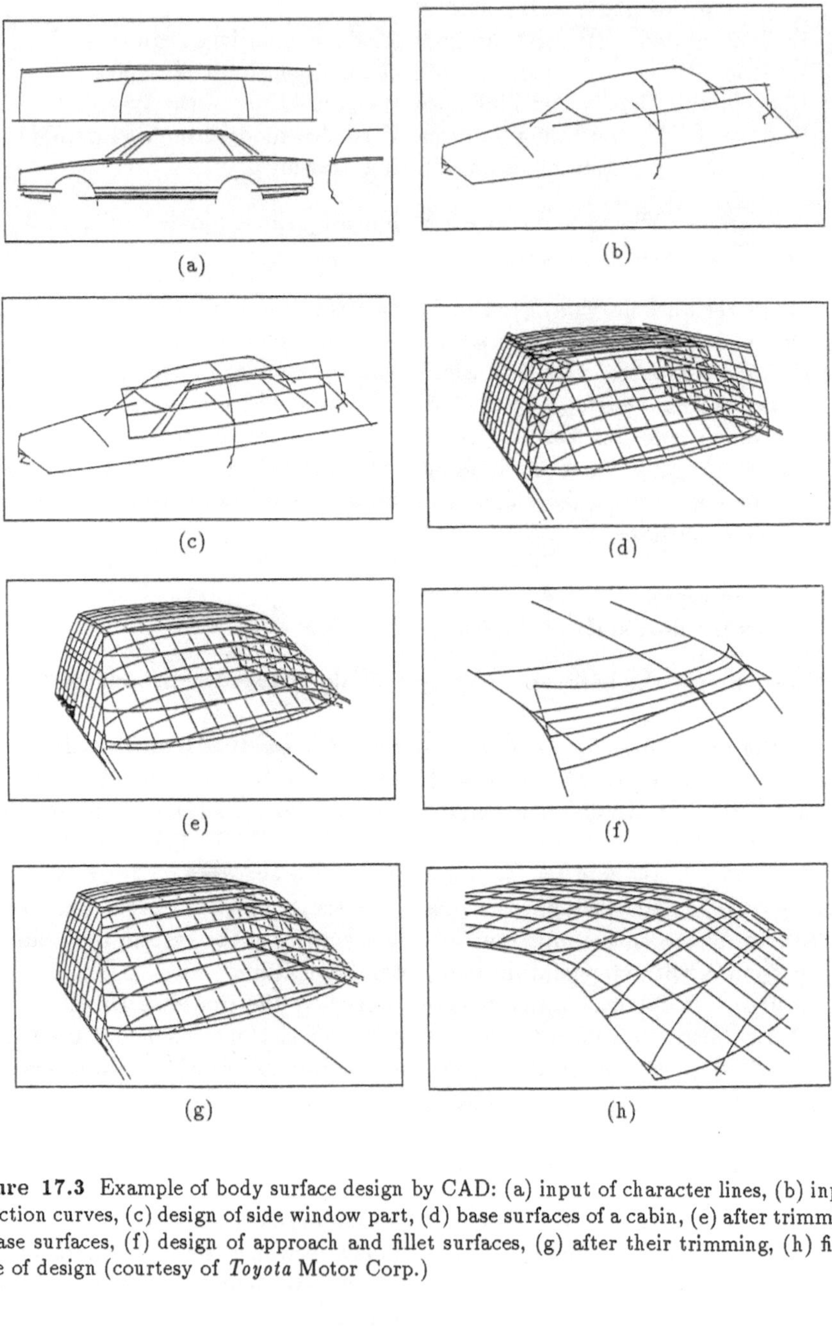

Figure 17.3 Example of body surface design by CAD: (a) input of character lines, (b) input of section curves, (c) design of side window part, (d) base surfaces of a cabin, (e) after trimming of base surfaces, (f) design of approach and fillet surfaces, (g) after their trimming, (h) final state of design (courtesy of *Toyota* Motor Corp.)

design. These are not precise drawings in a strict sense, but they contain the main features of the designers' images of the car. If these curves are to be input to the computer for style design, the characters of the curves have to be transferred and if any geometric contradictions exist in the drawings, they have to be removed. So far, to input a curve to the computer, the curve is traced as a point sequence by a person using a cursor and a digitizing tablet. From these data, a curve-fitting method of least squares is applied to obtain a smooth curve, but with this approach a desired result which conveys the character of the original curve is not always obtained. To improve the situation, a new method was developed.

Since designers create curve segments whose curvature profile varies monotonically by using appropriate curved rulers, besides least-square fitting, information on the curved rulers used by a designer at certain stages of drawing can be transferred to the computer. These curved rulers have been systematically prepared or sometimes newly created by designers according to their needs. The ruler information consists of its identifier, its used side, and its location on the drawing sheet. From this information, the mathematical expression of a single curve segment can be determined in the computer, and from the geometrical properties of the pair of projected curves, the corresponding space curve is obtained by a calculation which was explained in Sect. 5.5. Curve segments are connected to make their curvature distribution as smooth as possible; see Sect. 10.5. Methods of modifying input curves, without violating the original characters of the curve, are provided. They consist of the following functions:

(a) Shape unmodified: translation, rotation, sliding, extension.
(b) Shape modified: change of tangent angles, change of tangent magnitudes.
(c) Shape transformed: one end-point moving with the tangent free, one end-point moving with the tangent constrained, transforming in offset direction.

The basic data are manipulated, evaluated and modified by the designers using interactive graphic displays to construct a wire-frame model of the car.

17.4.3.2 Classes and Types of Surfaces in Style Design of Motor Cars

To facilitate construction of a high-quality surface, surface segments used in a car body are classified by their methods of creation and the locations at which they are used. The types of surfaces are called (quasi) cylindrical, conical and spherical or symbolically +, H and □ types by analogy with their generating elements. Each one can be used in various places as a class of base or approach or fillet surfaces.

The cylindrical (+ type) surface is created by sweeping a certain (variable) curve, which called a basis curve, along one guide curve. The conical (H type) surface uses two guide curves along which a basis curve moves. For the spherical (□ type) surface, four basis curves which make closed area move toward the center of the surface with blending of the sweeping surfaces. These construction

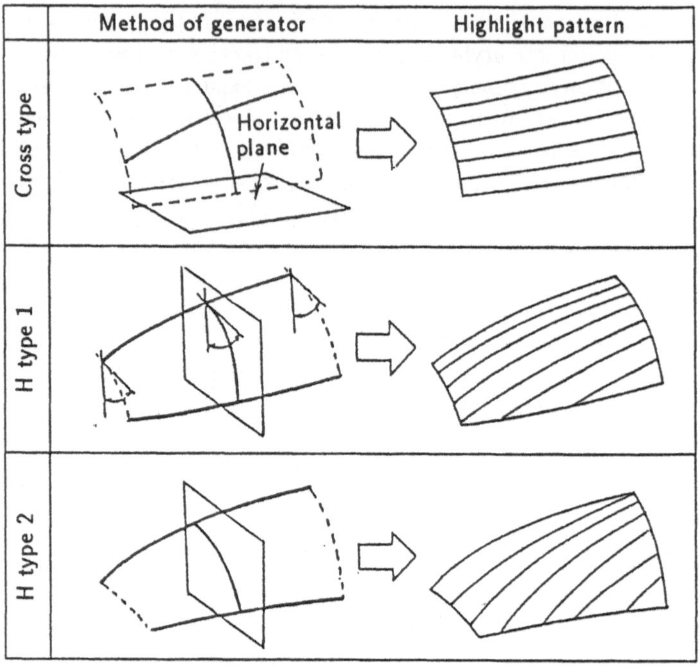

Figure 17.4 Types of surfaces and their highlight patterns and their generating methods (courtesy of *Toyota* Motor Corp.)

processes correspond to features of the highlight lines of desired surface. See Fig. 17.4.

While the base surface determines the skeleton of the whole shape, the approach surface modifies the parts of the base surface to make variations in the surface being designed and to make transient regions within the base surface. The fillet surfaces are rounding parts along the connection with the other surfaces to reduce abrupt changes of curvature. With these standardized surface construction methods, together with connection, interfering and trimming of surfaces, a basic internal model is constructed. At this stage, the quality of designed surfaces is evaluated by the highlight check. And then detailed surface design follows. The number of the commands which create and process curves and surfaces including their evaluation is as many as three hundred. They are classified into nineteen categories shown in Table 17.1.

17.4.3.3 Evaluation of Curve and Surface Quality

The next important function in a style CAD system is evaluation of the quality of the constructed curves and surfaces. In the manual process, the curve quality is checked by style designers observing from various directions the shape of the

Table 17.1 Nineteen categories of commands

	Category	Related items
1.	digitizer input	point sequence, straight line, circle/circular arc, tangent, etc.
2.	point	coordinate values, drafter input, point on line, point on surface, intersection of two lines, line and surface, etc.
3.	line	one, two points, angle, tangent, normal, etc.
4.	curve	circle, circular arc, curved ruler, fitting, etc.
5.	adjustment	connection, correction, generating line, integration, area definition, etc.
6.	2D to 3D	two projection curves, family curves, additional curve, rotation, etc.
7.	line modify	extension, rotation, slide, end-point displacement, offset, etc.
8.	control	control polygon or net manipulation.
9.	check	connection, tangent, curvature, highlight, hard point, etc.
10.	surface	plane, patch generation of three kinds, sweeping, rotation, symmetrical surface, etc.
11.	section	intersection, projection, etc.
12.	trim, delete	surface, line, figure, etc.
13.	math. op.	coordinate values, distance, angle, area, etc.
14.	std. parts	tire, lamp, etc.
15.	drawings	block registration, layout, size, line attributes, etc.
16.	NC data	tools, machining area, interference, cutter location path, cutter location control, etc.
17.	data trans.	to and from other systems.
18.	system	
19.	help/test	

curve drawn in a large scale on a sheet. The style designers' keen observing eyes can point out regions of a curve as good, uneven, extending, bulged and so on. To interpret their words, logarithmic values of radii of curvature along the inspected curve were plotted and the regions where the designers pointed were found to correspond to non-smoothly varying regions of the plotted curve. This plot is now called the test of profile or distribution of the radius of curvature.

The surface quality is checked in the manual process by the designers observing a real surface of the clay model, but with the computer method, a CRT surface is too small and has inadequate shape and its display quality is not good enough to evaluate the quality of designed lines and surfaces in space.

Even if the expressions of the surfaces to be designed satisfy conditions on patch connection and constraints specified on the shape, the preliminary designed surfaces do not always have characteristics the designers want to have. As their mathematical expressions are nothing but applications of interpolation formulas, it is not necessarily guaranteed that the expressions generate the favorable properties all over the designed surfaces.

On the other hand, the favorable properties themselves are not clearly defined mathematically, and they depend on the designers' subjective or intuitive recognition and interpretation. There are no general methodologies for the evaluation of the aesthetic aspect of shapes, but for the manual design of free-form surfaces of motor cars, a set of designers' and the modelers' empirical rules for constructing surfaces of good quality seems to have been implicitly established. Accordingly, to introduce a CAD system in this field, it must have features the experts in this field can accept.

A CAD system must provide its users not only with various surface definition methods, but also with surface quality evaluation techniques. Otherwise users cannot proceed their design process further with confidence. Since the current manual design methods and evaluation criteria cannot be applied directly to their computerization, it is important to systematize the procedures the style designers engage and to clarify what they observe for their evaluation of the quality of surface in terms of engineering and mathematical formulations. Then an acceptable CAD system for shape design can be constructed.

One possible method of checking the quality of a designed surface is to simulate the highlight of the surface. The highlight check for a manually made clay model is as follows: a model with smoothly finished surfaces is placed in a room with many parallel fluorescent lamps attached to its ceiling and walls. Images of these parallel lamps on the surfaces are observed to check where disorder occurs.

As precise simulation of this process is very difficult and time-consuming even with a very high-speed computer and the purpose is to check the quality of a surface and to locate regions of low quality, a similar but simplified method may be applied. The assumption is that the observing eye is at infinity, the line of sight is parallel and silhouette curves of the object surface are obtained, and then changing the direction of the parallel lines of sight, we obtain the silhouette pattern whose equation is given in Sect. 7.3. This pattern of the silhouette loci is called the highlight pattern in the factory. If a real highlight pattern is to be calculated, this is also possible, but with high cost of calculation. See Chap. 7.

Even with this simplified assumption, the calculation is rather complicated and the loci are very sensitive to the variation of the direction of the normal vector on each point on the surface. Disorder of the silhouette pattern corresponds to irregularity of the curvature distribution on the surface. On a boundary of $C^{(1)}$ connection, the highlight pattern has break points and for $C^{(0)}$ connection it has discontinuity of the highlight lines.

Another method of shape evaluation is to observe the radius of curvature

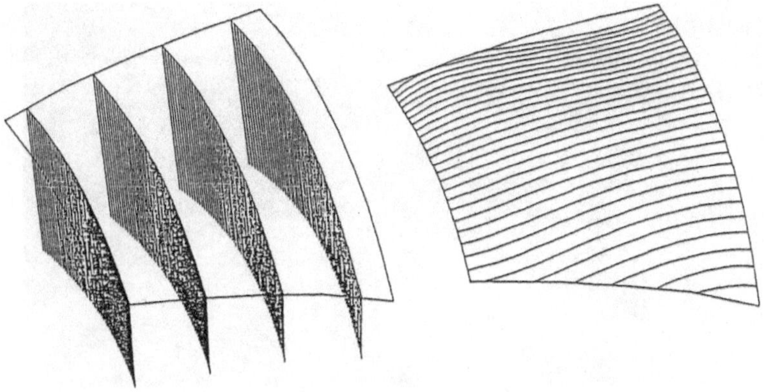

Figure 17.5 Profile of radius of curvature of section curves and highlight pattern of a surface (courtesy of *Toyota* Motor Corp.)

profiles of curves on the surface or section curves of the surface. Irregularity of the profiles indicates the existence of unfavorable regions on the surface. The radius of curvature distribution check can be performed in less time than the highlight check, so it is used frequently during the design stage. This test for a curve is rather severe and it corresponds well to the inspection test by the observing eyes of proficient designers. However, the curvature profile check is not sufficient for the surface quality evaluation. Figure 17.5 shows both the curvature profile and the highlight distribution. The curvature profile along the longitudinal sections are deemed good, but the highlight pattern has wavy disturbances. The surface construction must be modified. Figure 17.6 shows highlight lines of a model on a display and Fig. 17.7 is a clay model made from the same data as in Fig. 17.6 and shows the real highlight test. Figure 17.8 is a drawing of a completed body generated from an internal model. Figure 17.9 is a photograph of the real car developed from the internal model.

Usually, graphic display with color shading is not so effective, because the purpose of the shading display is to make pictures of good quality, but not to show the quality of the surface of the products. But recently, with greatly increased computing power, it has become possible to evaluate viewing effects of color paints on a car under various environmental light conditions. In order to do so, accurate shape data of a geometric model and data of real light reflection properties of paints on metal sheets are supplied for ray tracing calculation. Though it takes rather a long time for the calculation, the evaluation of a color simulation of a car and that of color paints is possible long before the real model is produced. Figure 17.10 is an example of a display of this kind.

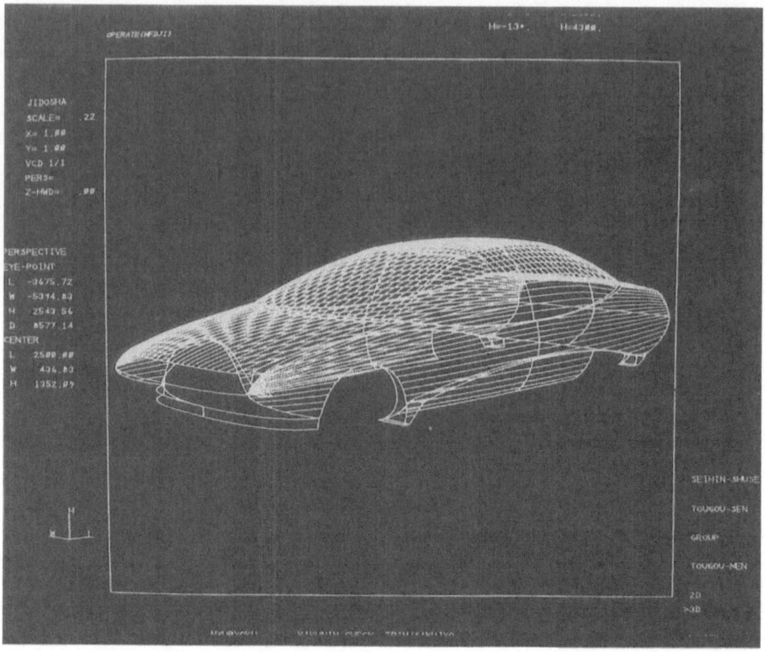

Figure 17.6 CRT display of highlight pattern of a car (courtesy of *Toyota* Motor Corp.)

17.4.4 Die-Face Design System

A blank steel sheet is deformed into the final shape of a smooth outer panel of a car by the stamping press. A sectional view of a stamping die and stamping action is schematically shown in Fig. 17.11. First a sheet of metal is placed on the lower die and the periphery of the sheet is clamped and then the stamping punch presses the sheet to deform it to the final shape. Refer to Fig.17.12. Since manufacture of a product by press forming requires a number of processes such as drawing (elongation of a steel plate by pressing), trimming, flanging, and so on, and the forming is a complicated process, the designer must be able to visualize all stages, from the blank steel sheet to the final product. Deformation of the sheet is produced by pressing the stamping punch, but the quality of deformed sheet is strongly affected by the shape of the holding surface and the surrounding part called the 'step draw' around the shape of the product part surface of the die. See Fig. 17.12. These two surfaces are called the die-face and its determination in the stamping die design is critical for the production of a car body surface of good quality.

A theoretical analysis of press deformation which is applicable to the real problem is very difficult, because it has to treat a 3D elasto-plastic large deformation problem, but with simplified FEM analysis the initial form of the steel plate at its boundary holding state is approximately determined to give the least

Figure 17.7 Highlight test of a clay model (courtesy of *Toyota* Motor Corp.)

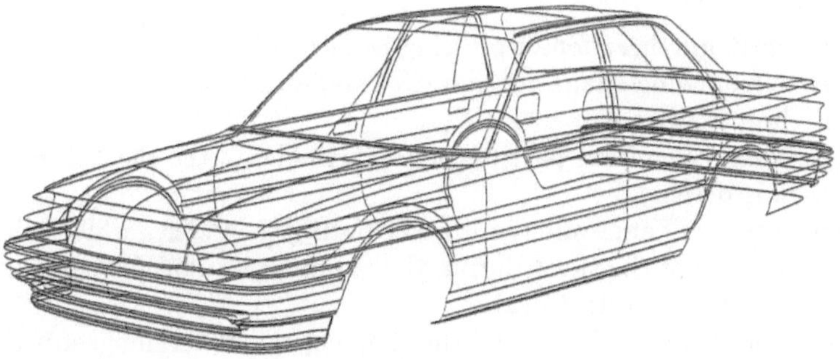

Figure 17.8 Line drawing from an internal model(courtesy of *Toyota* Motor Corp.)

unnatural deformation by the initial holding.

Since a considerable amount of knowledge on the behavior of steel sheet during stamping concerning the occurrence of fracture, surface distortion and other imperfections has been accumulated over the years, the designer of a die-face with this knowledge can evaluate the quality of its shape appropriately if he can visualize the deforming state of the sheet during pressing and calculate the indices of its local deformation. Accordingly, Computer Aided Die-face Design resembles Computer Aided Style Design in the formation of surfaces and the

Figure 17.9 Photograph of the real car of Fig. 17.8 (courtesy of *Toyota* Motor Corp.)

need to evaluate them, though the evaluation criteria are quite different and the sources of input base data are different. From the body style and structure databases, the shape information of the final product is obtained which defines the shape of the central region of the die. Around the central region, the die-face has to be determined to guarantee to deform the steel sheet, which is set in the cavity of the die, into the predefined shape by the stamping. Refer to Fig. 17.12. The design procedures are:

(1) The optimum direction of pressing is decided.
(2) The surrounding shape consisting of step-draw parts to prevent fracture of the sheet is designed.
(3) The clamping surface part to prevent wrinkles of the sheet during stamping action is designed.
(4) These surfaces are evaluated by calculating the indices which relate to local deformation of the sheet during the stamping action. If the criterion is not satisfied, then return to (2).
(5) Draw beads are designed and the shape of the material sheet is determined.

The indices calculation in item (4) consist of that of geometric displacement during the stamping punch movement, which is precisely obtained, and that of material deformation, which is performed by using FEM on the approximated shape. Figure 17.13(a) shows the initial blank steel sheet shape held by the holder around the peripheral region, and (b) shows diagram of variation of drawing depth during the press punching from the initial clamped position of the steel sheet. The shape of the die-face affects this distribution, from which values of the indices to estimate formability are obtained. Just like the style CAD subsystem, the die-face CAD subsystem has the following capabilities:

- Smooth reception of data from the other subsystem to construct an internal model for designing the die-face.

Figure 17.10 Shading display for evaluation of color paint on a car, (courtesy of *Toyota* Motor Corp.)

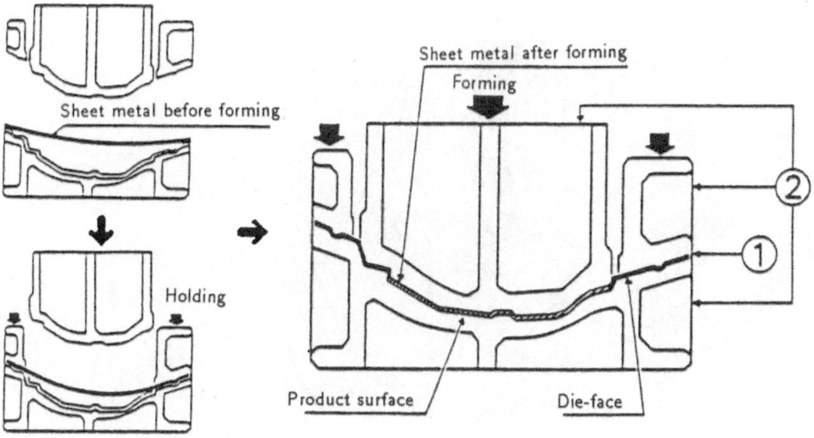

Figure 17.11 Sections of a stamping die (courtesy of *Toyota* Motor Corp.)

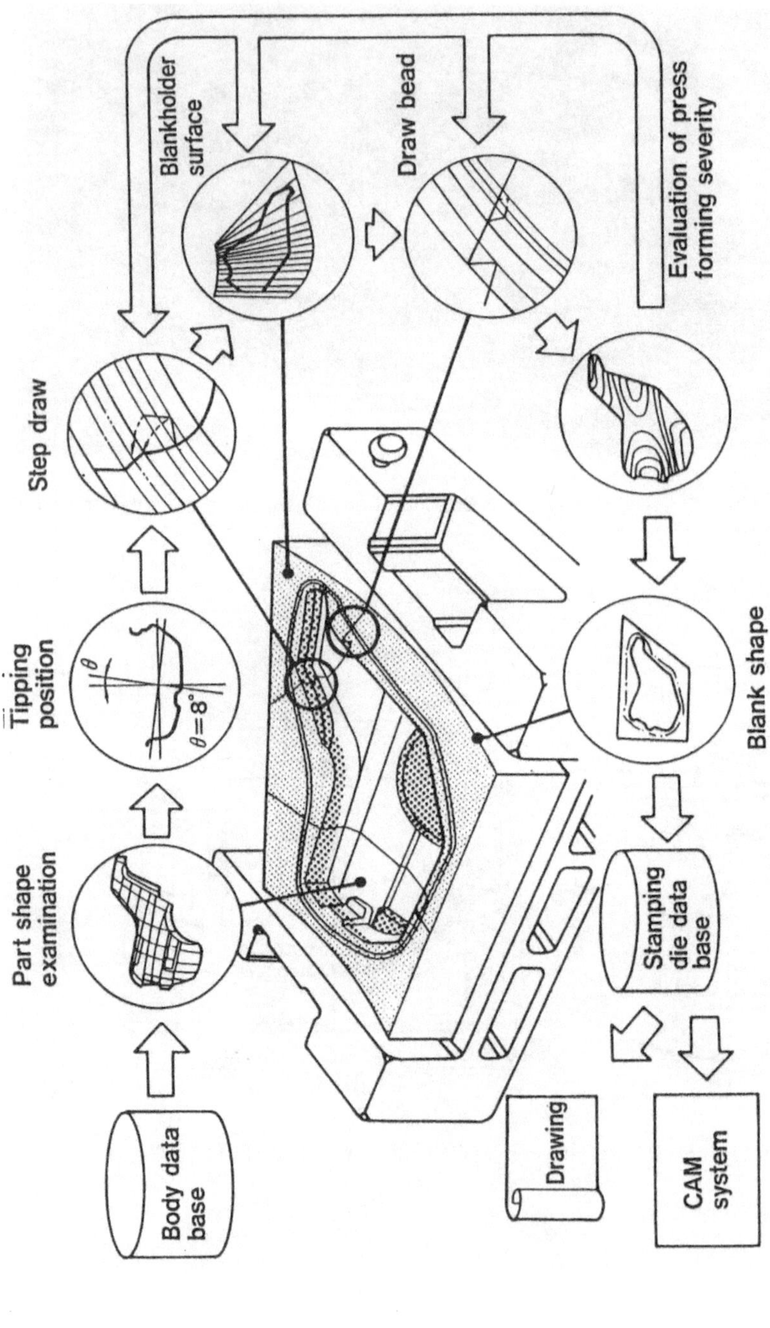

Figure 17.12 Design processes of a die-face (courtesy of *Toyota* Motor Corp.)

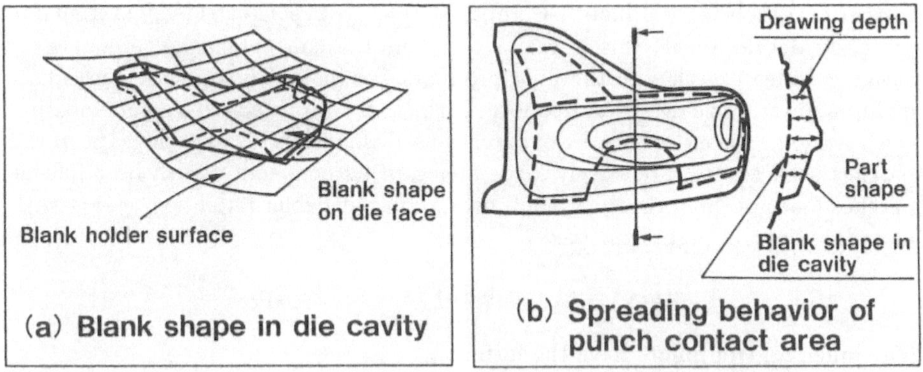

Figure 17.13 Evaluation of die-face design (courtesy of *Toyota* Motor Corp.)

- Interactive creation, alteration of die-face shapes.
- Generation of displays of various diagrams and charts for evaluating the formability of the steel sheet which is set in the die cavity.

Following the success of this system, real models made of plaster to investigate the die-face shapes in the conventional methods have been eliminated and the number of real try-outs of stamping using the newly manufactured dies has decreased considerably. Manufacturing and checking measurements of the stamping dies of car bodies and those of the fixtures for welding and inspecting are also controlled by the computer using the information in the internal model which are constructed in the previous stages. With this integration of CAD and CAM, the productivity and quality of new car development have been enhanced greatly in the company.

17.5 CAD/CAM of Free-Form Injection-Mold Products

In commercial products such as audio and video equipment, conventional design procedures can easily produce simple shapes, but currently industrial designers have to design aesthetically pleasing free-form shapes which have great commercial appeal even in a highly competitive and technically well-developed market.

The conventional procedures for design and manufacture of aesthetically pleasing products take a long time. Because mock-ups, molding dies and prototypes of the products are made based on two-dimensional drawings, each of these steps results in a slightly different version of the designers' ideas. Accordingly, negotiations between the designers and makers of mock-ups and molding dies take a long time to make them remodel or modify the objects until the designer are satisfied. But since most current commercial CAD/CAM systems are more concerned with the productivity of products than their aesthetic aspects, they are rarely helpful.

Sony's CAD/CAM system called FRESDAM is the one which is aimed at aesthetic aspects of products. Compared to Toyota's CAD/CAM system for car body development, this is very small, but the concept is the same. From designers' idea sketches to the final products, for instance, to dies for injection molding, interactive design of free-form surfaces is supported with an engineering work station. Since the number of patches used is not large and each shape of the patches used covers a relatively large area and its connection with its adjacent patches must be smooth, the system uses a kind of Bézier patch expression with abundant control points such as

$$\mathbf{s}(u,v) = (1-u+\mathsf{E}u)^3(1-v+\mathsf{F}v)^3\mathbf{p}_{00}$$

The inner control points have the form:

$$\mathbf{p}_{ij} = \frac{u^2(1-u)^2\mathbf{p}_{ij}^v + 2u(1-u)v(1-v)\mathbf{q}_{ij} + v^2(1-v)^2\mathbf{p}_{ij}^u}{\{u(1-u)+v(1-v)\}^2}, \quad i,j = 1,2.$$

The control points $\mathbf{q}_{ij}$ are used for modifying the shape of a patch without affecting its boundary conditions of $C^{(1)}$ connection. The control points $\mathbf{p}_{ij}^v$ and $\mathbf{p}_{ij}^u$ are used for modifying the cross-boundary tangents in u and v directions independently. The above expression is a modification of that explained in Sect. 14.4.

Tool path calculation for a ball-end mill is performed by intersection of the offset surface of the original patches with the tool driving planes. Calculation for collision detection and avoidance, compensation for heat shrinkage of the material used, and allowance for the electrode gap in an electro-discharge machine must all be performed. And since separately produced molded parts of a product must be glued on their parting surfaces to become a complete product, determination of the parting curves and parting surfaces becomes indispensable. The parting curve on the surface is a space curve which has the same geometric characteristic as a silhouette curve from a viewpoint at infinity .

The design and manufacturing process are shown in Fig. 17.14(a)–(f) for an in-the-ear headphone with a large resonant tube and a nice fit to an ear: (a) shows the designer's final idea sketches with rendering, (b) is the defined free-form shape, (c) is its shading display for evaluation, (d) is a photograph of the produced model at five times actual size, (e) shows the electrodes of the discharge machine, and (f) is a shading display for a case for the device, which is also aesthetically designed.

Since this example is on a small scale, all the information is generated, stored and transferred without converting to engineering drawings. Evaluation of designed objects is performed on CRT in a shading display.

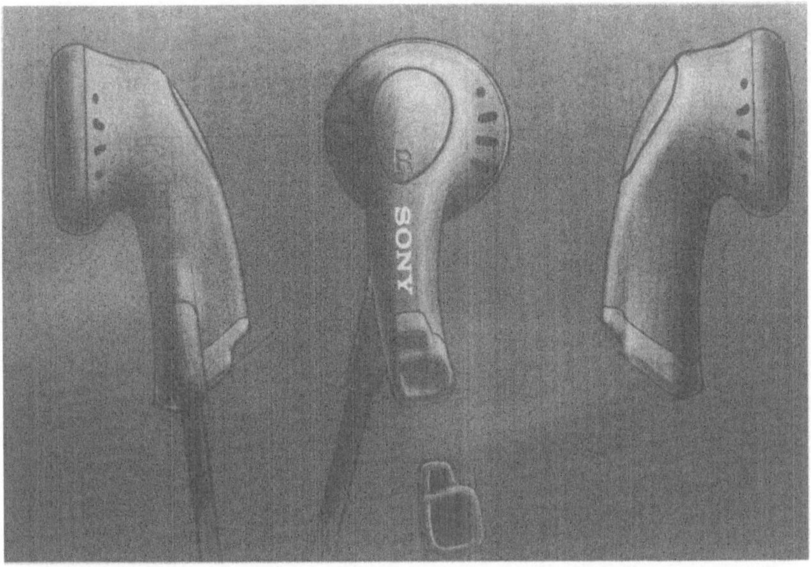

(a)

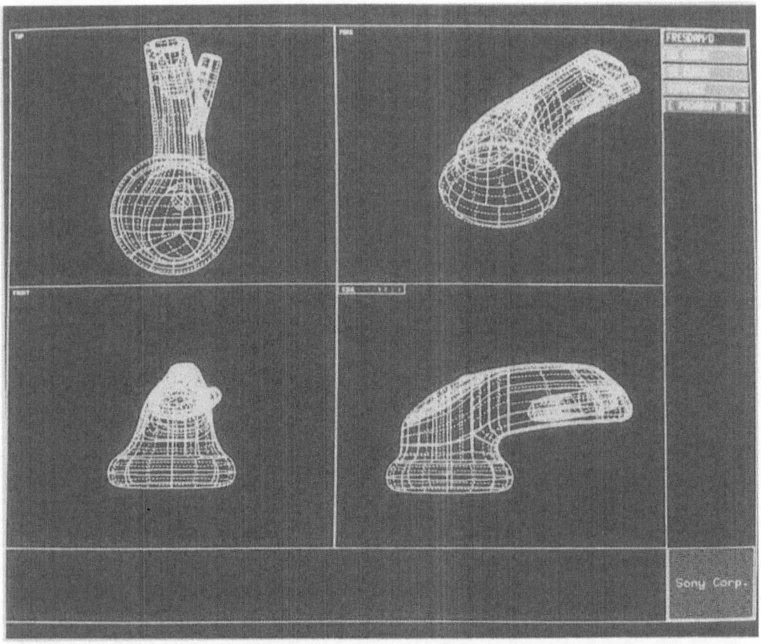

(b)

Figure 17.14 Design and manufacturing of in-the-ear head-phone: (a) final idea sketch, (b) model output, (c) shading display, (d) machined model for evaluation, (e) produced electrodes, (f) display for its case design (courtesy of *Sony* Corp.)

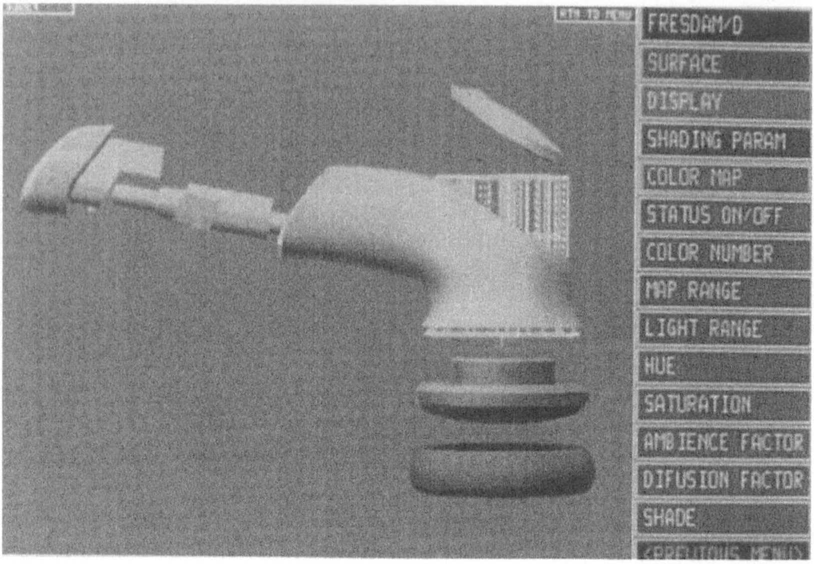

(c)

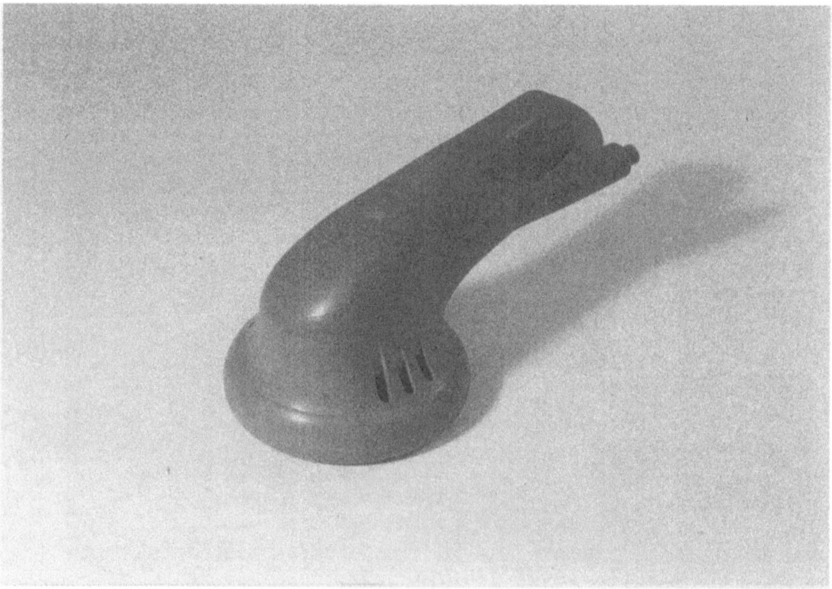

(d)

Figure 17.14 (continued)

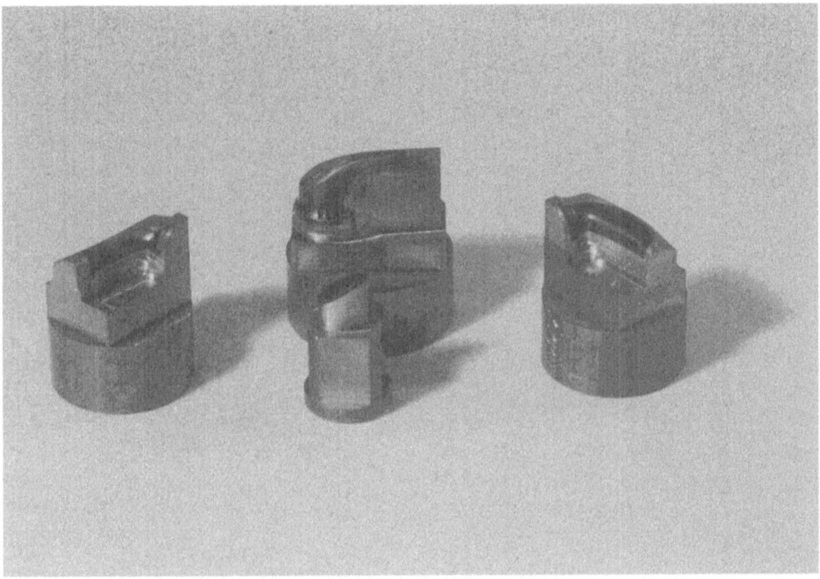

(e)

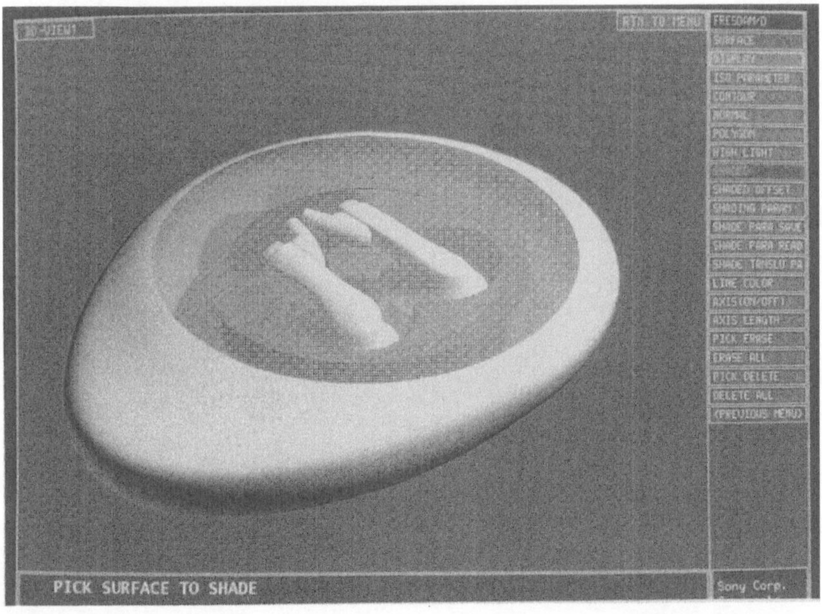

(f)

Figure 17.14 (continued)

Appendix Numerical Methods of Differential Equation Solving

A.1 Introduction

In this book, we have used the numerical methods of solving differential equations for tracing curves on surfaces which show their special features or intersection with other surfaces. Though some of these surfaces are represented by implicit equations, we convert them into the corresponding differential equations to trace the curves, because it is easier to trace a curve with step-by-step solution of its differential equation than to calculate a great many of the coordinate values of discrete points on the curve by solving its nonlinear equations. Since usually these curves are determined uniquely by given geometric conditions, their tracing by the differential equations does not become unstable, so long as truncation errors do not accumulate. In rare cases at singular points on the curves the validity of the differential equations fails. In such cases, we have to take the omitted higher-order terms of the differential equation into consideration.

There are several differential equation solvers available for our problems. Among them, those methods equipped with adaptive stepsize control show good results. This enables us not only to avoid the problem of determining the integration stepsizes, but also provides the optimum ones for tracing the curves. We explain the methods we have used in order to give an understanding of their principles and uses. We do not intend to introduce their detailed theoretical aspects, which, if needed, should be consulted in those references treating numerical methods of computation [14][15][16]. Since the equations for the intersection curves have been given, we can check the accuracy of the results obtained from the integration process at any stage of the curve tracing. But we can keep expected errors of the integration process within the given tolerance, hence we need not perform extra checks of accuracy in the real space using the original equations. The tracing curve passes through the points on the intersection curve, which are obtained at the initial stage, and the loop is closed.

By our experience on problems treated in Chap. 7 and Chap. 16, the fourth-order Runge-Kutta method with adaptive stepsize control is most reliable though its speed is low. It is faster than the Runge-Kutta-Fehlberg method, which uses terms up to the fifth order, but the Bulirsch-Stoer method is good and relatively fast. Predictor-corrector methods are fast, but frequently they fail in critical regions. If such a method is equipped with a mechanism of variable stepsize control, then together with the Runge-Kutta method it gives good results and is the fastest, but not always reliable. The Bulirsch-Stoer method with rational extrapolation works well, though that with polynomial extrapolation is a little faster and less reliable.

We explain these methods in this chapter. For simplicity, we treat a differential equation with a single dependent variable, but the solving methods can easily be extended to cases of multiple dependent variables.

Given a differential equation of the form

$$\frac{dy}{dt} = f(y,t) \tag{A.1}$$

with an initial condition of the value y_i at the start $t = t_i$, our objective is to deduce the value y_n at $t = t_i + H$ as a good estimate of $y(t_i + H)$ with a given tolerance.

A.2 Adaptive Runge-Kutta Method

A.2.1 Runge-Kutta Step

In this case, we set $H = h$ and $y_n = y_{i+1}$ at $t = t_{i+1} = t_i + h$ and we obtain y_{i+1} by the following procedure:
at $t = t_i$, $t_i + h/2$, and $t_i + h$, calculate k_1, k_2, k_3 and k_4 by the following formulas:

$$\left.\begin{array}{lcl} k_1 &=& hf(t_i, y_i) \\ k_2 &=& hf(t_i + h/2, y_i + k_1/2) \\ k_3 &=& hf(t_i + h/2, y_i + k_2/2) \\ k_4 &=& hf(t_i + h, y_i + k_3). \end{array}\right\} \tag{A.2}$$

Then the value y_{i+1} is given by

$$y_{i+1} = y_i + (k_1 + 2k_2 + 2k_3 + k_4)/6 + O(h^5). \tag{A.3}$$

Four evaluations of the right-hand side of the differential equation (A.1) are required per step and the error term is $O(h^5)$. This formula gives good results for most differential equations if a proper stepsize is chosen. Since it is difficult to give a proper stepsize in advance, we are liable to take a smaller stepsize. This brings long computation time, but still the fixed stepsize does not always guarantee precise results in critical cases. Accordingly, adaptive stepsize methods are preferable.

A.2.2 Runge-Kutta with Quality Control

This is a stepsize doubling method. Let us integrate the same equation with stepsizes $2h$ and h using the Runge-Kutta step and call the results y_a and y_b. Then we can assume that two equations:

$$\left.\begin{array}{lcl} y(t+2h) &=& y_a + (2h)^5\phi + O(h^6) \\ y(t+2h) &=& y_b + 2h^5\phi + O(h^6) \end{array}\right\} \tag{A.4}$$

hold for the correct value $y(t+2h)$, where ϕ is considered a constant independent of the stepsize adopted so long as h is small. From the above two, we know that the difference $\Delta = y_b - y_a$ of the two results is proportional to h^5:

$$\Delta = y_a - y_b = 30h^5\phi. \tag{A.5}$$

We use this relation for adjusting the stepsize of the Runge-Kutta integration. Let Δ_0 denote the desired accuracy with the stepsize h_0 and Δ_1 be a value obtained with a stepsize h_1 for the same equation. There is a relation which is obtained from eq.(A.5):

$$\frac{h_0}{h_1} = |\frac{\Delta_0}{\Delta_1}|^{0.2}. \tag{A.6}$$

This can be used to adjust h_1 for the next try of the integration: if Δ_1 is larger than Δ_0 in magnitude in the first try, the value of the right-hand side indicates how much the stepsize should be decreased in the second try of the same interval, and if it is smaller than Δ_0, the right-hand side value shows the degree of increase of the stepsize in the next section.

The magnitude of Δ_0 for the desired accuracy depends of course on the problems we treat and the maximum magnitude of y and h. Therefore, for safety, with introduction of a safety factor $S(\approx 0.95)$, eq.(A.6) is changed to a slightly different criterion:

$$\begin{aligned}\frac{h_0}{h_1} &= S|\frac{\Delta_0}{\Delta_1}|^{0.2} \text{ for } \Delta_0 \geq \Delta_1,\\ \frac{h_0}{h_1} &= S|\frac{\Delta_0}{\Delta_1}|^{0.25} \text{ for } \Delta_0 < \Delta_1.\end{aligned}$$

In the examples used in this book, value of Δ_0 was taken as $10^{-6} \sim 10^{-8}$ or $\epsilon = \Delta_0/h = 10^{-4} \sim 10^{-6}$, and Δ_1 was the maximum one chosen among the differences Δ for several dependent variables.

A.2.3 Runge-Kutta-Fehlberg Method

Both the fourth-order and the fifth-order Runge-Kutta formulas are applied to approximate y in the given interval of t. The number of subintervals is increased step-by-step until the fractional difference between the results of the fourth-order and the fifth-order formulas (which give a truncation error h^5 and h^6, respectively) in the same interval is less than the specified tolerance. This method gives good results, but compared to the Runge-Kutta method with adaptive stepsize control it is slow and its accuracy seems the same order as that of the adaptive Runge-Kutta method for our problems.

The procedure is given below. At $t = t_i$, $t_i+h/4$, $t_i+3h/8$, $t_i+12h/13$, t_i+h and $t_i + h/2$, calculate k_1, k_2, k_3, k_4, k_5 and k_6 by the following formulas:

$$k_1 = hf(t_i, y_i),$$

$$\begin{aligned}
k_2 &= hf(t_i + h/4, y_i + k_1/4), \\
k_3 &= hf(t_i + 3h/8, y_i + 3k_1/32 + 9k_2/32), \\
k_4 &= hf(t_i + 12h/13, y_i + 1932k_1/2197 - \\
&\quad 7200k_2/2197 + 7296k_3/2197), \\
k_5 &= hf((t_i + h, y_i + 429k_1/216 - 8k_2 + \\
&\quad 3680k_3/513 - 845k_4/4104), \\
k_6 &= hf(t_i + h/2, y_i - 8k_1/27 + 2k_2 - \\
&\quad 3544k_3/2565 + 1859k_4 - 11k_5/40).
\end{aligned}$$

Then calculate r for an error estimate,

$$r = (\frac{k_1}{360} - \frac{128k_3}{4275} - \frac{2197k_4}{75240} + \frac{k_5}{50} + \frac{2k_6}{55}), \tag{A.7}$$

if $r \le \epsilon$, the value y_{i+1} is given by

$$y_{i+1} = y_i + \frac{25k_1}{216} + \frac{1408k_3}{2565} + \frac{2197k_4}{4104} - \frac{k_5}{5}. \tag{A.8}$$

For the next stepsize h', calculate the following criterion:

$$q = 0.84(\frac{\epsilon}{r})^{1/4}. \tag{A.9}$$

If $q \le 0.1$, then $h' = 0.1h$, if $q \ge 4$, then $h' = 4h$, otherwise $h' = qh$.

A.3 Variable Stepsize Predictor-Corrector Method

Starting from (t_{i-3}, y_{i-3}) for the next three points, the Runge-Kutta method with a stepsize h is used to obtain $(y_{i-2},\ y_{i-1},\ y_i)$. After these points, the following explicit Adams-Bashforth formula for determining a predicted next value $\tilde{y}_{i+1}$:

$$\tilde{y}_{i+1} = y_i + (h/24)\{55f(t_i, y_i) - 59f(t_{i-1}, y_{i-1}) + 37f(t_{i-2}, y_{i-2}) - 9f(t_{i-3}, y_{i-3})\} \tag{A.10}$$

is applied, then the following implicit Adams-Moulton formula for its corrected value:

$$y_{i+1} = y_i + (h/24)\{9f(t_{i+1}, \tilde{y}_{i+1}) + 19f(t_i, y_i) - 5f(t_{i-1}, y_{i-1}) + f(t_{i-2}, y_{i-2})\} \tag{A.11}$$

is used for a more precise y_{i+1}. This method is fast compared to the Runge-Kutta method and the accuracy is the same. Accordingly if it is reasonably certain that the differential equation is well-behaved, this is one of the most desirable methods. But when a curve to be traced has regions of great variation of curvature and torsion, the tracing often fails.

Accordingly, if the method with adaptive stepsize control is introduced, the situation is greatly improved, but since the predictor-corrector method always requires four starting points, the change of stepsize must be kept minimal. As a consequence, we use it only when the estimated truncation error

$$e = \frac{|y(t_{i+1}) - y_{i+1}|}{h} \approx |y_{i+1} - \tilde{y}_{i+1}| \frac{19}{270h} \tag{A.12}$$

becomes smaller than $(\epsilon/10)$ or greater than ϵ, that is,

$$(\epsilon/10) > e \text{ } or \text{ } e > \epsilon.$$

The stepsize h is changed to qh, where

$$q = 1.5\left(\frac{h\epsilon}{|y_{i+1} - \tilde{y}_{i+1}|}\right)^{1/4}. \tag{A.13}$$

If $q > 4$ or $q < 0.1$, set $q = 4$ or $q = 0.1$ to prevent abrupt change of the interval.

If h becomes smaller than a predefined minimum value h_m, switch the integration process to the Runge-Kutta method with stepsize h_m, proceed for several steps and then return to the predictor-corrector method.

A.4 Bulirsch-Stoer Method

A.4.1 Principle of the Method

This usually gives good results with automatic adjustment of stepsizes and its speed is relatively high. After many trials of various differential equation solvers, we have relied on this method for application in geometric computation.

For the integration procedure of this solver, the modified midpoint method is applied, whose error is a function of even powers of the stepsize h. Since in this method integration is performed several times for the same interval $(t,\ t + H)$ with different stepsizes, we use an index i to indicate its integration number. Hence the symbol t_i used in the previous sections is abandoned hereafter. Let the correct value of y at $t + H$ be $y(t + H)$ and the obtained value there be y_n after $n = H/h$ steps starting from $y_0 = y(t)$. The error at $t = t + H$ is given by

$$y_n - y(t + H) = \sum_{j=1} \alpha_j h^{2j}, \tag{A.14}$$

where α_j are coefficients independent of the stepsize h. Let $w_1, w_2, w_3 \cdots$ be values of the integration at $t = t + H$ obtained by the modified midpoint method with the stepsizes $h_1, h_2, h_3 \cdots$ $(h_1 > h_2 > h_3 \cdots)$. We consider these values $w_1, w_2, w_3 \cdots$ to be a function of the square of the stepsize: let this function pass through the points $(h_1^2, w_1), (h_2^2, w_2), (h_3^2, w_3) \cdots$. The estimate of $y(ti + H)$ is considered as the extrapolated value of this function at stepsize zero.

This is the principle of extrapolation methods of differential equation solvers. The Bulirsch-Stoer method adopts a rational extrapolation, for it gives usually better results than a polynomial extrapolation does.

A.4.2 Outline of the Procedures

Next we explain the procedures of the method. We use the stepsizes of decreasing order in a section between t and $t+H$. The section is divided into eleven sets of subsections whose numbers are $(2,4,6,8,12,16,24,32,48,64,96)$, each one of which is indicated by the index k.

1. For each stepsize h_i, we calculate the integration value w_i and produce a current extrapolated estimate y at $h=0$ using the current and previously obtained values w_i's.
2. If the difference between the current estimate and the last obtained estimate becomes negligible, further decrease of the substeps is unnecessary. When the estimates do not converge even at the eleventh stepsize, the length H of this section is made a quarter and we start again from the beginning.
3. When the estimate converges at an intermediate step, the length of the next section should be adjusted to keep the index value for subsections nearly constant. Let the desirable index of subsections be k_d and that of the current stepsize be k_c. If $k_c = k_d - 1$, the next section length is extended by 1.2 times and if $k_c = k_p$, it is shortened by 0.95 times.
4. Otherwise it is given by the formula $H \times$ (number of subsections of the index k_d-1)/(number of current subsections). The index $k_d = 7$ is said to be good.

A.4.3 Integration Procedure

In the Bulirsch-Stoer method, the modified midpoint method is used for the integration procedure, because its error of integration is expressed by a function of the square of the stepsize h and the calculation is simple. The integration procedure between t and $t+H$ is given by:

$$\left.\begin{aligned} h &= H/n, \\ z_0 &= y(t), \\ z_1 &= z_0 + hf(t, z_0), \\ z_{i+1} &= z_{i-1} + 2hf(t+ih, z_i) \quad for\ i = 1 \ldots n-1, \\ y(t+H) &\approx y_n = (1/2)\{z_n + z_{n-1} + hf(t+H, z_n)\}. \end{aligned}\right\} \tag{A.15}$$

There are $n+1$ evaluations of $f(t,y)$ in the interval H which comprises n subdivisions, whereas the Runge-Kutta method requires $4n$ evaluations. In Bulirsch-Stoer, the number of evaluations is at most 73 compared to 96 in Runge-Kutta for the same interval H. But since in Bulirsch-Stoer H is adjusted automatically to be taken much larger, it is more suitable for fast tracing of curves. When H is too large for drawing a curve, the intermediate points can be obtained from

$$y_i \approx (1/4)\{z_{i-1} + 2z_i + z_{i+1}\} \approx z_i. \tag{A.16}$$

A.4.4 Polynomial Extrapolation

Polynomial and rational interpolations and extrapolations are treated in Sect. 8.2. Our object here is to estimate a value y at $x = 0$ from a given sequence of points $(x_1, y_1), (x_2, y_2), \cdots (x_n, y_n)$, $\{x_1 > x_2 > \cdots > x_n > 0\}$, by extrapolation, where terms x_i represent the square of the sub-interval in the integration processes and terms y_i the corresponding result values of the integration. Use of polynomial extrapolation does not seem to provide good results because of its oscillatory behavior, but usually it gives good results in non-critical cases. The reason may be an extrapolation of a very short distance. Calculation methods of obtaining extrapolated values by polynomial or rational functions are almost the same, of course, the polynomial calculation is faster. We first explain the method of polynomial functions, because it is easy to understand and can be extended to the case of the rational functions.

Let $P_{i(i+1)\cdots(i+m)}$ represent the extrapolated value of y at $x = 0$ from a given sequence of $m + 1$ points $(x_i, y_i), (x_{i+1}, y_{i+1}), \cdots (x_i + m, y_i + m)$ and let it be called the value on level m . We want to obtain $P_{12\ldots n}$ for the estimate of y at $x = 0$, but at the same time we want to make n as small as possible within a given tolerance of the estimated value y.

We assume that from $y_1, \cdots y_{n-1}$, an estimated value $y = P_{1\ldots n-1}$ has already been determined and now a new point y_n at x_n has been obtained. Since we can make two estimates $P_{23\ldots n}$ from the sequence of y values $(y_2 \cdots y_n)$, and $P_{12\ldots n}$ from $(y_1 \cdots y_n)$, if their difference is smaller than the tolerance, we take $P_{12\ldots n}$ as the estimate of y at $x = 0$, otherwise we proceed to x_{n+1} for the next finer sub-interval and obtain the integration result y_{n+1}. Then make $P_{23\ldots n,(n+1)}$ and $P_{12\ldots n,(n+1)}$, and test if it is final.

To make the above stated process systematic, at first we define the differences between $P_{i(i+1)\ldots(i+m)}$ and its lower neighboring ones :

$$\left.\begin{array}{rcl} C_{i,m} & = & P_{i\ldots(i+m)} - P_{i\ldots(i+m-1)}, \\ D_{i,m} & = & P_{i\ldots(i+m)} - P_{(i+1)\ldots(i+m)}, \\ C_{i,0} & = & P_i = y_i, \qquad D_{i,0} = P_i = y_i. \end{array}\right\} \tag{A.17}$$

We make two triangular tables whose entries contain $C_{i,m}$ or $D_{i,m}$. The first index i indicates the i-th row of the tables of the difference and the second index m the m-th column of the tables. According to the following formulas:

$$\left.\begin{array}{rcl} C_{i,m+1} & = & \dfrac{x_i(C_{i+1,m} - D_{i,m})}{(x_i - x_{i+m+1})}, \\ D_{i,m+1} & = & \dfrac{x_{i+m+1}(C_{i+1,m} - D_{i,m})}{(x_i - x_{i+m+1})}, \end{array}\right\} \tag{A.18}$$

entries of the tables are filled step by step when a new value y_i for x_i is added in the entries $(i, 0)$ of both tables.

The above tables are constructed and the intermediate estimates are determined in the following way.

Table A.1

$$\begin{array}{lllll} y_1 = P_1 & C_{1,1} & C_{1,2} & C_{1,3} & C_{1,4} \\ y_2 = P_2 & C_{2,1} & C_{2,2} & C_{2,3} & \\ y_3 = P_3 & C_{3,1} & C_{3,2} & & \\ y_4 = P_4 & C_{4,1} & & & \\ y_5 = P_5 & & & & \end{array} \tag{A.19}$$

Table A.2

$$\begin{array}{lllll} y_1 = P_1 & D_{1,1} & D_{1,2} & D_{1,3} & D_{1,4} \\ y_2 = P_2 & D_{2,1} & D_{2,2} & D_{2,3} & \\ y_3 = P_3 & D_{3,1} & D_{3,2} & & \\ y_4 = P_4 & D_{4,1} & & & \\ y_5 = P_5 & & & & \end{array} \tag{A.20}$$

1. Let $P_1(= y_1)$ already have been obtained. Then $P_2(= y_2)$ is added. $D_{1,1}$ and $C_{1,1}$ are calculated and stored, but $C_{1,1}$ is not used. They are given from equations (A.17) and (A.18) by

$$\left.\begin{aligned} C_{1,1} &= P_{12} - P_1 = \frac{x_1(P_2 - P_1)}{(x_1 - x_2)} = \frac{x_1(C_{2,0} - D_{1,0})}{(x_1 - x_2)}, \\ D_{1,1} &= P_{12} - P_2 = \frac{x_2(P_2 - P_1)}{(x_1 - x_2)} = \frac{x_2(C_{2,0} - D_{1,0})}{(x_1 - x_2)}. \end{aligned}\right\} \tag{A.21}$$

   An estimated value at level two is obtained from

$$P_{12} = P_2 + D_{1,1}. \tag{A.22}$$

2. P_3 is added. The level-one values $C_{2,1}$ and $D_{2,1}$ are calculated and stored using formulas similar to the above. From $D_{1,1}$ and $C_{2,1}$, new entries $C_{1,2}$ and $D_{1,2}$ of level two are calculated by using (A.18) and stored,

$$\left.\begin{aligned} C_{1,2} &= P_{123} - P_{12} = \frac{x_1(C_{2,1} - D_{1,1})}{(x_1 - x_3)}, \\ D_{1,2} &= P_{123} - P_{23} = \frac{x_3(C_{2,1} - D_{1,1})}{(x_1 - x_3)}. \end{aligned}\right\} \tag{A.23}$$

   An estimated value at level three is $P_{123} = P_{23} + D_{1,2}$. Since $P_{23} = P_3 + D_{2,1}$, we have

$$P_{123} = P_3 + D_{2,1} + D_{1,2}. \tag{A.24}$$

3. P_4 is added. $C_{3,1}$ and $D_{3,1}$ are calculated and stored. From $D_{2,1}$ and $C_{3,1}$, the level-two values $C_{2,2}$ and $D_{2,2}$ are obtained. From $D_{1,2}$ and $C_{2,2}$ the final entries $C_{1,3}$ and $D_{1,3}$ of level three are filled. We obtain

$$P_{1234} = P_4 + D_{3,1} + D_{2,2} + D_{1,3}. \tag{A.25}$$

4. P_5 is added. $C_{4,1}$ and $D_{4,1}$ are obtained. From $D_{3,1}$ and $C_{4,1}$, $C_{3,2}$ and $D_{3,2}$ are determined. From $D_{2,2}$ and $C_{3,2}$, $C_{2,3}$ and $D_{2,3}$ are obtained. From $C_{1,3}$ and $D_{2,3}$, the final entries $C_{1,4}$ and $D_{1,4}$ are filled.

$$P_{12345} = P_5 + D_{4,1} + D_{3,2} + D_{2,3} + D_{1,4}. \tag{A.26}$$

Generally the estimate of level n is given by

$$P_{12} \dots n = P_n + D_{n-1,1} + D_{n-2,2} + \dots + D_{1,n-1}. \tag{A.27}$$

The last term of the estimate of level n is considered the last correction value for its final estimate. If $D_{1,n-1}$ is smaller than a given tolerance, the estimate value of y is given by $P_{12} \dots n$.

A.4.5 Rational Extrapolation

The Bulirsch-Stoer algorithm uses a rational function for the interpolation instead of the polynomial one.

Let $R_{i(i+1)\dots(i+m)}$, which is the rational function equivalent of $P_{i(i+1)\dots(i+m)}$, be the value obtained by extrapolation at $x = 0$. We want to find it by constructing two triangular tables similar to Tables A.1 and Tables A.2 of the polynomial case. Let us define the differences of adjacent rational functions similarly to eq.(A.17):

$$\left.\begin{aligned} C_{i,m} &= R_{i\dots(i+m)} - R_{i\dots(i+m-1)}, \\ D_{i,m} &= R_{i\dots(i+m)} - R_{(i+1)\dots(i+m)}, \\ C_{i,0} &= R_i = y_i, \; D_{i,0} = R_i = y_i. \end{aligned}\right\} \tag{A.28}$$

Instead of eq.(A.18), we have the recurrence formulas for $C_{i,m}$ and $D_{i,m}$:

$$\left.\begin{aligned} C_{i,m+1} &= \frac{(x_i/x_{i+m+1})D_{i,m}(C_{i+1,m} - D_{i,m})}{(x_i/x_{i+m+1})D_{i,m} - C_{i+1,m}}, \\ D_{i,m+1} &= \frac{C_{i,m}(C_{i+1,m} - D_{i,m})}{(x_i/x_{i+m+1})D_{i,m} - C_{i+1,m}}. \end{aligned}\right\} \tag{A.29}$$

Then we can construct tables similar to Tables A.1 and A.2 each time a new pair (x_i, y_i) is obtained. Similarly to eq.(A.27), the estimate $R_{12\dots n}$ of the level n is given by

$$R_{12\dots n} = R_n + D_{n-1,1} + D_{n-2,2} + \dots + D_{1,n-1}. \tag{A.30}$$

A.5 Examples and Evaluation

Various figures of curves shown in Chapters 7 and 16 are results obtained from differential equation solvers explained in this Appendix. We add two examples for quantitative evaluation of some of the methods.

Example 1. Contour curves of a very flat surface which is synthesized from 3×3 patches and has a flat line segment contacting with a horizontal plane are shown in Fig. A.1(b). The length of the contact line segment is 0.5. Its contour curves were produced by intersection with the horizontal cutting planes whose heights are 2×10^{-8}, 2×10^{-7}, $2 \times 10^{-6}, \cdots 2 \times 10^{-2}$, 2×10^{-1}. They are shown in Fig. A.1(a). The differential equation method was used for tracing contour curves. The Runge-Kutta method with adaptive stepsize control gave good results: a contour curve whose height is as small as 2×10^{-8} was obtained. With a value smaller than this, the procedure did not work with double precision calculation.

Example 2. Contour curves of a surface which has a subtle flat part are shown in Fig. A.2, where the global shape is shown by the contour curves with each height step 0.01 of the cutting plane. In Fig. A.3, the height step is 0.0002 and its shape around the peak is shown, whereas in Fig. A.4 it is 0.00001, in which clearly a local hump and two straight lines can be observed.

In tracing the curves near the critical points, a predictor-corrector method fails in the early stage, and so does Runge-Kutta without adaptive stepsize. The Bulirsch-Stoer method with the polynomial extrapolation fails in the last critical stage. Bulirsch-Stoer with rational extrapolation is very fast in the non-critical region, but becomes very slow at the critical region and past it, while Runge-Kutta with adaptive stepsize is not fast, but at the critical region it is faster than Bulirsch-Stoer with rational extrapolation and successful. Runge-Kutta-Fehlberg is slow but succeeds.

Figure A.5 shows variation of the stepsize in the Bulirsch-Stoer method, though there occur very large stepsizes, the calculated points are accurate. The smooth curves are drawn by extracting the intermediate integration values given by eq.(A.16) in the Bulirsch-Stoer method. We can recognize in these examples that the truncation errors are very small. Otherwise, we cannot discriminate the intersection curves with very small height differences.

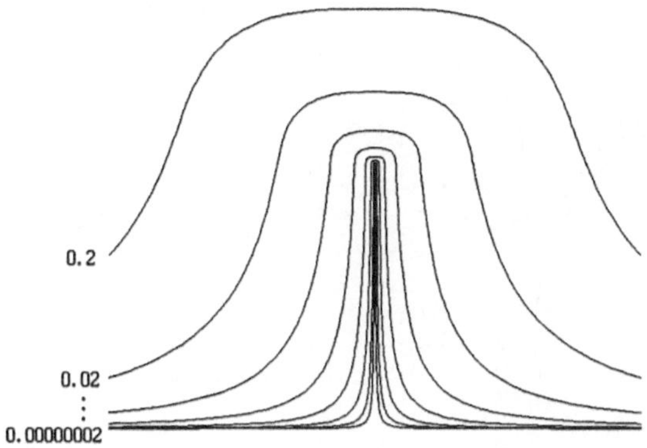

(a)

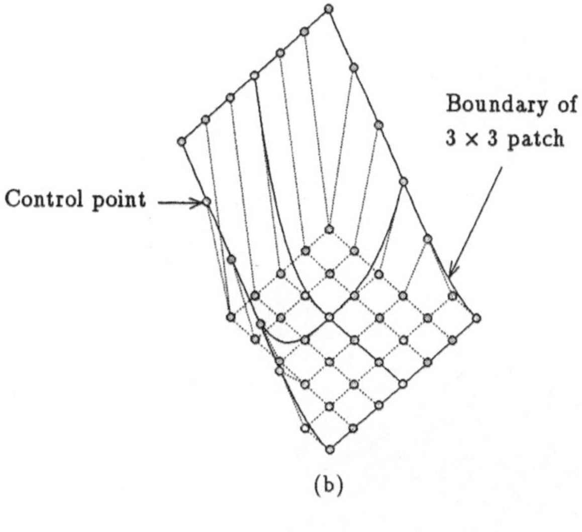

(b)

Figure A.1 Contour curves at very small heights obtained by Runge-Kutta method with quality control: (a) contour curves, (b) locations of control points of connected patches

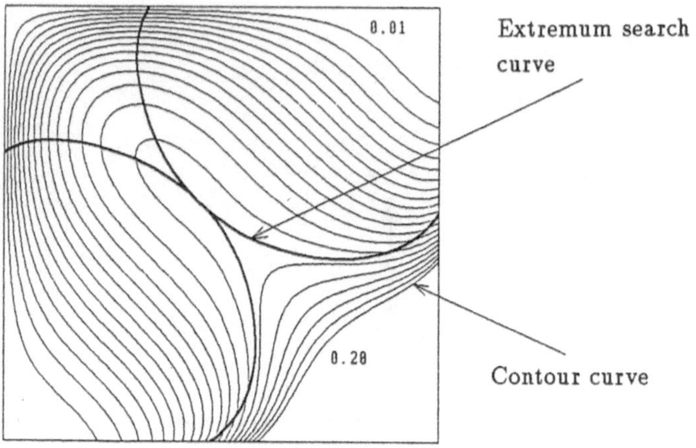

Figure A.2 Global shape with contour curves and extremum search curves

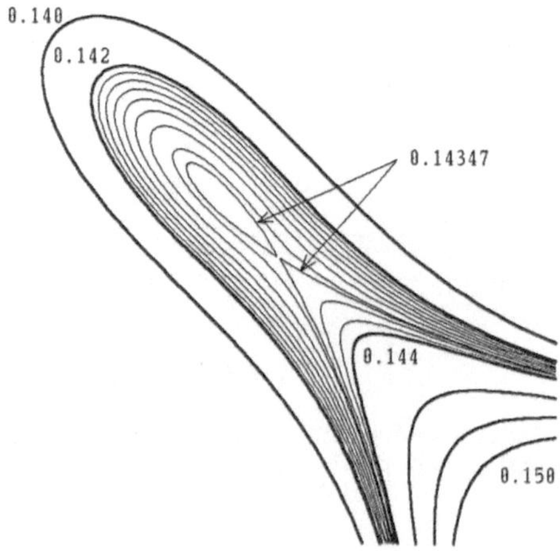

Figure A.3 Magnified figure around the critical region. Contour curves are shown

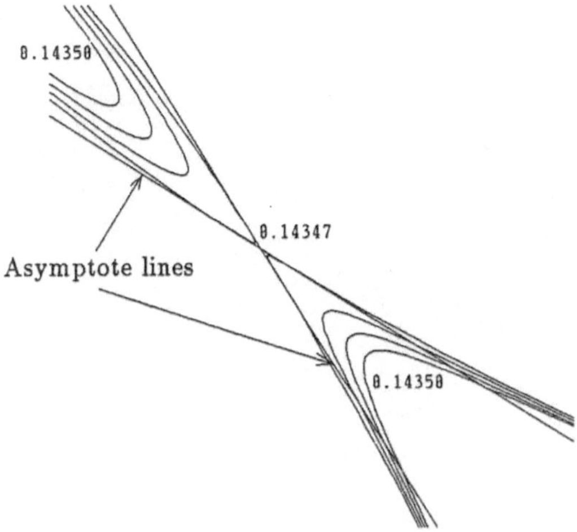

Figure A.4 More magnified figure of the contour curves

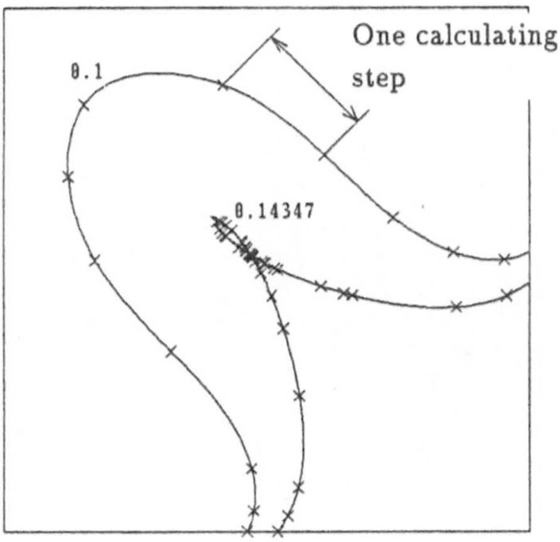

Figure A.5 Example of automatic stepsize adjustment by Bulirsch-Stoer method

Bibliography

Since a comprehensive bibliography in the field of *Geometric modelling and CAD/CAM* is to be included in Kimura's book (see Preface), only literature and references related to the contents of this book are shown separately for each part. An asterisk indicates the papers written in Japanese.

Abbreviations for Japanese journals and academic societies cited in this book are as follows:

- IPS Japan: the Information Processing Society of Japan. Its publications are:
- J. IPS Japan: Journal of the Information Processing Society of Japan (in Japanese).
- Trans. IPS Japan: Transactions of the Information Processing Society of Japan (in Japanese).
- J. Inf. Process.: Journal of Information Processing (in English)
- Information Processing in Japan. Annual publication before 1978 (in English).
- Proc. -th Annual Convention IPS Japan (in Japanese).
- IPSJ SIG reports (in Japanese).
- Proc. IPSJ SIG Symposium (in Japanese).

Journals of other academic societies.

- J. ICSJ: Journal of the Society of Instrument and Control Engineers of Japan (in Japanese).
- J.JSPE: Journal of the Japan Society of Precision Engineering (in Japanese).
- J.JSME: Journal of the Japan Society of Mechanical Engineers (in Japanese).
- Trans.JSME: Transactions of the JSME (in Japanese).

Preface

[1] Farin, G.: Curves and surfaces for computer aided geometric design: a practical guide. 2nd ed. San Diego CA: Academic Press 1990

[2] Hoffmann, C.M.: Geometric and solid modeling. San Mateo CA: Morgan Kaufmann 1989

[3] Hosaka, M.: MARS-1. Seat Reservation System of Japanese National Railways. In: History Committee of IPS Japan (ed.): *History of Early Japanese Computers. Tokyo: Ohmsha 1985, pp. 155–172

[4] Hosaka, M.: *Digital differential analyzer. Hitachi Hyoron (J. of R & D of Hitachi Ltd.) 45(10): 61–65, 1963
Hosaka, M.: *Generation of figures by DDA. J.ICSJ 3(7): 422–452, 1965
Hosaka, M.: *On generation, storage and processing of graphics. J.IPS

Japan 6(3): 129–139, 1965. English abstract: Information Processing in Japan 6: 29-40, 1967

[5] Sutherland, I.E.: Sketchpad – A man machine graphical communication system. In: Proc. SJCC. 1963, pp. 329–346
Coons, S.: An Outline of the requirements for the computer aided design. In: Proc. SJCC: 1963, pp. 299–304

[6] Hosaka, M.: *Man-machine interaction and computer display. Monthly Memoranda of IPS Japan 22, Feb. 1967
Hosaka, M.: *Design methods by computer assistance. J.JSME 71(590): 118–127, 1967

[7] Hosaka, M.: *Theory and design of free-form surface. J.IPS Japan 8(2): 65-72, 1967. English abstract: Information Processing in Japan 7: 54–61, 1968
Hosaka, M.:Design of free-form surface and computer. J.JSPE 34(1): 4–10, 1968
Hosaka, M.: Theory of curve and surface synthesis and their smooth fitting. J.IPS 10(3): 121–131, 1969 (English abstract, Information Processing in Japan 9: 60–68, 1970)

[8] Ishimatu, Y. et al.: Computer-aided generation of sculptured surfaces. J. Numerical Control: 33–39, April, 1974

[9] Matsushita, T., Hosaka, M.: *Shape processing by a new interactive language GIL. In: Proc. 11th annual convention IPS Japan: 65–66, 1970
Kimura, F., Hosaka, M.: *On processing of solid body. In: Proc.13th annual convention IPS Japan: 271-272, 1972
Kimura, F., Hosaka, M.: *Extension of interactive language GIL. In: Proc 14th annual convention IPS Japan: 275–276, 1973
Kimura, F., Hosaka, M.: *Input-output control of GIL. In: Proc. 14th annual convention IPS Japan: 277–278, 1973

[10] Braid, I. Lang, C. A.: Computer-aided design of mechanical components with volume building blocks. In: Proc. Prolamat '73, Amsterdam: North-Holland 1973
Engli, M. E.: Language for 3D graphics application. In: A. Gunther et al.(eds.): Proc. European Computing symp., Amsterdam: North-Holland, 1973

[11] Hosaka, M., Kimura, F. et al.: A unified method for processing polyhedra. In: Rosenfeld, J.R. (ed.): Proc.IFIP Congress '74, Amsterdam: North-Holland 1974, pp. 768–772
Hosaka, M., Kimura, F. et al.: GIL- A Language for Interactive Graphics, In: Proc. 2nd US-Japan computer conference: Montvale NJ: AFIPS Press 1975, pp. 546–550
Hosaka, M., Kimura, F., Kakishita, N.: A software system for computer-aided activities. In: J.J. Allan (ed.): CAD Systems. Proc. IFIP working conference : Amsterdam: North-Holland 1977, pp. 169–198

[12] Hosaka, M.: *Computer Graphics: Tokyo: Sangyo-Tosho Jan. 1974, Many references for Hosaka's research work during 1963–1972 are included

[13] Hosaka, M., Kimura, F.: An interactive geometric design system with handwriting input. In: Gilchrist, B.(ed.): Proc. IFIP Congress '77: Amsterdam: North-Holland 1977, pp. 167–172

[14] Kimura, F.: Geomap-III: Desining solids with free-from surfaces. IEEE Computer Graphics and Applications 4: 58–72, 1984

[15] Hosaka, M., Kuroda, M.: *Generation of curves and surfaces in CAD, J. IPS Japan 17(12): 1120–1127, 1976. English abstract, Information Processing in Japan (17): 75–79, 1976

[16] Hosaka, M., Kimura, F.: Synthesis method of curves and surfaces in interactive CAD. In: Proc. Intn'l conference on interactive technology (IEEE 78 CH 289-8C) 1978, pp. 151–156

[17] Hosaka, M., Kimura, F.: A mathematical theory and design methodolgy for free-form shapes. J. IPS Japan 13(3): 1–12, 1980

[18] Ohara, M.: CAD/CAM at Toyota Motor Corporation. In: Kitagawa, T. (ed): JARECT, Computer science and technologies 18, Tokyo: Ohmsha and North-Holland 1988, pp. 191–210

[19] Hosaka, M., Kimura, F.: Interactive input methods for free-form shape design. In: Sata, T., Warman, E. (eds.): Proc. IFIP working conference on man-machine communication, Amsterdam: North-Holland 1981, pp. 103–115

[20] Hosaka, M.: *Theory of shape connection. In: Proc. symp. Graphics and CAD '84 IPS Japan 1984, pp. 51–62
Hosaka, M.: *Intersction and connection of surfaces 1. In: Proc. symp. Graphics and CAD '87 IPS Japan 1987, pp. 57–66
Hosaka, M.: *Intersection and connection of surfaces 2. In: Proc. symp. Graphics and CAD '88 IPS Japan 1988, pp. 121–132

Chapters 1–4

[1] Hosaka, M., Kimura, F.: A unified method of processing polyhedra. In: Rosenfeld, J.R. (ed.): Information Processing '74 Proc. IFIP Congress '74, Amsterdam: North-Holland 1974, pp. 768–772

[2] Hosaka, M., Kimura, F.: An interactive geometric design system with handwriting input. In: Gilchrist, B. (ed.): Information Processing '77 Proc. IFIP Congress '77, Amsterdam: North-Holland 1977, pp. 167–172

[3] Levin, J.: A parametric algorithm for drawing pictures of solid objects of quadric surfaces. Comm. ACM (10): 555–563, 1976

[4] Levin, J.: Mathematical models for detecting the intersection of quadric surfaces. Computer Graphics and Image Processing 11: 73–87, 1979

[5] Sarraga, R.F.: Algebraic methods for intersections of quadric surfaces in GMSOLID. Computer Vision, Graphics and Image Processing 22: 222–328, 1983

[6] Mäntylä, M.: An introduction to Solid Modeling. Rockville MD: Computer Science Press Inc. 1988
[7] Miller, J.R.: Analysis of quadric-surface based solid models. IEEE Computer Graphics and Applications 1: 28–41, 1988
[8] Saitoh, T.: *Interference and display of quadrics. In: Proc. JSPE '90 JSIP 1990, pp. 1037–1038

Chapters 5–8

[1] Coxeter, H.S.M.: Introduction to geometry. 2nd edition, John Wiley 1969
[2] Hosaka, M.: *Theory of curve and surface synthesis and their smooth fitting. J. IPS Japan 10(3): 121–131, 1969 (English abstract, Information Processing in Japan 9: 60–68, 1969)
[3] Enomoto, H. et al.: *Computer experiment on global properties of structure lines of images using graphic display and its consideration). J. IPS Japan 17(7): 641–649, 1976
[4] Kajiya, J.T.: Ray tracing parametric patches. Computer Graphics (Proc. Siggraph '82) 16(3): 224-254, 1984
[5] Sederberg, T.W., Anderson, D.C., Goldman, R.N.: Implicit representation of parametric curves and surfaces. Computer Vision, Graphics and Image Processing 28: 72–84, 1984
[6] Goldman, R.N., Sederberg, T.W., Anderson, D.C.: Vector elimination: a technique for implicitization, inversion and intersection of planar parametric rational polynomial curves. Computer Aided Geometric Design 1(4): 327–356, 1984
[7] Sederberg, T.W.: Planar piecewise algebraic curves. Computer Aided Geometric Design 1(3): 241–255, 1984
[8] Poeschle, R.: Detecting surface irregularities using isophotes. Computer Aided Geometric Design 1(2): 163–168, 1984
[9] Tiller, W., Hanson, E.G.: Offsets of two-dimensional profiles. IEEE Computer Graphics and Applications 9: 36–46, 1984
[10] Satterfield, S.D., Rogers, D.F.: A procedure for generating contour lines from a B spline surface. IEEE Computer Graphics and Applications 4: 71–75, 1985
[11] Farouki, R., Rajan, V.: Exact offset procedures for simple solids. Computer Aided Geometric Design 2(4): 257–294, 1985
[12] Beck, J.M., Farouki, R.T., Hind, J.K.: Surface analysis methods. IEEE Computer Graphics and Applications 6(12): 18–36, 1986
[13] Klok, F.: Two moving coordinate frames for sweeping along 3D trajectory. Computer Aided Geometric Design 3(3): 217-229, 1986
[14] Farouki, R.: The approximation of non-degenerate offset surface. Computer Aided Geometric Design 2(1): 15–44, 1986
[15] Rossignac, J.R., Requicha, A.A.G.: Offsetting operations in solid modelling. Computer Aided Geometric Design 3(2): 129-148, 1986

[16] Stoer, J., Bulirsch, R.: Introduction to numerical analysis. New York: Springer-Verlag 1980

[17] Press, W. H., Flannery, B.P., Teukolsky, S.A., Vetterling W.T.: Numerical Recipies in C Cambridge: Cambridge University Press 1988

[18] Higashi, M., Kushimoto, T., Hosaka, M.: On formulation and display for visualizing features and evaluating quality of free-form surfaces. In: Vandoni, C.E., Duce, D.A.(eds): Proc. Eurographics '90 1990, pp. 299–309

[19] Love, A.E.H.: Treatise on the mathematical theory of elasticity. Cambridge: Cambridge University Press 1934, pp. 401–410

Chapters 9–12

[1] Hosaka, M.: *Theory and design of free-form surface. J. IPS Japan 8(2): 65–72, 1967

[2] Hosaka, M.: Synthesis and smoothing of free-form curves and surfaces. J. IPS Japan 10(3): 121–131, 1969

[3] Forrest, A.: Interactive interpolation and approximation by Bézier polynomials. Computer J. 15: 71–79, 1972

[4] Riesenfeld, R.F.: Non-uniform spline curves. In: Proc. 2nd USA-Japan computer conference IPS Japan 1975, pp. 551–555

[5] Hosaka, M., Kuroda, M.: *Generation of curves and surfaces in CAD. J. IPS Japan 17(12): 1120–1127, 1976. (English abstract, Information Processing in Japan (17): 75–79, 1976)

[6] Hosaka, M., Kimura, F.: Synthesis methods of curves and surfaces in interactive CAD. In: Proc. intn'l conference on interactive techniques in CAD, IEEE 1978, pp. 151–156

[7] Faux, I.D., Pratt, M.J.: Computational geometry for design and manufacture. London: Ellis Horwood 1979

[8] Hosaka, M., Kimura, F.: A theory and methods for three dimensional free-form shape construction. J. Inf. Process 3(3): 140–151, 1980

[9] Hosaka, M., Kimura, F.: Interactive input methods for free-form shape design. In: Sata,T., Warman, E. (eds.) Proc. IFIP working conference on man-machine communication. IFIP 1980, pp. 103–115

[10] Boehm, W.: Generating the Bézier points of B-splines. Computer Aided Design 13(6): 365–366, 1981

[11] Barsky B.A., Beatty, J.C.: Local control of bias and tension in Beta-splines. Proc. Siggraph '83 17(3): 193–218, 1983

[12] Hosaka, M.: Theory of shape connection. In: Proc. symp. Graphic and CAD '84 IPS Japan 1984, pp. 51–62

[13] Boehm, W., Farin, G., Kahmann, J.: A survey of curve and surface in computer aided geometric design. Computer Aided Geometric Design 1(1): 1–60, 1984

[14] Varady, T., Pratt, M.J.: Design techniques for the definition of solid objects with free-form geometry. Computer Aided Geometric Design 1(3): 207–225, 1984
[15] Boehm, W.: Curvature continuous curves and surfaces. Computer Aided Geometric Design 2(2): 313–323, 1985
[16] Piegl, L.: The sphere as a rational Bézier surface. Computer Aided Geometric Design 3(1): 45-52, 1986
[17] Piegl, L.: Infinite control points – a method for representing surface of revolution using boundary data. IEEE Computer Graphics and Applications 3: 45–55, 1987
[18] Piegl, L.: Interactive data interpolation by rational Bézier curves, IEEE Computer Graphics and Applications 4: 45–53, 1987
[19] Sarraga, R.F.: G^1 interpolation of generally unrestricted cubic Bézier curves. Computer Aided Geometric Design 4(1-2): 23–40, 1987
[20] Boehm, W.: Rational geometric splines. Computer Aided Geometric Design 4(1-2): 67–78, 1987
[21] Farin, G., Rein, G. et al.: Fairing cubic B-spline curves. Computer Aided Geometric Design 4(1-2): 91–104, 1987
[22] Bartels, R. H., Beatty, J.C, Barsky, B.A.: An introduction to splines for use in computer graphics and geometric modeling. San Mateo CA: Morgan Kaufman 1987
[23] Farouki, R., Rajan, V.: On the numerical condition of polynomials in Berstein form. Computer Aided Geometric Design 4(3): 191–216, 1987
[24] Hagen, H., Schultz, G.: Automatic smoothing with geometric surface patches. Computer Aided Geometric Design 4(3): 231–236, 1987
[25] Meier, H., Nowacki, H.: Interpolating curves with gradual change in curvature. Computer Aided Geometric Design 4(4): 297–306, 1987
[26] Farouki, R., Rajan, V.: Algorithm for polynomials in Berstein form. Computer Aided Geometric Design 5(1): 1–26, 1988
[27] Hosaka, M.: *Intersection and connection of surfaces-2. In: Proc. Symp. graphics and CAD IPS Japan 1988, pp. 121–132
[28] Higashi, M., Kaneko, K., Hosaka, M.: Generation of high-quality curve and surface with smoothly varying curvature. In: Duce, D.A., Jancene, P. (eds.): Proc. Eurographics '88 Amsterdam: North-Holland, 1988, pp. 79–92
[29] Fergson, D.R., et al.: Surface shape control using constrained optimization on the B-spline representation. Computer Aided Geometric Design 5, 1989
[30] Piegl, L.: Modifying the shape of rational B-splines. 1: curves. Computer Aided Design 21(8): 509–518, 1989
[31] Piegl, L.: Modifying the shape of rational B-splines. 2: surfaces. Computer Aided Design 21(9): 538–545, 1889
[32] Dokken, T. et al.: The role of NURBS in geometric modeling and CAD/CAM. In: Krause, F-L., Jansen, H. (eds.): Advanced geometric modeling in engineering application, Berlin: APZ 1989, pp. 95–102

[33] Farin, G.: Curves and surfaces for computer aided geometric design: a practical guide. 2nd ed. San Diego CA: Academic Press 1990

[34] Saitoh, T., Hosaka, M.: *On the extended rational Bézier curve and its use for curve fitting methods. Trans.IPS Japan 31(1): 33–41, 1990

[35] Saitoh, T. Hosaka, M.: *High quality outline font by the extended rational Bézier curve. Trans.IPS Japan 31(4): 562–570, 1990

Chapters 13–15

[1] Coons S.A.: Surfaces for computer aided design of space forms. MIT Project MAC TR-41 1967

[2] Gregory J.A.: Smooth interpolation without twist constaints. In: Burnhill, R.E., Riesenfeld, R.F.(eds.): Computer aided geometrric design, New York: Academic Press 1974, pp. 71–87

[3] Sabin, M.A.: The use of piecewise forms for the numerical representation of shape. Computer and Automation Inst. Hungarian Academy of Sci. 1977

[4] Barnhill, R.E., Brown, J.H., Klucewicz I.M.: A new twist in computer aided geometric design. Computer Graphics and Image Processing 8: 78–91, 1978

[5] Higashi, M. et al.: An interactive CAD systems for construction of shapes with high quality surface. In: Warman, E.A. (ed.): Proc. CAPE'83 IFIP 1983, pp. 371–390

[6] Chiyokura, H., Kimura, F.: Design of solids with free-form surfaces. Computer Graphics 17(3): 289–298, 1983

[7] Hosaka, M., Kimura, F.: Non-four-sided patch expressions with control points. Computer Aided Geometric Design 1(1): 75–86, 1984

[8] Charrot, P., Gregory, J.A.: A pentagonal surface patch for CAGD. Computer Aided Geometric Design 1(1): 87–94, 1984

[9] Herron, G.: Smooth closed surface with triangular interpolants. Computer Aided Geometric Design 2(4): 297–306, 1985

[10] Farin G. Triangular Bernstein-Bézier patches. Computer Aided Geometric Design 3(2): 83–127, 1986

[11] Sabin, M. A.: Some negtive results in N sided patches. Computer Aided Design 18(1): 38–44, 1986

[12] Chiyokura, H.: Localized surface interpolation method for irregular meshes. In: Kunii, T. (ed.): Computer Graphics. Berlin: Springer-Verlag, 1986

[13] Holmstroem, L.: Piecewise quadric blending of implicitly defined surface. Computer Aided Geometric Design 4: 171–189, 1987

[14] Shirman, L.A., Sequin, C.H.: Local surface interpolation with Bézier patches. Computer Aided Geometric Design 4(4): 279–295, 1987

[15] van Wijk J.J.: Bicubic patch for approximating non-rectangular meshes. Computer Aided Geometric Design 3(1): 1–14, 1987

[16] Farin, G. et al.: The octant of a sphere as a non-degenerate triangular patch. Computer Aided Geometric Design 4(4): 329–332, 1987

[17] Storry, D.J.T., Ball, A.A.: Design of an n-sided surface patch from Hermite boundary data. Computer Aided Geometric Design 6(2): 111–120, 1989

[18] Kushimoto, T., Hosaka, M.: *Design methods of surface in non-four-sided regions. 1, J.JSPE 55(08): 55-60, 1989. 2, J.JSPE 55(11): 119-124, 1989

[19] Kushimoto, T., Hosaka, M.: *Connection of triangular patches and its application. J.JSPE 55(10): 67–72, 1989

[20] Saitoh, T., Hosaka, M.: Interpolating curve networks with new blending patches. In: Vandoni, C.E., Duce, D.A. (eds.): Proc. Eurographics '90, pp. 137–146

[21] Hosaka, M.: *New solution of connection problem in free-form surfaces. J. IPS Japan 31(5): 612–622, 1990

[22] Peters, J.: Local cubic and bicubic C^1 surface interpolation with linearly varying boundary normal. Computer Aided Geometric Design 7: 499–516, 1990

Chapter 16 and Appendix

[1] Phillips, M.B., Odel, G.M.: An algorithm for locating and displaying the intersection of two arbitrary surfaces. IEEE Computer Graphics and Applications 9: 48–55, 1984

[2] Dokken, T.: Finding intersections of B-spline represented geometies using recursive subdivision techniques. Computer Aided Geometric Design 2(1–3): 37–48, 1985

[3] Pfeifer, H-U.: Methods used for intersecting geometrical entities in the GPM module for volume geometry. Computer Aided Design 17(7): 311–318, 1985

[4] Houghton, E.G.: Implementation of a divide-and-conquer method for intersection of parametic surfaces. Computer Aided Geometric Design 2(1-3): 173–183, 1985

[5] Miller, J.R.: Sculptured surfaces in solid models: issues and alternative approaches. IEEE Computer Graphics and Applications 12: 37–48, 1986

[6] Barnhill, R.E., Farin, G., Jordan, M., Piper, R.R.: Surface/suface intersection. Computer Aided Geometric Design 4(1–2): 3–16, 1987

[7] Bajaj, C.L., Hoffmann, C.M., Lynch, R.E.: Tracing surface intersections. Computer Aided Geometric Design 5(4): 295–307, 1988

[8] Hosaka, M.: *Connection and interference of surfaces. In: Proc. symp. on graphics and CAD '87: IPS Japan 1988, pp. 57–66

[9] Hosaka, M., Kushimoto, T., Gonda, H.: *Connection and interference of surfaces. In: Proc. symp. on graphics and CAD '88 IPS Japan: 1988, pp. 121–131

[10] Hoffmann, C.M.: Geometric and solid modeling, San Mateo CA: Morgan Kaufmann 1989

[11] Higashi, M, Mori, T., Hosaka, M.: Interference calculation of surfaces based on their geometric properties. In: Kimura, F., Rolstadas, A.(eds.): Computer Applications in Production and Engineering CAPE'89 : Amsterdam: Elsevier 1989, pp. 275-282
[12] Aziz, M.A., Bata, R., Bhat. S.: Bézier surface/surface intersection IEEE Computer Graphics and Applications 1: 50–58, 1990
[13] Hosaka, H., Kushimoto, T.:*New methods of intersection tracing of curves surfaces. In: Proc. symp. on design JSPE: 1990, pp. 1-3, pp. 7-9
[14] Stoer, J., Bulirsch, R.: Introduction to numerical analysis. New York: Springer-Verlag 1980
[15] Press, W.H., Flannery, B.P., Teukolsky, S. A., Vetterling, W.T. : Numerical recipies in C. Cambridge: Cambridge University Press 1988
[16] Burden, R.L., Faires, J.D.: Numerical analysis-4th edition. PWS-KENT, 1989

Chapter 17

[1] Sakai, Y., Kuranaga, Y.: Development of CADDETT system. In: Proc. ISATA'81, Croydon: Automotive Automation Ltd. 1981
[2] Takahashi, A., Inoue, M., Okamaoto, I., Mori, T.,: Stamping die CAD at Toyota. In: Proc.CAMP '83 Berlin: AMK , 1983, pp. 623–636
[3] Higashi, M., Kohzen, I., Nagasaka, J.: An interactive CAD system for construction of shapes with high-quality surface. In: Warman, E.A. (ed.): Proc. CAPE'83 Amsterdam: North-Holland 1983, pp. 371–390
[4]. Nagasaka, J., Higashi, M. Sannokyou, H., Aoyama, N.: High speed machining system for styling design. In: Proc. ISATA 84, Automotive Automation Ltd., Croydon: 1984, pp. 1083–1098
[5] Higashi, M., Mori T., Taniguchi, H., Yoshimi, J.: Geometric modelling for efficient evaluation of press forming severity. In: Wang, M.M., Fang, F.C.(eds.): Proc. TMS-AIME symposium on computer modelling for sheet metal forming process, Detroit: AIME, 1985, pp. 21–35
[6] Takahashi, A., Okamoto, I., Hiramatsu, T.,Yamada, N.: Evaluation methods of press forming severity in CAD application. In: Wang, M.M., Fang, F.C.(eds.): Proc. TMS-AIME symposium on computer modelling of sheet metal forming process, Detroit: AIME, 1985, pp. 37–50
[7] Araki, H., Kawai, M., Amano, Y., Nonaka, M.: Replacement of physical models by accurate numerical models for manufacturing stamping dies in the integrated CAD/CAM system. In: Crestin, J.P., MacWaters, J.F.(eds.) Proc. Prolamat: Amsterdam: North-Holland 1985, pp. 47–60
[8] Araki, H., Niwa, S, Makino, T., Yoshinaka, T.: Computer aided finishing system for stamping die manufacturing. In: Proc. CAMP'85, Berlin: AMK, 1985, pp. 203-208
[9] Kobayashi, S.: Calculation of free-form surface intersection and its application to evaluation of mass-property. In: Proc. Intn'l conf. computers

in engineering. ASME: 1985, pp. 325–330

[10] Izumi, M., Kobayashi, S.: Development of a solid modeller for FEM data generation. In: Proc. ISATA '85, Croydon: Automotive Automation Ltd. 1985, pp. 535–570

[11] Kuragano T. et al.: The FRESDAM system for generating tool paths for manufacturing aesthetically pleasing injection mold products. In: Kimura, F., Rolstadas, A. (eds.) Proc. CAPE '89, Amsterdam: Elsevier 1989, pp. 411–418

Subject Index

Springer Series
Computer Graphics – Systems and Applications

Former Subseries of SYMBOLIC COMPUTATION

J. L. Encarnação, R. Schuster, E. Vöge (eds.):
Product Data Interfaces in CAD/CAM Applications.
Design, Implementation and Experiences.
IX, 270 pages, 147 figs., 1986

U. Rembold, R. Dillmann (eds.):
Computer-Aided Design and Manufacturing.
Methods and Tools. Second, revised and enlarged edition.
XIV; 458 pages, 304 figs., 1986

G. Enderle, K. Kansy, G. Pfaff:
Computer Graphics Programming. GKS – The Graphics
Standard. Second, revised and enlarged edition.
XXIII, 651 pages, 100 figs., 1987

Y. Shirai:
Three-Dimensional Computer Vision.
XII, 297 pages, 313 figs., 1987

D. B. Arnold, P. R. Bono:
CGM and CGI. Metafile and Interface Standards
for Computer Graphics.
XXIII, 279 pages, 103 figs., 1988

J. L. Encarnação, P. C. Lockemann (eds.):
Engineering Databases. XII, 229 pages, 152 figs., 1990

P. Wisskirchen:
Object-Oriented Graphics. From GKS and PHIGS
to Object-Oriented Systems.
XIII, 236 pages, 83 figs., 1990

J. L. Encarnação, R. Lindner, E. G. Schlechtendahl:
Computer Aided Design. Fundamentals and System
Architectures. Second, revised and extended edition.
XII, 432 pages, 240 figs., 1990

H. Hagen, D. Roller (eds.):
Geometric Modeling. Methods and Applications.
VII, 286 pages, 140 figs., 1991

T. Yagiu:
Modeling Design Objects and Processes
X, 327 pages, 82 figs., 1991

M. Hosaka:
Modeling of Curves and Surfaces in CAD/CAM.
XXI, 350 pages, 90 figs., 1992